| 高等学校计算机专业规划教材 |

Android
开发与应用

马玉春 著

DEVELOPMENT AND
APPLICATION
ON ANDROID PLATFORM

机械工业出版社
China Machine Press

图书在版编目（CIP）数据

Android 开发与应用 / 马玉春著. —北京：机械工业出版社，2019.9
（高等学校计算机专业规划教材）

ISBN 978-7-111-63700-4

I. A… II. 马… III. 移动终端 – 应用程序 – 程序设计 – 高等学校 – 教材　IV. TN929.53

中国版本图书馆 CIP 数据核字（2019）第 198986 号

本书基本涵盖了 Android 编程的所有技术，首先介绍了开发环境，然后在第一部分介绍编程基础与技巧，包括常用组件与技术及 Intent 的综合应用。第二部分为实用案例分析，是第一部分的扩展与应用，重点展示了个性化 ListView 组件、SQLite 数据库、文件读写、通讯录操作、广播接收器、后台服务、消息机制及数据传递和百度地图等的灵活应用。第三部分讲解了 Android 应用于远程温度监测的一个综合案例。

本书可以作为大专院校计算机相关专业学生的 Android 程序设计教材和毕业设计的参考书，也可为计算机与自动控制专业相关的工程技术人员及硕士研究生从事项目研发提供技术方案参考。

出版发行：机械工业出版社（北京市西城区百万庄大街 22 号　邮政编码：100037）
责任编辑：佘　洁　　　　　　　　　　　　责任校对：李秋荣
印　　刷：北京瑞德印刷有限公司　　　　　版　　次：2019 年 10 月第 1 版第 1 次印刷
开　　本：185mm×260mm　1/16　　　　　印　　张：20.75
书　　号：ISBN 978-7-111-63700-4　　　　定　　价：59.00 元

客服电话：（010）88361066　88379833　68326294　　投稿热线：（010）88379604
华章网站：www.hzbook.com　　　　　　　　　　　　读者信箱：hzjsj@hzbook.com

版权所有 • 侵权必究
封底无防伪标均为盗版
本书法律顾问：北京大成律师事务所　韩光／邹晓东

前　　言

　　Android是以Google为首的Open Handset Alliance（OHA，开放手机联盟）推出的一款开放的嵌入式操作系统平台，2007年11月推出Android SDK 1.0版，现已推出9.0版。Android应用越来越广泛，已经席卷整个智能手机产业和移动互联网行业，至2017年年底，其市场占有率已经达到85.9%。面对这种形势，软件从业者纷纷转向Android应用的开发。目前市场上有很多关于Android基础及技术的书籍，但是其中大部分主要讲解Android系统中各组件的使用及零散的应用技术，读完这些书读者难以掌握研发Android应用的关键和主流技术。

本书的主要内容

　　本书作者长期从事软件工程的科研与教学工作，获得国家软件著作权11项。本书是作者多年来从事Android应用研究和教学实践的结晶，并融入作者20年计算机监控系统的研发经验。本书首先从Android开发环境入手，介绍了常用组件与技术及Intent的综合应用，然后在此基础之上完成了6个实用案例，并详细分析了代码实现，最后介绍了一个远程温度监测实例，涉及数据处理技术、TCP客户机与服务器编程及温度监测的最终实现。各章内容具体安排如下所示。

　　第1章"开发环境搭建与应用入门"主要介绍了开发环境的搭建方法、工作空间与相关文件，并结合第一个应用程序详细介绍了程序的调试方法、项目的导入方法、调试设备的选择等。

　　第一部分（第2~4章）介绍编程基础与技巧，包括常用组件与技术及Intent的综合应用。第2章"常用开发组件"首先介绍了各组件通用的常见属性，然后分别介绍各组件的重要属性，通过实例从多角度展示组件事件的处理方法，并将框架布局知识融入其中。

　　第3章"常用技术"开始引入Library类库的概念，将常用技术和具有共性的代码放入类库中进行维护，以Sound类为例，所有进行情景模式操作的软件只需引用Library类库即可调用相关功能，对Sound类的改进只需在Library类库项目中进行，其他导入Sound类的项目无须更改代码，只需重新编译。该章依次介绍了Activity的生命周期、绘图方法、情景模式与音频播放、消息机制、多线程处理、定时功能、SQLite数据库、查询联系人、用户偏好和内部文件存取方法以及手机地图的开发方法，所有这些技术都通过详细实例进行讲解，并将通用技术融入Library类库中。

　　第4章"Intent的综合应用"首先介绍了Intent的基础知识，然后介绍了利用Intent打开Activity并传递数据、状态栏通知的实现及通过状态栏打开Activity，接着采用配置文件注册或软件注册的方法，从开机自动启动、来电、短信发送与接收等方面讲解广播接收器的应用，最后介绍了服务的基础知识、启动服务和绑定服务的实现，并介绍了使服务长期运行而不被系统杀死的方法。第3章与第4章的内容是本书的技术基础，也是Android应用研发的技术基础。

第二部分（第5～10章）为实用案例分析。第5章"课堂随机点名软件"结合3.8节的SQLite数据库技术，实现了多门课程多班级不重复点名；第6章"简易英语学习软件"利用Library类库中的文件存取类FileProcess和多线程文本读取类ThreadReadText，实现自行设置学习内容，既可以用来学习英语对话，也可以用来背单词；第7章"通讯录备份与恢复软件"在Library类库中创建了PhoneBook类，集成了通讯录的查找与更新功能，可以一键完成所有通讯录的备份，以及一键将备份全部恢复（合并）到通讯录。

第8章"服务账号登记软件"主要利用SQLite数据库技术与文本文件存取技术分类保存、检索各种账号，并将敏感数据加密保存到外部文件，以及从外部文件解密恢复到数据库中。第8章的列表选项采用了多种适配器并实现了拖曳技术。本章内容是前面基本技术的具体应用，并利用第11章的数据编码与处理技术对文本数据进行加密和解密。

第9章"地址定位及辅助服务软件"实现的是一个启动服务，结合3.12节的百度地图实现了手机定位功能，并利用第11章的编码技术将定位信息加密后发送到目标手机；利用4.5节的显示来电和接收短信技术，以及3.9节的查询联系人技术确保白名单来电响铃，还可实现开启"WiFi关闭"提醒，向目标手机发送余额不足信息。第10章的"地址查询与地图打点软件"向第9章的"地址定位及辅助服务软件"发送地址查询命令，利用第11章的编码技术将收到的地址信息解密后在百度地图上标注出来。

第三部分（第11～15章）为基于互联网的远程温度监测案例，包括数据处理、网络编程与具体实现三个主题。第11章"数据编码与处理技术"实现了字节（数组）、字符（串）和汉字等的相互转换及简单的编码功能，并实现了随机字节的生成与字节的位操作技术。第12章"数据包的校验技术"实现了多种校验码的计算和结尾码的处理，可通过一个函数为原始数据包添加校验码和结尾码，也可通过一个函数检验综合数据包是否正确并删除尾部的结尾码与校验码，留下有效数据。这两章内容广泛应用于数据传输与处理，也是计算机监控的核心技术。

第13章"通用TCP客户机与服务器测试软件"在Library类库中创建了通用网络处理类NetworkProcess及TCP客户机与服务器类TcpClientServer，密切结合第11章的数据编码技术及第12章的数据包校验技术，实现了通用TCP客户机和服务器测试软件，可以以字节或普通字符串收发和显示数据，根据要求添加校验码或结尾码并进行相关检验，还可测试客户机与服务器之间的时间间隔。

第14章"I-7013D模块仿真软件"（简称仿真模块）是第13章的通用TCP服务器测试软件的一个特例，它以手动、自动和锁定三种方式提供正弦波形式的温度数据，按照I-7013D协议的要求响应客户机的温度查询。

第15章"I-7013D模块监测软件"（简称监测软件）则是第13章的通用TCP客户机测试软件的一个特例，它根据测试出的时间间隔对仿真模块进行查询，并对所收到的温度数据以趋势线的形式实时显示。仿真模块与监测软件构成一个简单的以物联网为基础的计算机监控系统。

本书特色

- 共享类库：创建的Library类库项目包含数据库、TCP、通讯录等共享类。

- 经验丰富：将作者多年的研发经验列入书中，让读者少走弯路。
- 编排合理：先介绍基本概念与简单例程、常用技术，最后介绍工程应用。
- 作品实用：通用 TCP 客户机和服务器可直接应用于测试工程项目。
- 实践性强：将知识点融入实例，且解释详尽，通俗易懂，便于模仿与应用。
- 自定义数据库类：方便用户快速创建 SQLite 数据库软件，快捷检索和更新。
- 自定义 TCP 类：可以用来快速构建稳定可靠的 TCP 服务器或客户机应用。
- 完善的数据编码与数据包校验技术：可用于快速构建计算机监控系统。

本书的学习方法与约定

本书的每章（节）都有对应的源代码实例，而 Library 类库项目中包含通用核心共享类，首先按照 1.4 节介绍的方法将其导入并复制到工作空间，然后根据章节安排或需要导入相关的其他项目。阅读章节内容时，必须打开相应的例程，一边操作例程，一边学习书本知识。对于源代码讲解部分所涉及的变量或函数名，请通过快捷键【Ctrl+F】查找它们出现的位置，理解其含义。遇到有疑问的地方，则设置断点跟踪程序的运行，以弄清程序的逻辑。第 5 章及以后的实例都列有主要知识点，与第 2～4 章的编程基础和技巧相关，读者在学习综合案例前需要了解对应的主要知识点。本书面向编程人员，因而软件界面一般采用英文格式，Android SDK 中的子程序一般称为方法，本书创建的子程序一般称为函数，需要读者注意的关键代码用粗体加以强调。

致谢

本书得到海南省高等学校教育教学改革研究项目（批准号：Hnjg2018-57）、海南省高等学校科学研究项目（批准号：Hnky2019-57）及海南热带海洋学院 2019 年校级教材建设项目（批准号：RHYJC2019-03）的资助，在此表示衷心的感谢。

声明

本书中的所有应用程序或软件工具都由作者独立开发，已经或正在申报软件著作权，软件的使用仅限于购买了本书的读者本人或者已经取得作者或出版社授权的单位，未经许可不得以任何形式复制、传播。

由于学识有限，书中不足和疏漏之处在所难免，请读者不吝赐教，以便作者进一步完善（walker_ma@163.com）。

马玉春
2019 年 8 月 8 日于三亚

目　录

前言

第1章　开发环境搭建与应用入门……1
- 1.1　搭建开发环境……1
- 1.2　创建虚拟设备……2
- 1.3　第一个应用程序……3
- 1.4　工作空间与相关文件……4
- 1.5　程序的调试方法……8
- 1.6　本章小结……10

第一部分　编程基础与技巧

第2章　常用开发组件……12
- 2.1　常见属性……12
- 2.2　EditText 组件与菜单……13
- 2.3　Button 组件……17
- 2.4　ToggleButton 组件……20
- 2.5　CheckBox 组件……22
- 2.6　RadioButton 组件……24
- 2.7　Spinner 组件……26
- 2.8　ListView 组件……29
- 2.9　Switch 组件……33
- 2.10　DatePicker 组件……36
- 2.11　AlertDialog 组件……37
- 2.12　本章小结……45

第3章　常用技术……46
- 3.1　进一步了解 Activity……46
- 3.2　绘图……47
- 3.3　用静态库函数设置手机情景模式和音量……51
- 3.4　播放音频……55
- 3.5　利用消息机制处理后退键……56
- 3.6　利用多线程和消息机制获取 IP 地址……59
- 3.7　定时功能的实现……62
- 3.8　SQLite 与自定义 ListView……64
- 3.9　查询联系人……74
- 3.10　使用 SharedPreferences 对象存储数据……80
- 3.11　内部文本文件存取……85
- 3.12　百度地图……86
- 3.13　本章小结……91

第4章　Intent 的综合应用……92
- 4.1　Intent 的基础知识……92
- 4.2　在 Activity 之间传递数据……94
- 4.3　状态栏通知……98
- 4.4　广播接收器与开机自动启动……99
- 4.5　显示来电和接收短信……101
- 4.6　带回执的短信发送……105
- 4.7　服务的基础知识……108
- 4.8　启动服务的实现……110
- 4.9　绑定服务的实现……115
- 4.10　本章小结……117

第二部分　实用案例分析

第5章　课堂随机点名软件……120
- 5.1　主要功能和技术特点……120
- 5.2　软件操作……120
- 5.3　界面布局与资源说明……121
  - 5.3.1　字符串定义文件……121
  - 5.3.2　菜单项定义文件……121
  - 5.3.3　颜色定义文件……122
  - 5.3.4　自定义对话框布局文件……122
  - 5.3.5　ListView 列表布局文件……123
  - 5.3.6　版权窗体布局文件……124
  - 5.3.7　主窗体布局文件……124

5.4	配置文件 …………………………… 126	8.6.1	适配器布局文件 …………… 185
5.5	主窗体源代码 ……………………… 127	8.6.2	窗体布局文件 ……………… 186
5.6	本章小结 …………………………… 135	8.6.3	拖放阴影源代码 …………… 187
		8.6.4	适配器源代码 ……………… 188
第6章	简易英语学习软件 ……………… 136	8.6.5	窗体源代码 ………………… 190
6.1	主要功能和技术特点 ……………… 136	8.7	单位详细信息窗体 ………………… 199
6.2	软件操作 …………………………… 136	8.7.1	适配器布局文件 …………… 199
6.3	界面布局与资源说明 ……………… 137	8.7.2	窗体布局文件 ……………… 200
	6.3.1 适应多屏幕的 dimens 文件 …… 137	8.7.3	适配器源代码 ……………… 201
	6.3.2 菜单项定义文件 …………… 138	8.7.4	窗体源代码 ………………… 202
	6.3.3 主窗体布局文件 …………… 138	8.8	单位搜索窗体 ……………………… 207
	6.3.4 主题设置文件 ……………… 140	8.8.1	布局文件 …………………… 207
6.4	配置文件 …………………………… 141	8.8.2	源代码 ……………………… 208
6.5	目录与文件处理类源代码 ………… 141	8.9	本章小结 …………………………… 211
6.6	文本读取类源代码 ………………… 142		
6.7	主窗体源代码 ……………………… 144	第9章	地址定位及辅助服务
6.8	本章小结 …………………………… 151		软件 ………………………………… 212
		9.1	主要功能和技术特点 ……………… 212
第7章	通讯录备份与恢复软件 ………… 152	9.2	软件操作 …………………………… 212
7.1	主要功能和技术特点 ……………… 152	9.3	配置文件 …………………………… 213
7.2	软件操作 …………………………… 152	9.4	广播接收器源代码 ………………… 215
7.3	界面布局 …………………………… 153		9.4.1 启动完成 …………………… 215
7.4	配置文件 …………………………… 153		9.4.2 来电处理 …………………… 216
7.5	通讯录操作源代码 ………………… 154		9.4.3 情景模式改变 ……………… 217
7.6	外部文本写入源代码 ……………… 156		9.4.4 屏幕状态变化 ……………… 217
7.7	主窗体源代码 ……………………… 156		9.4.5 短信接收 …………………… 218
7.8	本章小结 …………………………… 161		9.4.6 WiFi 设置变化 ……………… 221
		9.5	服务源代码 ………………………… 222
第8章	服务账号登记软件 ……………… 162	9.6	适配器源代码 ……………………… 223
8.1	主要功能和技术特点 ……………… 162	9.7	窗体源代码 ………………………… 226
8.2	软件操作 …………………………… 163	9.8	本章小结 …………………………… 229
8.3	配置文件 …………………………… 164		
8.4	登录窗体 …………………………… 165	第10章	地址查询与地图打点
	8.4.1 布局文件 …………………… 165		软件 ………………………………… 230
	8.4.2 源代码 ……………………… 167	10.1	主要功能和技术特点 ……………… 230
8.5	服务浏览窗体 ……………………… 172	10.2	软件操作 …………………………… 230
	8.5.1 适配器布局文件 …………… 172	10.3	配置文件 …………………………… 231
	8.5.2 窗体布局文件 ……………… 173	10.4	短信接收与处理源代码 …………… 232
	8.5.3 适配器源代码 ……………… 174	10.5	窗体源代码 ………………………… 233
	8.5.4 窗体源代码 ………………… 177	10.6	本章小结 …………………………… 234
8.6	单位浏览窗体 ……………………… 185		

第三部分 基于互联网的远程温度监测案例

第 11 章 数据编码与处理技术 ……… 236
- 11.1 十六进制字符串的预处理 ……… 236
- 11.2 字节与两个十六进制字符相互转换 ……… 237
- 11.3 字与十六进制字符串相互转换 ……… 238
- 11.4 字节数组与十六进制字符串相互转换 ……… 238
- 11.5 字节数组与 ByteBuffer 对象相互转换 ……… 239
- 11.6 英文字符串的多种编码方法 ……… 239
- 11.7 适用于汉字的 Unicode 编码 ……… 240
- 11.8 随机字节的生成与数字至字节数组的转换 ……… 243
- 11.9 字节的位操作技术 ……… 243
- 11.10 本章小结 ……… 244

第 12 章 数据包的校验技术 ……… 245
- 12.1 枚举类型的定义与说明 ……… 245
- 12.2 累加和校验码的生成与检验 ……… 246
- 12.3 异或校验码的生成与检验 ……… 247
- 12.4 循环冗余校验码的生成与检验 ……… 249
- 12.5 累加求补校验码的生成与检验 ……… 250
- 12.6 结尾码的处理 ……… 252
- 12.7 数据包的综合处理 ……… 254
- 12.8 应用实例 ……… 257
- 12.9 本章小结 ……… 257

第 13 章 通用 TCP 客户机与服务器测试软件 ……… 258
- 13.1 主要功能和技术特点 ……… 258
- 13.2 软件操作 ……… 258
- 13.3 界面布局 ……… 259
- 13.4 配置文件 ……… 264
- 13.5 网络处理类 ……… 265
- 13.6 通用 TCP 客户机与服务器类 ……… 268
  - 13.6.1 各种声明的说明 ……… 268
  - 13.6.2 构造函数 ……… 269
  - 13.6.3 获取 Socket 对象与多线程的启动 ……… 269
  - 13.6.4 数据接收与发送 ……… 269
  - 13.6.5 TcpClientServer 源代码 ……… 270
- 13.7 窗体源代码 ……… 275
- 13.8 TCP 服务器的关键代码 ……… 286
- 13.9 本章小结 ……… 286

第 14 章 I-7013D 模块仿真软件 ……… 287
- 14.1 主要功能和技术特点 ……… 287
- 14.2 软件操作 ……… 287
- 14.3 界面布局 ……… 288
- 14.4 配置文件 ……… 291
- 14.5 窗体源代码 ……… 292
- 14.6 本章小结 ……… 298

第 15 章 I-7013D 模块监测软件 ……… 299
- 15.1 主要功能和技术特点 ……… 299
- 15.2 软件操作 ……… 299
- 15.3 配置文件 ……… 300
- 15.4 参数设置窗体 ……… 301
  - 15.4.1 界面布局 ……… 302
  - 15.4.2 源代码 ……… 304
- 15.5 主窗体 ……… 307
  - 15.5.1 实时温度显示组件 ……… 307
  - 15.5.2 portrait 布局 ……… 309
  - 15.5.3 landscape 布局 ……… 311
  - 15.5.4 源代码 ……… 311
- 15.6 对实物模块的监控 ……… 320
- 15.7 本章小结 ……… 321

参考文献 ……… 322

第 1 章　开发环境搭建与应用入门

Android 是一种基于 Linux 的自由及开放源代码的操作系统，主要用于移动设备，如智能手机和平板电脑，由 Google 公司和开放手机联盟领导及开发。2007 年 11 月，Google 与 84 家硬件制造商、软件开发商及电信营运商组建开放手机联盟，共同研发、改良 Android 系统。随后 Google 以 Apache 开源许可证的授权方式，发布了 Android 的源代码。第一部 Android 智能手机发布于 2008 年 10 月，随后 Android 逐渐扩展到平板电脑及其他领域，如电视、数码相机、游戏机等。2011 年第一季度，Android 在全球的市场份额首次超过塞班系统，跃居全球第一；2013 年第四季度，Android 平台手机的全球市场份额达到 78.1%，2017 年则增至 85.9%。Android 移动应用行业前景好，人才需求大，发展潜力大，学好 Android 应用程序开发可以快速提升自身价值。

1.1　搭建开发环境

Android 开发环境需要 Java SE 的 JDK、Android SDK 和 SDK Platform 三部分。Java SE（Java Platform, Standard Edition）是 Java 平台标准版的简称，用于开发和部署桌面、服务器以及嵌入式设备和实时环境中的 Java 应用程序。Android SDK（Software Development Kit）是 Android 软件开发工具包，是开发 Android 应用所需要的特定的软件包、软件框架、硬件平台、操作系统等建立应用软件的开发工具的集合。

Java SE 可从 https://developer.oracle.com/java 站点根据提示下载，本书下载的是 Windows 32 位版本的 jdk-8u20-windows-i586.exe，安装后 JDK 所在目录为 C:\Program Files\Java\jdk1.8.0_20，通过"控制面板→系统和安全→系统→高级系统设置→高级选项卡→环境变量"设置系统变量名 JAVA_HOME 为 C:\Program Files\Java\jdk1.8.0_20，然后在系统变量 Path 前面添加"%JAVA_HOME%\bin;"，这样就可以在任何路径下调用 JDK 中的可运行程序。

Android SDK 可从 Android 的官方站点 http://developer.android.com/sdk/index.html 下载，本书下载的版本为 adt-bundle-windows-x86-20140702.zip（对应 64 位版本的压缩包名为 adt-bundle-windows-x86_64-20140702.zip），直接解压缩即可使用，其中还包括 Eclipse 工具。解压缩后的目录太长，本书将该目录修改为 Android_ADT32，可在该目录下发现 eclipse 与 sdk 子目录和"SDK Manager.exe"文件。设置环境变量名 ANDROID 为 D:\Android_ADT32\sdk\platform-tools;D:\Android_ADT32\sdk\tools，然后，在系统变量 Path 前添加"%ANDROID%;"即可。

运行 eclipse.exe，在 Eclipse 的 Windows 主菜单上配置 Android SDK，即通过 Windows→Preferences→Android→【Browse】按钮在 SDK Location 处填写 D:\Android_ADT32\sdk，这样就可以使用 Eclipse 开发 Android 应用程序了。

开发 Android 应用程序的 SDK Platform 提供了编写、调试 Android 应用程序所需要的专用类和虚拟设备，而下载的 Android SDK 只包含基本的 SDK、编译工具和例程等，为了开发新版本的 Android 应用程序，需要再通过 Eclipse 环境中的"Android SDK Manager"快捷方式启动"SDK Manager.exe"以下载新版本的 SDK，如图 1-1 所示。通过 SDK 管理器可以更新或安装包，也可以删除已经安装或过时的包。下载齐全后，sdk 目录中将有各个版本的例程、平台工具、编译工具和系统映像等。

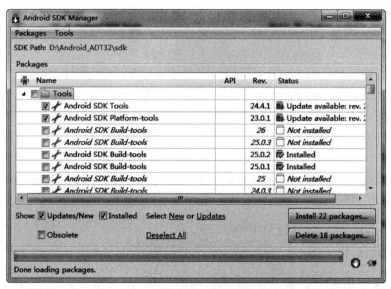

图 1-1　Android SDK 管理器

1.2　创建虚拟设备

Android 应用程序最终是在移动设备上运行的，但一般在 PC 上编写和调试，因而需要一个手机模拟器，即 Android 虚拟设备（Android Virtual Device，AVD）。一般在 Eclipse 环境中点击"Android Virtual Device Manager"（Android 虚拟设备管理器）来创建 AVD，Create 为创建，Edit 为修改 AVD 参数。如图 1-2 所示，AVD Name 中输入 AVD 的名称，可以以操作系统的版本号作为后缀名；Device 栏中可选择设备型号，主要考虑屏幕大小和分辨率；Target 栏中选择 Android 操作系统的版本号；CPU/ABI 栏中选择 ARM 或者 Intel 系列；Skin 栏中选择 AVD 的分辨率，选用动态分辨率即可。另外，Memory Options 设置太大会显示"Cannot set up guest memory"，一般设置为 512MB 即可。

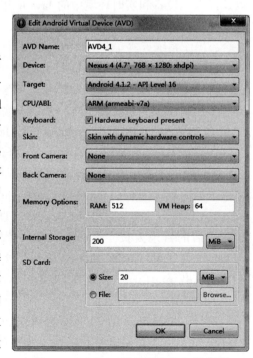

图 1-2　Android 虚拟设备的参数说明

只有下载了相应的 Android SDK 版本，才能在 Target 栏中做出对应选择；同理，只有下载了该版本下的 ARM 或 Intel 系统映像文件，才能在 CPU/ABI 栏中选择其映像文件。采用 ARM 映像文件创建的 AVD，启动速度很慢，通常为 2 分钟以上，不适合课堂或实验室教学演示。一般使用 Intel 映像文件创建 AVD，这样 AVD 的启动速度大约为 30 秒，但要求计算机的 CPU 为 Intel 公司的，并且支持虚拟技术（Virtual Technology），还需要将 CMOS 的"CPU Setup"或 Advanced 栏中的"Intel（R）Virtualization Technology"项设置为 Enabled。这样即可从图 1-1 所示的 Packages 中找到"Intel x86 Emulator Accelerator"（模拟器加速器）安装包，下载安装即可。如果存在无法下载的情况，可以到 Intel 公司的官方网站下载，解压后找到可运行程序 haxm_check，运行后显示"VT support – yes"，表示可以安装使用，点击批处理文件 silent_install 完成安装。

AVD 在调试程序的过程中启动较慢，比较耗时，因而，可以使用实际设备来调试应用程序。在使用 Android 手机调试程序时，确保 PC 端和手机端不要安装"手机助手"，因为该工具会占用调试端口；另外，还需要通过手机的"设置"功能打开"开发者选项"功能，打开"USB 调试"，有的手机还需要将手机的"USB 配置"设置为 RNDIS（USB Ethernet）。部分手机没有"开发者选项"，通过"设置→关于手机→版本号"，连续点击"版本号"7 次即可出现"开发者选项"，不同的手机可能有差异。

1.3　第一个应用程序

搭建好开发环境，设置好 AVD 并准备好 Android 手机后，即可开发第一个应用程序。通过 Eclipse 的 File→New→Android Application Project，打开如图 1-3 所示的对话框，在 Application Name 中输入 Hello，Project Name 和 Package Name 中的内容也将自动变化，将 Package Name 中默认的 example 替换为 ch01。在 Minimum Required SDK 一栏选择 API 14（Android 4.0），使得应用程序兼容的最低版本为 4.0；Target SDK 为 5.1.1 版，这个版本最好不要超过 5.1.1，否则应用程序在安装时，虽然用户已经授权，但是在运行时有的权限仍然每次都需要用户授权，这是 5.1.1 以上版本安全性设置所致。Compile With 一栏使用默认值即可。

图 1-3　创建应用程序

在图 1-3 中点击【Next】按钮，在 Configure Project 对话框中选择"Create Project in Workspace"，使用默认工作空间来保存项目。如果需要更换新的路径，可以清除该选项，通过【Browse】按钮重新选择合适的工作空间。

在 Configure Project 对话框中点击【Next】按钮，进入 Configure Launcher Icon（图标设置）对话框，Foreground 的【Image】按钮用于让用户选择自定义图标，【Clipart】按钮用于选择系统自带的剪贴画作为应用程序图标，【Text】按钮用于使用用户自定义文本作为图标。

在图标设置对话框中点击【Next】按钮进入 Create Activity 对话框，使用默认值。继续点击【Next】按钮，在新的对话框中点击【Finish】完成项目的创建。

在 Package Explorer 窗口中，右击刚刚创建的 Hello 项目→Run As→Android Application，在 AVD 中显示如图 1-4 所示的运行效果。

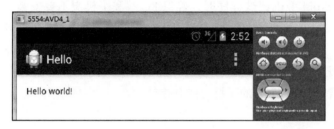

图 1-4　Hello 项目的运行效果

1.4　工作空间与相关文件

通过 Eclipse 的 Window→Preferences→General→Startup and Shutdown→Workspaces，可以发现 Hello 项目所在的目录，如图 1-5 所示，bin 子目录下的 Hello.apk 即编译完成的 Android 应用程序。在图 1-5 中，如果选择"Prompt for workspace on startup"，则可以在启动 Eclipse 时让用户重新选择工作空间。对于多个工作空间的情况，可以通过【Apply】按钮选择一个工作空间，也可以通过【Remove】按钮删除不需要的工作空间。

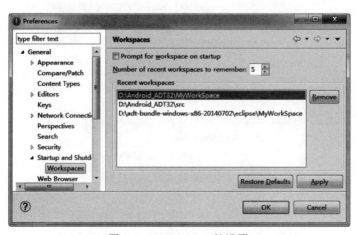

图 1-5　Workspaces 的设置

（1）项目删除与导入

在 Eclipse 环境中右击 Hello 项目可以选择删除（Delete）项目，如果选择"Delete Project…"，还将删除硬盘中的对应文件（不可恢复）。在 Windows 的资源管理器中，可以将 Hello 项目直接复制到 U 盘上，需要时右击 Eclipse 的 Package Explorer 窗口的空白处，选择 Import 导入此项目以便进一步完善。在打开的窗口中依次选择 Android→Existing Android Code Into Workspace，然后在下一个窗口中浏览选择 U 盘上的 Hello 文件夹，选择"Copy projects into workspace"，这样，在导入项目的同时将项目复制到工作空间。

（2）布局文件 activity_main.xml

Hello 项目中的所有文件都是自动生成的。res 目录中存放应用程序的资源文件，其中，layout 子目录下的 activity_main.xml 文件是用于界面设计的布局文件，如图 1-6 所示为图形方式（Graphical Layout）显示的界面布局。

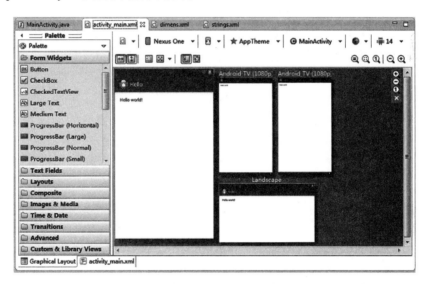

图 1-6　用于界面设计的布局文件

点击图 1-6 底端的 activity_main.xml 标签，则显示文本形式的布局文件：

```
<RelativeLayout xmlns:android="http://schemas.android.com/apk/res/android"
    xmlns:tools="http://schemas.android.com/tools"
    android:layout_width="match_parent"
    android:layout_height="match_parent"
    android:paddingBottom="@dimen/activity_vertical_margin"
    android:paddingLeft="@dimen/activity_horizontal_margin"
    android:paddingRight="@dimen/activity_horizontal_margin"
    android:paddingTop="@dimen/activity_vertical_margin"
    tools:context="com.ch01.hello.MainActivity" >
    <TextView
        android:layout_width="wrap_content"
        android:layout_height="wrap_content"
        android:text="@string/hello_world" />
</RelativeLayout>
```

RelativeLayout 标记表示相对布局方式，这种方式下的组件以上下左右的相对方式排列。android:layout_width 属性表示布局的宽度（下文省略"android:"），属性值 match_parent 让组件充满父容器的其他空间。padding 是当前组件中的内容相对于组件本身边界的上下左右的距离，在这里，TextView 组件就是 RelativeLayout 布局中的内容，paddingLeft 表示 TextView 左边界与 RelativeLayout 左边界的距离。同理，如果 paddingLeft 属性放在 TextView 标记中，则表示文本内容的左边界与文本框的左边界之间的距离。padding 是一种内部距离，而 margin 表示外部距离，如果在 TextView 标记中添加 layout_marginLeft 属性，则表示该组件左侧需要保留的距离。

TextView 标记中的属性值 wrap_content 表示布局元素将根据内容更改大小，对文本框来说，文本多大，文本框就有多大。一个属性可能对应多个属性值，写完"="以后，可以按组合键【Alt + = + ?】来显示所有属性值，从列表中选择一项即可。

（3）距离定义文件 dimens.xml

RelativeLayout 标记的 "@dimen/activity_vertical_margin" 中使用了一个变量 activity_vertical_margin，其数值的大小在 dimens.xml 文件（位于 values 子目录）中定义，这里为 16dp。dp 是一个屏幕分辨率相关的像素单位，是 density-independent pixel 的简称，又称为 dip。Android 鼓励用变量来替代具体的数值，这样可以增加代码的可读性。可以修改这里的具体数值，也可以在 TextView 标记中添加 layout_marginLeft 属性，并通过图 1-6 的预览来观察和理解对应的属性。从 dimens.xml 文件中可以看出，XML 使用左标记"<!--"和右标记"-->"作为注释标记，凡是在这两个标记之间的内容都是注释。

```
<resources>
    <!-- Default screen margins, per the Android Design guidelines. -->
    <dimen name="activity_horizontal_margin">16dp</dimen>
    <dimen name="activity_vertical_margin">16dp</dimen>
</resources>
```

（4）字符串定义文件 strings.xml

TextView 标记中的 Text 属性值为 "@string/hello_world"，hello_world 是一个字符串变量，定义在 strings.xml 文件（位于 values 子目录）中，这里为"Hello world!"，其中，"Hello"为应用程序的名称，"Settings"为菜单项。与图 1-6 一样，strings.xml 也有可视化标签"Resources"，此时可分别使用【Add】、【Remove】、【Up】和【Down】按钮添加、删除、上移和下移字符串定义。

```
<?xml version="1.0" encoding="utf-8"?>
<resources>
    <string name="app_name">Hello</string>
    <string name="hello_world">Hello world!</string>
    <string name="action_settings">Settings</string>
</resources>
```

（5）菜单定义文件 main.xml

menu 子目录下的 main.xml 是一个菜单文件，其内容如下所示。其中的"item"标记表示一个菜单项，每个菜单项都有一个独立的"item"标记。id 属性是对象的名称，程序根据这个 id 号处理菜单请求。orderInCategory 属性表示菜单顺序，数字小的排在数字大的菜单项的前面。showAsAction 属性表示菜单项的显示方式，"never"值表示只在溢出列表中显示菜单，"always"值表示总是在应用程序的标题栏显示菜单，"ifRoom"值则表示如果有空间就在标题栏显示菜单。title 属性规定菜单项的显示内容，由字符串变量 action_settings 表示，在 string.xml 中定义。

```xml
<menu xmlns:android="http://schemas.android.com/apk/res/android"
    xmlns:tools="http://schemas.android.com/tools"
    tools:context="com.ch01.hello.MainActivity" >
    <item
        android:id="@+id/action_settings"
        android:orderInCategory="100"
        android:showAsAction="never"
        android:title="@string/action_settings"/>
</menu>
```

（6）图标与图片文件夹

res 下的 drawable-hdpi、drawable-ldpi 等目录中存放的是图标文件，其他图片文件也可以存放在这些目录中。

（7）项目配置文件 AndroidManifest.xml

AndroidManifest.xml 文件包含了 Android 程序中所使用的 Activity（窗体）、Permission（权限）、Service（服务）和 Receiver（广播接收器）等，在该文件中，可以通过 <intent-filter> 设置默认启动的 Activity。"uses-sdk"标记规定了本程序适应的最小和最大 SDK 版本，可以在创建应用程序的时候进行设置（如图 1-3 所示），也可以在这里进行调整。

```xml
<?xml version="1.0" encoding="utf-8"?>
<manifest xmlns:android="http://schemas.android.com/apk/res/android"
    package="com.ch01.hello"
    android:versionCode="1"
    android:versionName="1.0" >
    <uses-sdk
        android:minSdkVersion="14"
        android:targetSdkVersion="22" />
    <application
        android:allowBackup="true"
        android:icon="@drawable/ic_launcher"
        android:label="@string/app_name"
        android:theme="@style/AppTheme" >
        <activity
            android:name=".MainActivity"
            android:label="@string/app_name" >
            <intent-filter>
                <action android:name="android.intent.action.MAIN" />
                <category android:name="android.intent.category.LAUNCHER" />
            </intent-filter>
        </activity>
    </application>
</manifest>
```

有时该文件中会出现提示性警告，如目标版本 targetSdkVersion 没有采用当前的最新版本，如图 1-7 第 7 行所示。这时可以将光标移动到提示处，按组合键【Ctrl+1】，在菜单中选择"Disable Check in This File Only"即可清除提示性警告。

（8）资源索引文件与源代码文件

gen 目录中的 R.java 文件是创建 Android 程序时自动生成的，用来定义 Android 程序中所有资源的索引，在 Java 源文件中编写代码时可以直接通过该索引访问各种资源。

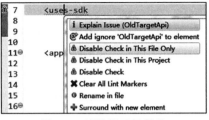

图 1-7　清除提示性警告

src 目录下的 MainActivity.java 就是 Android 应用程序的源代码文件，程序启动时，首先执行其中的 onCreate 方法，调用 setContentView 方法完成布局的设置，然后调用 onCreateOptionsMenu 方法初始化菜单。onOptionsItemSelected 是一个选择菜单项事件的侦听方法，需要通过菜单项执行的代码可在这里完成。

1.5　程序的调试方法

Android 应用程序发生错误时会直接崩溃，而编写代码与运行程序又处于两个不同的空间，因而，调试 Android 应用程序相对于可视化桌面应用程序来说比较困难。但是，以下方法可以用来方便地调试 Android 应用程序。

（1）使用 Toast 类显示浮动消息

对于可能发生错误的地方，可以调用 Toast 类的静态方法 makeText 来显示一个浮动消息，输出需要观察的信息。Toast 类的 makeText 方法有三个参数，第一个参数为上下文相关环境，第二个参数是需要输出的信息，第三个参数是浮动消息显示的时间长度。如下代码显示短时浮动消息"OK"，可将此代码写入菜单项中观察运行效果。

```
Toast.makeText(this, "OK", Toast.LENGTH_SHORT).show();
```

（2）使用 Log 类输出调试信息

也可以采用 Log 类来获取程序运行时的日志信息，该类位于 android.util 包，提供了一些方法以输出日志信息。以 d 方法为例，其使用方式一般为 Log.d(tag, msg)，第一个参数是日志标签，第二个参数为输出的具体信息。在菜单事件处理函数中输入代码，如图 1-8 所示，此时，Log 类下有红色波浪线，将鼠标移动到该处，显示不能解析 Log 类，选择导入 android.util 包，红色波浪线自动消失。程序运行时，可以在 LogCat 窗口观察到输出的调试信息。

```
public boolean onOptionsItemSelected(MenuItem item) {
    // Handle action bar item clicks here. The action bar will
    // automatically handle clicks on the Home/Up button, so long
    // as you specify a parent activity in AndroidManifest.xml.
    int id = item.getItemId();
    if (id == R.id.action_settings) {
        Toast.makeText(this, "OK", Toast.LENGTH_SHORT).show();
        Log.d("Debug", "Info");
        return true;
    }
    return super.onOptionsItemSelected(item);
}
```

图 1-8　调试信息的输出

(3) 使用断点观察变量

在图 1-8 的第 32 行左边条双击，可以设置一个圆形断点标志。如果按照 1.3 节中的 "Run As" 方法运行程序，断点处不会暂停。为了观察断点处的数据变化情况，应该选择 "Debug As"，则程序运行到第 32 行时暂停，图 1-9 的标签将自动显示为 "Debug" 有效，将鼠标移动到变量 id 处，可以观察其数值。需要继续运行程序，可以点击主菜单 Run 下的 Resume 子菜单（对应功能键【F8】）。点击 Java 标签，将显示 MainActivity 类中的方法与事件等，以用于在代码中快速定位。DDMS（Dalvik Debug Monitor Server）标签是 Android 开发平台自带的一个调试工具，用于提供端口传输服务、在当前连接调试设备上的截图、设备的线程和堆信息、LogCat 信息、进程状态信息和广播信息等。

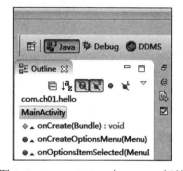

图 1-9　Java、Debug 与 DDMS 标签

(4) 设置调试设备

Android 应用程序最终是要部署在 Android 设备上的，而且，AVD 启动比较耗时，需要生成系统映像，再加载应用程序；有的程序在 AVD 上不方便调试，如 TCP 服务器程序。这时，可以使用 Android 设备来调试程序，因为只需要安装应用程序，因而速度很快，而且能够完成所有的功能。通过主菜单 Run→Debug Configuration，打开如图 1-10 所示的对话框，左侧选择需要调试的项目，这里为 "Hello"，右侧选择 "Target" 标签，可以点击第一个单选按钮 "Always prompt to pick device" 或者第三个单选按钮，以选择 Android 设备。

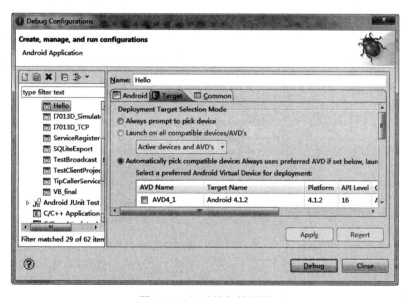

图 1-10　调试设备的设置

(5) 选择 Android 设备

在图 1-10 中点击【Debug】按钮，显示如图 1-11 所示的对话框，上半部分为 Android

设备列表，下半部分为 AVD 列表，点击上面运行的 Android 设备，再点击【OK】按钮，即可利用 Android 设备进行调试。在调试过程中，在原 Android 设备或 AVD 中运行的以前版本的软件可能会影响调试并导致出现错误，应关闭以前版本软件后再进行调试。另外，如果设备中已经安装了高版本编译器编译的软件，而调试软件是低版本编译器编译的，则不能安装，应该先卸载高版本软件，才能调试此低版本软件。

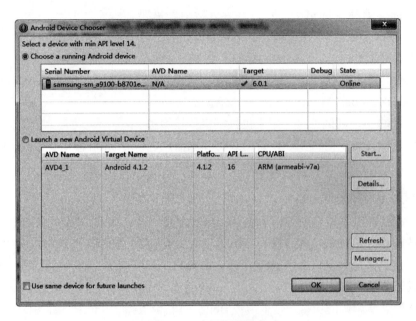

图 1-11　选择调试设备

在对项目进行任何修改（包括代码、res 目录下的资源文件或 AndroidManifest.xml 配置文件）后，再次运行还有可能显示原有的资源文件甚至代码逻辑，这时应该尝试 Clean 操作，即通过主菜单 Project 下的"Clean..."子菜单，选择"Clean projects selected below"单选项，再在列表中选择需要清除的项目，最后点击【OK】按钮以清除项目中 bin 目录下先前编译好的 apk 和 dex 等文件。这样再次运行软件，即可生成新的 apk 文件。还可以删除项目中 gen 目录下的 R.java 文件，让系统重新自动创建，以达到刷新资源（图片或字符串定义等）的目的。

1.6　本章小结

开发 Android 应用程序首先需要根据需求分析构建界面，完成布局，然后编写代码以实现相关功能，最后进行调试完善。相对于 Android Studio 开发环境来说，Eclipse 开发环境更容易入门，提示更好，也更方便使用。本章介绍了 Android Eclipse 开发环境的搭建，创建调试用的 AVD，然后通过一个 Hello 项目介绍了如何开发应用程序，对于所涉及的文件进行了详细的讲解，最后介绍了程序调试的多种方法，特别是使用 Android 设备的详细方法。在 Hello 项目中只对菜单项进行了简单的处理，尚未进行真正意义上的编程。下一章将介绍应用程序的常用开发组件。

第一部分

编程基础与技巧

第 2 章 常用开发组件

第 3 章 常用技术

第 4 章 Intent 的综合应用

第 2 章 常用开发组件

与可视化桌面应用程序一样，Android 应用程序也有各种常用的组件和容器。常用容器主要包括相对布局管理器 RelativeLayout 和线性布局管理器 LinearLayout，其间可以存放文本框 TextView、可编辑文本框 EditText 和按钮 Button 等，也可放置子布局管理器；单选按钮容器 RadioGroup 则用来存放单选按钮 RadioButton，使得该容器中的单选按钮互斥。

2.1 常见属性

组件的属性采用"android: 属性名 =" 属性值 ""的形式，1.4 节已经介绍了几个最基本的属性，layout_width 用于设置组件在父组件中的宽度，取值"match_parent"表示当前组件的宽度填满父组件的剩余空间，"wrap_content"表示当前组件的宽度正好能够显示其中的文本或子组件，这里也可以使用以"dp"为单位的具体数值，如"16dp"。"fill_parent"目前仍在使用，但是，API level 8 以后使用"match_parent"来替代"fill_parent"。layout_height 属性用来设置组件在父组件中的高度。

id 属性用于唯一地标识组件，因而不能重复，Java 源代码中使用该 id 标识处理事件或动态更新该组件的其他属性信息。id 属性值的形式为"@+id/id 值"。第 1 章中的 Hello 项目中的 TextView 组件没有 id 标识，因为其文本内容是静态指定的，不需要在源代码中动态变化。

textSize 属性用来指定 TextView 相关组件中的文本字号大小，单位为 sp（scaled pixel），如"24sp"。正如距离可以在 dimens.xml 中定义，字体大小也可以在其中定义。

textStyle 属性用来表示字体显示的方式，normal 表示正常显示，bold 表示加粗显示，italic 表示倾斜显示，如果既要加粗又要倾斜，可以使用"|"标识进行叠加。

gravity 属性用来表示组件或布局内部内容的对齐方式，比如一个文本框的长度是固定的，属性值"center"表示无论数据长短都在此文本框里居中显示。此外，还有"top""bottom""left""right"等属性值。而 layout_gravity 表示组件本身在其容器中的对齐方式。

background 属性用来设置组件的背景颜色或背景图像。如果采用颜色来表示，使用 RGB 格式，1 字节（2 位十六进制数）表示 1 种颜色，依次表示 R（Red，红色）、G（Green，绿色）和 B（Blue，蓝色），前面用前导字符"#"。例如，纯蓝色对应的颜色值为"#0000FF"。颜色也可采用 8 位十六进制数表示，最左边两位表示透明度，从"00"（完全透明）到"FF"（完全不透明）。如果采用图像，使用"@drawable/ 图像名"，图像存于 res 目录下的某个 drawable 子目录下，图像名一般采用小写，系统将自动查找图像位置并识别扩展名。

  layout_below 是相对布局 RelativeLayout 中组件（或子布局）的属性，以属性值"@id/layoutTitle"为例，表示本组件（或子布局）位于 layoutTitle 组件（或子布局）的下方，同理，还有 layout_top、layout_left 和 layout_right 等。

  orientation 是线性布局 LinearLayout 中的属性，属性值"horizontal"表示该布局中的组件或子布局是以水平即横向从左到右排列的，而属性值"vertical"表示它们从上到下垂直排列。

  layout_weight 属性只有在 LinearLayout 中才有效，表示其中的各个组件所占的比例。假如 LinearLayout 是横向排列的（orientation 属性值为"horizontal"），其间有三个按钮，按钮中的 layout_weight 属性值都为 1，表示这三个按钮横向均匀排列，但是要求 layout_width 属性值为 0dp。同理，对于纵向排列，要求 layout_height 的属性值为 0dp，这时可以使用 layout_weight 属性值来指定纵向排列的组件或子布局所占空间比例。

  对于一个应用程序来说，窗体中出现的字体常常需要统一规划，如果在每个窗体布局文件中重复书写相关属性，则不便管理和维护。这时，可以使用存放 Android 主题与样式的 styles.xml 文件（位于 values 目录下）进行统一定义。例如，以下对文本进行了统一定义，组件大小为刚好包含文本，居中显示，其中"?android:attr/textAppearanceMedium"表示引用主题属性 textAppearanceMedium，即采用中号字体。

```
<style name="TextStyle">
    <item name="android:layout_width">wrap_content</item>
    <item name="android:layout_height">wrap_content</item>
    <item name="android:textAppearance">?android:attr/textAppearanceMedium </item>
    <item name="android:gravity">center</item>
</style>
```

  则在布局文件中，以 TextView 为例，只要设置 id 属性即可，其他属性统一用定义好的 style 来引用，"@style/TextStyle"表示引用 styles.xml 文件中的 TextStyle 样式。

```
<TextView
    android:id="@+id/txtPort"
    style="@style/TextStyle" />
```

  上文介绍的 background 属性可以用来在布局文件中设置窗体背景，但是如果一个应用程序中有多个窗体使用同一背景，这种方式就不太合适。这时，可以在 styles.xml 文件的应用程序主题 AppTheme 中进行统一设置，这里 background 是背景文件名，这样所有窗体背景都被统一，布局文件中将不再需要进行任何设置。

```
<style name="AppTheme" parent="AppBaseTheme">
    <item name="android:windowBackground">@drawable/background</item>
</style>
```

2.2 EditText 组件与菜单

  EditText 组件允许用户输入文本，主要用于输入用户名、账号、密码以及运行参数等。当该组件获得焦点时，下方会显示一个虚拟键盘，为了更加高效地输入数据，又经常需要对键盘进行限制，如只能输入整数、字母或时间等，这些都可以通过属性进行设置。

（1）EditText 的常用属性

在 xml 布局文件中，<EditText> 标记用于可编辑文本框，在这个标记之内，可以根据需要设置各种属性，其中常用属性如表 2-1 所示。

表 2-1 EditText 组件的常用属性

属　　性	取　　值	说　　明
inputType	"number"	仅允许输入数字
	"numberDecimal""	仅允许输入十进制数值
	"numberPassword"	仅允许输入数字密码
	"text"	可以输入任何文本
	"textPassword"	可以输入任何文本密码
	"phone"	仅允许输入电话号码相关字符
	"datetime"	仅允许输入日期和时间
digits	允许的字符集合	设置允许输入的字符集合，如"YyNn"
hint	提示字符串	如"######"可提示输入 6 位整数
maxLength	正整数	设置可输入的最大字符数
lines	正整数	设置输入文本框的行高
minLines	正整数	设置输入文本框的最小行高
maxLines	正整数	设置输入文本框的最大行高

（2）处理 TextChanged 事件

完整地掌握一个组件的使用，一般需要掌握其属性的设置方法及其事件处理。为了操作 EditText 对象，获取 EditText 对象中的数据，处理 EditText 事件时必须声明全局 EditText 对象，在 Activity 对象的 onCreate 事件中对其初始化，并监控文本变化（详见示例代码及注释）。

侦听数据变化时，需要通过 EditText 对象的 addTextChangedListener 方法实现 TextWatcher 接口中的三个虚拟函数，当文本发生变化时，beforeTextChanged 为文本变化之前的事件，s 表示变化之前的文本，start 表示开始位置，将有 after 个字符替换原有的 count 个字符。

```
public void beforeTextChanged(CharSequence s, int start, int count, int after)
```

onTextChanged 为文本变化中的事件，在 beforeTextChanged 之后执行，s 表示替换完成后的文本，start 为开始位置，count 长度的字符将先前 before 长度的字符替换掉。

```
public void onTextChanged(CharSequence s, int start, int before, int count)
```

afterTextChanged 为文本变化后的事件，s 中存有变化后的最终文本。

```
public void afterTextChanged(Editable s)
```

（3）使用菜单动态选择输入类型

创建一个 EditText 项目，将菜单 main.xml 文件修改为如下三个菜单项：电话号码、日期时间和普通文本。菜单项的 title 定义详见 strings.xml 文件，这里不再列出。菜单出现的先后顺序通过 orderInCategory 属性进行控制。

```
<menu xmlns:android="http://schemas.android.com/apk/res/android"
    xmlns:tools="http://schemas.android.com/tools"
    tools:context="com.ch02.edittext.MainActivity" >
    <item
        android:id="@+id/action_phone"
        android:orderInCategory="100"
        android:showAsAction="never"
        android:title="@string/action_phone"/>
     <item
        android:id="@+id/action_datetime"
        android:orderInCategory="101"
        android:showAsAction="never"
        android:title="@string/action_datetime"/>
    <item
        android:id="@+id/action_text"
        android:orderInCategory="102"
        android:showAsAction="never"
        android:title="@string/action_text"/>
</menu>
```

（4）主窗体布局文件 activity_main.xml

利用上一章 Hello 项目的框架，在 TextView 前面添加一个 EditText 标记，通过 layout_alignBaseline 属性设置 EditText 与 TextView 对象显示在同一水平线上，且 EditText 对象最多容纳 16 个字符。TextView 标记通过 layout_toRightOf 属性使得该对象居于可编辑文本框的右侧，并采用中等字号动态显示输入的字符数。

```
<RelativeLayout xmlns:android="http://schemas.android.com/apk/res/android"
    xmlns:tools="http://schemas.android.com/tools"
    android:layout_width="match_parent"
    android:layout_height="match_parent"
    android:paddingBottom="@dimen/activity_vertical_margin"
    android:paddingLeft="@dimen/activity_horizontal_margin"
    android:paddingRight="@dimen/activity_horizontal_margin"
    android:paddingTop="@dimen/activity_vertical_margin"
    tools:context="com.ch02.edittext.MainActivity" >
    <EditText
        android:id="@+id/edInput"
        android:layout_width="wrap_content"
        android:layout_height="wrap_content"
        android:hint="@string/strHint"
        android:layout_alignBaseline="@+id/txtCount"
        android:maxLength="16"
        android:inputType="text" />
    <TextView
        android:id="@+id/txtCount"
        android:layout_width="wrap_content"
        android:layout_height="wrap_content"
        android:layout_toRightOf="@+id/edInput"
        android:textAppearance="?android:attr/textAppearanceMedium"
        android:hint="@string/strCount"/>
</RelativeLayout>
```

（5）源代码 MainActivity.java

本程序在窗体的 onCreate 事件中注册事件变化侦听器，在 afterTextChanged 事件中获取 EditText 对象中的有效文本长度。字符的输入类型通过菜单选择，由于有三项菜单，因而在 onOptionsItemSelected 事件中采用 switch 语句来处理分支。

```java
package com.ch02.edittext;
import android.app.Activity;
import android.os.Bundle;
import android.text.Editable;
import android.text.InputType;
import android.text.TextWatcher;
import android.view.Menu;
import android.view.MenuItem;
import android.widget.EditText;
import android.widget.TextView;
public class MainActivity extends Activity {
    EditText edInput;                   //声明EditText对象
    TextView txtCount;                  //声明TextView对象，对输入的文本动态计数
    @Override
    protected void onCreate(Bundle savedInstanceState) {
        super.onCreate(savedInstanceState);
        setContentView(R.layout.activity_main);
        txtCount = (TextView)findViewById(R.id.txtCount);      //初始化
        edInput = (EditText)findViewById(R.id.edInput);         //初始化
        edInput.addTextChangedListener(new TextWatcher(){
            @Override
            public void beforeTextChanged(CharSequence s, int start, int count,
                    int after) {
                return;
            }
            @Override
            public void onTextChanged(CharSequence s, int start, int before,
                    int count) {
                return;
            }
            @Override
            public void afterTextChanged(Editable s) {
                //更新当前字符长度
                txtCount.setText(Integer.toString(s.length()));
            }
        });
    }
    @Override
    public boolean onCreateOptionsMenu(Menu menu) {
        getMenuInflater().inflate(R.menu.main, menu);
        return true;
    }
    @Override
    public boolean onOptionsItemSelected(MenuItem item) {
        int id = item.getItemId();
        switch(id){
        case R.id.action_datetime:           //动态设置日期时间输入类型
            edInput.setInputType(InputType.TYPE_CLASS_DATETIME);
            return true;
```

```
        case R.id.action_phone:    //动态设置电话号码输入类型
            edInput.setInputType(InputType.TYPE_CLASS_PHONE);
            return true;
        case R.id.action_text:      //动态设置文本输入类型
            edInput.setInputType(InputType.TYPE_CLASS_TEXT);
            return true;
        }
        return super.onOptionsItemSelected(item);
    }
}
```

（6）运行效果

在菜单项中选择"Date Time"，虚拟键盘中只能输入数字与"-/:"日期或时间分隔符，且文本的长度在编辑框右侧动态显示，超过 16 个字符将不再接收输入，运行效果如图 2-1 所示。

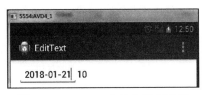

图 2-1　EditText 中输入日期效果

2.3　Button 组件

Button 组件直接接收用户指令并做出响应。Button 注册用户的 Click 事件主要有两种方法，即直接在布局文件中指定 Button 组件的 onClick 事件处理函数，或者采用 EditText 组件的方式并利用代码进行注册，该方法有若干变体。

（1）在布局文件中指定事件处理函数

创建项目 ButtonXml，布局文件 activity_main.xml 中的 Button 标记如下所示，这里指定了对象名称为 btXml，Click 事件的处理函数为 onClickXml，该函数名为用户自定义的，对于多个按钮，可以使用相同的处理函数，根据唯一的 id 进行区分处理。

```
<Button
    android:id="@+id/btXml"
    android:layout_width="wrap_content"
    android:layout_height="wrap_content"
    android:onClick="onClickXml"
    android:text="@string/strXml" />
```

主窗体源代码文件中的 onClickXml 函数定义如下，btXml 作全局声明，并在 onCreate 中进行初始化，这里利用 switch 分支来处理多个按钮的情况，用布局文件中定义的 id 号进行区分。这种处理 Click 事件的方法简单方便。

```
public void onClickXml(View v){
    int id = v.getId();
    switch(id){
    case R.id.btXml:
        btXml.setText("btXml");
        break;
    default:
        break;
    }
}
```

（2）在 MainActivity 类中继承 OnClickListener 接口

创建 ButtonInterface 项目，在主窗体源代码 MainActivity.java 中实现 OnClickListener 接口，如图 2-2 所示，在第 11 行的 Activity 后面写入"implements OnClickListener"，这时，OnClickListener 下面出现提示错误的红色波浪线，将鼠标移至此处，在出现的浮动窗口选择导入"android.view.View.OnClickListener"，MainActivity 下又出现提示错误的红色波浪线，选择第一项修复方案，MainActivity 中出现一个 onClick 事件处理函数，可以将上文 onClickXml 中的代码直接复制过来。另外，还需要在 onCreate 方法中注册 OnClick 侦听器，传入 this 作为参数，用来表示 OnClick 事件处理方法在 Activity 对象中。如果不进行设置，则点击按钮无效。

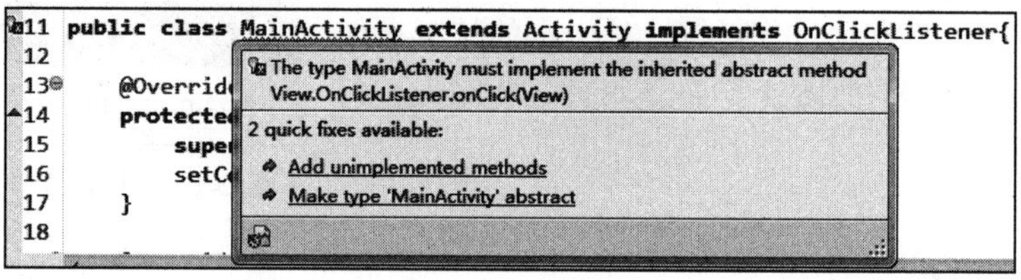

图 2-2　直接实现 OnClickListener 接口

```
protected void onCreate(Bundle savedInstanceState) {
    super.onCreate(savedInstanceState);
    setContentView(R.layout.activity_main);
    btInterface = (Button)findViewById(R.id.btInterface);
    btInterface.setOnClickListener(this);
}
```

直接实现接口函数的方法也比较方便，不需要说明每个 Button 标记的 onClick 属性，也可针对多个按钮，用 id 进行区分即可。但是，对于 Activity 需要实现多个接口的情况，就显得比较凌乱。

（3）使用自定义嵌入类实现侦听

创建 ButtonClass 项目，用一个自定义嵌入类 ButtonListener 实现 OnClickListener 接口，然后在 onCreate 方法中声明 ButtonListener 对象 listener，最后通过 btClass 对象注册 OnClick 侦听器，用来表示 OnClick 事件处理方法在 listener 对象中。

```
protected void onCreate(Bundle savedInstanceState) {
    super.onCreate(savedInstanceState);
    setContentView(R.layout.activity_main);
    btClass = (Button)findViewById(R.id.btClass);
    ButtonListener listener = new ButtonListener();
    btClass.setOnClickListener(listener);      //多个按钮可使用一个侦听对象
}
class ButtonListener implements OnClickListener{
    @Override
    public void onClick(View v) {
        int id = v.getId();
        switch(id){
        case R.id.btClass:
```

```
            btClass.setText("btClass");
            break;
        default:
            break;
        }
    }
}
```

（4）直接实现侦听

对于窗口中只有单个命令按钮用于测试的情况，可以直接实现接口的侦听，其中 onClick 中的代码都相似，这里不再赘述。

```
btClass.setOnClickListener(new OnClickListener(){
    @Override
    public void onClick(View v) {
        // TODO Auto-generated method stub
    }});
```

（5）ButtonClass 项目的主要源代码

以 ButtonClass 项目为例，主窗体 MainActivity.java 中的主要源代码如下所示。

```
package com.ch02.buttonclass;
import android.app.Activity;
import android.os.Bundle;
import android.view.Menu;
import android.view.MenuItem;
import android.view.View;
import android.view.View.OnClickListener;
import android.widget.Button;
public class MainActivity extends Activity {
    Button btClass;
    @Override
    protected void onCreate(Bundle savedInstanceState) {
        super.onCreate(savedInstanceState);
        setContentView(R.layout.activity_main);
        btClass = (Button)findViewById(R.id.btClass);
        ButtonListener listener = new ButtonListener();
        btClass.setOnClickListener(listener);     //多个按钮可使用一个侦听对象
    }
    class ButtonListener implements OnClickListener{
        @Override
        public void onClick(View v) {
            // TODO Auto-generated method stub
            int id = v.getId();
            switch(id){
            case R.id.btClass:
                btClass.setText("btClass");
                break;
            default:
                break;
            }
        }
    }
}
```

（6）运行效果

以 ButtonClass 项目为例，其运行效果如图 2-3 所示，初始按钮上的文本为 Class，点击按钮以后，文本变为按钮的对象名称 btClass。

2.4 ToggleButton 组件

ToggleButton 组件具有命令按钮和状态表达两种功能，与开关一样有两种状态，可做出两种控制，同时在

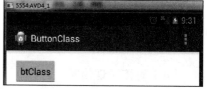

图 2-3　ButtonClass 运行效果

按钮表面还能做出对应的文字与图片提示。本节通过项目 ToggleButton 来举例说明。

（1）关键属性

关键属性见 activity_main.xml 布局文件所示，其中 checked 属性有"true"和"false"两种状态，如果不设置该属性值，则默认为 false。textOn 属性用来指定状态为 true 时的按钮文本，如"ON"；textOff 属性用来指定状态为 false 时的按钮文本，如"OFF"。button 属性用来设置按钮为 true 或 false 时的图片，此方式下默认图片靠左显示，与 drawableLeft 属性具有相同的功能，另外还可以设置属性 drawableRight，使得图片靠右显示，而 drawableTop 使得图片靠上显示，drawableBottom 使得图片靠下显示。

```
<RelativeLayout xmlns:android="http://schemas.android.com/apk/res/android"
    xmlns:tools="http://schemas.android.com/tools"
    android:layout_width="match_parent"
    android:layout_height="match_parent"
    android:paddingBottom="@dimen/activity_vertical_margin"
    android:paddingLeft="@dimen/activity_horizontal_margin"
    android:paddingRight="@dimen/activity_horizontal_margin"
    android:paddingTop="@dimen/activity_vertical_margin"
    tools:context="com.ch02.togglebutton.MainActivity" >
    <ToggleButton
        android:id="@+id/tb"
        android:layout_width="wrap_content"
        android:layout_height="wrap_content"
        android:button="@drawable/toggle_button"
        android:textOn="ON"
        android:textOff="OFF"
        android:background="@null"
        android:checked="false"/>
</RelativeLayout>
```

（2）用 XML 文件来表达图片

button 属性中的图片用 toggle_button.xml 文件来指定，其中的 selector 标记可以表示 state_checked 为 "true" 和 "false" 时的图片。将 red.jpg 和 green.jpg 复制到资源文件夹下的 drawable-hdpi 目录中，右击该目录并选择 "Refresh" 以刷新，这两个文件即可显示在目录中，本 XML 文件也位于此目录，系统将自动定位本文件及所使用的图片。

```
<selector xmlns:android="http://schemas.android.com/apk/res/android">
    <item android:state_checked="false"
        android:drawable="@drawable/red" />
```

```xml
        <item android:state_checked="true"
            android:drawable="@drawable/green" />
</selector>
```

(3) 主要源代码

ToggleButton 也有 Button 的特性，使用"直接实现侦听"的方法，主窗体的主要源代码如下所示。如果 ToggleButton 对象 tb 为 true，则通过 Toast 浮动消息输出"true state"；如果为 false，则输出"false state"，在实际工程项目中可以根据需要调用相关函数。

由于 Toast 的 makeText 的第一个参数为上下文环境，如果使用 this，则表示 tb 对象，显然，不能在 tb 对象里输出浮动消息。因而，需要在 onCreate 中保存 MainActivity 对象的 this 环境，然后在侦听处理事件中传给 makeText 函数。

```java
package com.ch02.togglebutton;
import android.app.Activity;
import android.content.Context;
import android.os.Bundle;
import android.view.Menu;
import android.view.MenuItem;
import android.view.View;
import android.view.View.OnClickListener;
import android.widget.Toast;
import android.widget.ToggleButton;
public class MainActivity extends Activity {
    ToggleButton tb;
    @Override
    protected void onCreate(Bundle savedInstanceState) {
        super.onCreate(savedInstanceState);
        setContentView(R.layout.activity_main);
        final Context context = this;
        tb = (ToggleButton)findViewById(R.id.tb);
        tb.setOnClickListener(new OnClickListener(){
            @Override
            public void onClick(View v) {
                // TODO Auto-generated method stub
                if(tb.isChecked())
                    Toast.makeText(context, "true state",
                        Toast.LENGTH_SHORT).show();
                else
                    Toast.makeText(context, "false state",
                        Toast.LENGTH_SHORT).show();
            }});
    }
}
```

(4) 运行效果

程序运行效果如图 2-4 所示，初始状态为"红灯 OFF"，点击按钮后显示为"绿灯 ON"，同时显示"true state"浮动消息。

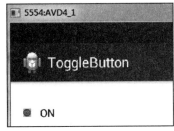

图 2-4　ToggleButton 运行效果

2.5 CheckBox 组件

CheckBox 组件又称复选框，通常附带一个文本标识。若干个该组件放在一个容器中，允许选择多项。CheckBox 组件一般有三种状态：选中、未选中和禁选。

（1）关键属性

CheckBox 组件也有 onClick 属性，可以指定一个事件侦听函数，这里的两个组件属性值都为 onClickCheckBox，用 id 区分事件对象。checked 属性用来设置初始布尔值，默认为 false。CheckBox 组件的标识文本默认居于右侧，可以通过设置 button 和 drawableRight 属性，使得复选框居于右侧，详见如下 activity_main.xml 文件。

```
<RelativeLayout xmlns:android="http://schemas.android.com/apk/res/android"
    xmlns:tools="http://schemas.android.com/tools"
    android:layout_width="match_parent"
    android:layout_height="match_parent"
    android:paddingBottom="@dimen/activity_vertical_margin"
    android:paddingLeft="@dimen/activity_horizontal_margin"
    android:paddingRight="@dimen/activity_horizontal_margin"
    android:paddingTop="@dimen/activity_vertical_margin"
    tools:context="com.ch02.checkbox.MainActivity" >
    <CheckBox
        android:id="@+id/chk1"
        android:onClick="onClickCheckBox"
        android:layout_width="wrap_content"
        android:layout_height="wrap_content"
        android:checked="false"
        android:text="Check1" />
    <ToggleButton
        android:id="@+id/tb"
        android:onClick="onClickToggle"
        android:layout_width="wrap_content"
        android:layout_height="wrap_content"
        android:layout_alignBaseline="@id/chk1"
        android:layout_marginLeft="10dp"
        android:background="@null"
        android:textOn="checked"
        android:textOff="unchecked"
        android:layout_toRightOf="@id/chk1" />
    <CheckBox
        android:id="@+id/chk2"
        android:onClick="onClickCheckBox"
        android:layout_width="wrap_content"
        android:layout_height="wrap_content"
        **android:button="@null"**
        **android:drawableRight="?android:attr/listChoiceIndicatorMultiple"**
        android:layout_below="@id/chk1"
        android:checked="false"
        android:text="check2" />
</RelativeLayout>
```

（2）常用方法

CheckBox 组件的 isChecked() 方法读取对象的当前布尔值，setChecked 方法设置布尔值，

setEnabled 方法设置是否禁选（false 表示禁选）。

（3）事件处理

事件处理函数除了通过 onClick 属性指定外，也可直接实现侦听，当选中状态发生变化即触发事件。与 Button 组件一样，还可在此基础之上做一些变形实现。其基本源代码框架如下所示，chk1 是 CheckBox 对象。

```
chk1.setOnClickListener(new OnClickListener (){
    @Override
    public void onClick(View v) {
        // TODO Auto-generated method stub
    }});
```

（4）主要源代码

MainActivity.java 中的源代码实现以上所有功能测试，两个 CheckBox 对象的文本标识左右各异，可以同时选中，由 chk1 对象实现事件响应，由 Toast 浮动消息显示选中的对象及其状态。ToggleButton 对象 tb 控制 chk1 是否选中，chk2 是否禁选：tb 为 true 时，chk1 选中，chk2 不禁选；tb 为 false 时，chk1 未选中，chk2 禁选。

```
package com.ch02.checkbox;
import android.app.Activity;
import android.content.Context;
import android.os.Bundle;
import android.view.Menu;
import android.view.MenuItem;
import android.view.View;
import android.widget.CheckBox;
import android.widget.Toast;
import android.widget.ToggleButton;
public class MainActivity extends Activity {
    CheckBox chk1;
    CheckBox chk2;
    ToggleButton tb;
    Context context;
    @Override
    protected void onCreate(Bundle savedInstanceState) {
        super.onCreate(savedInstanceState);
        setContentView(R.layout.activity_main);
        chk1 = (CheckBox)findViewById(R.id.chk1);
        chk2 = (CheckBox)findViewById(R.id.chk2);
        tb = (ToggleButton)findViewById(R.id.tb);
        chk2.setEnabled(false);            //初始设置
        context = this;                    //保存上下文环境
    }
    public void onClickCheckBox(View v){
        int id = v.getId();
        switch(id){
        case R.id.chk1:
            if(chk1.isChecked())
                Toast.makeText(context, "Check1 checked",
                    Toast.LENGTH_SHORT).show();
            else
```

```
                Toast.makeText(context, "Check1 unchecked",
                    Toast.LENGTH_SHORT).show();
                break;
            case R.id.chk2:
                //代码省略
            }
        }
        public void onClickToggle(View v){
            if(tb.isChecked()){
                chk1.setChecked(true);
                chk2.setEnabled(true);
            }
            else{
                chk1.setChecked(false);
                chk2.setEnabled(false);
            }
        }
    }
```

（5）运行效果

CheckBox 的运行效果如图 2-5 所示，ToggleButton 对象为 unchecked 状态时，Check2 禁选，Check1 可以得到响应。

2.6 RadioButton 组件

图 2-5 CheckBox 运行效果

RadioButton 为单选按钮，即一个容器或一个组中只能有一个被选中。在一个布局文件中，单选按钮一般被放置到 RadioGroup 容器中。

（1）主要相关属性

创建项目 RadioButton，activity_main.xml 文件如下所示。在作为 RadioButton 容器的 RadioGroup 标记中，需要指定布局方向，horizontal 表示容器中的 RadioButton 水平排列，vertical 表示垂直排列。RadioButton 的 checked 属性默认为 false，如果需要默认指定某个 RadioButton 有效，将其设置为 true 即可。事件处理函数可通过 onClick 属性设置。

```
<RelativeLayout xmlns:android="http://schemas.android.com/apk/res/android"
    xmlns:tools="http://schemas.android.com/tools"
    android:layout_width="match_parent"
    android:layout_height="match_parent"
    android:paddingBottom="@dimen/activity_vertical_margin"
    android:paddingLeft="@dimen/activity_horizontal_margin"
    android:paddingRight="@dimen/activity_horizontal_margin"
    android:paddingTop="@dimen/activity_vertical_margin"
    tools:context="com.ch02.radiobutton.MainActivity" >
    <RadioGroup
        android:id="@+id/radioGroup1"
        android:layout_width="wrap_content"
        android:layout_height="wrap_content"
        android:layout_gravity="center"
        android:orientation="horizontal" >
        <RadioButton
```

```xml
        android:id="@+id/rb1"
        android:layout_width="wrap_content"
        android:layout_height="wrap_content"
        android:onClick="onClick_RB"
        android:text="RB1" />
    <RadioButton
        android:id="@+id/rb2"
        android:layout_width="wrap_content"
        android:layout_height="wrap_content"
        android:onClick="onClick_RB"
        android:text="RB2" />
    </RadioGroup>
</RelativeLayout>
```

（2）主要源代码

MainActivity.java 中的源代码实现以上主要功能的测试，两个 RadioButton 对象实现互斥，由 Toast 浮动消息显示选中的对象。由于点中即选中，其 checked 属性就为 true，因而，一般不需要使用 RadioButton 对象的 isChecked() 方法来获取状态值。

```java
package com.ch02.radiobutton;
import android.app.Activity;
import android.content.Context;
import android.os.Bundle;
import android.view.Menu;
import android.view.MenuItem;
import android.view.View;
//import android.view.View.OnClickListener;
import android.widget.RadioButton;
import android.widget.Toast;
public class MainActivity extends Activity {
    Context context;
    RadioButton rb1;
    RadioButton rb2;
    @Override
    protected void onCreate(Bundle savedInstanceState) {
        super.onCreate(savedInstanceState);
        setContentView(R.layout.activity_main);
        rb1 = (RadioButton)findViewById(R.id.rb1);
        rb2 = (RadioButton)findViewById(R.id.rb2);
        rb1.setChecked(true);    //也可用代码设置初始值
        context = this;

/*        rb1.setOnClickListener(new OnClickListener(){
            @Override
            public void onClick(View v) {
                // TODO Auto-generated method stub

            }});*/
    }
    public void onClick_RB(View v){
        int id = v.getId();
        switch(id){
        case R.id.rb1:
            Toast.makeText(context, "RB1 checked",
```

```
                Toast.LENGTH_SHORT).show();
            break;
        case R.id.rb2:
            Toast.makeText(context, "RB2 checked",
                Toast.LENGTH_SHORT).show();
            break;
        default:
            //
    }
}
```

2.7 Spinner 组件

Spinner 组件又称下拉框，当点击该组件时将显示所有可选列表项，用户只能从中选择某一项，默认情况下显示当前项。这些列表项可使用与这个组件相关联的适配器 Adapter 来实现，即 Adapter 充当 Spinner 组件与列表项之间的桥梁，并确保每次从中选取某一项。Spinner 组件是 AdapterView 类的子类。

（1）数据项处理

对于一个应用程序来说，Spinner 中的列表项基本是固定的。创建 Spinner 项目，可以在资源目录的 strings.xml 文件中定义一个字符串数组，这里数组的名称为 items，可以在代码中通过 this.getResources().getStringArray 方法获取。另外，也可以直接定义并初始化一个字符串数组。

```xml
<?xml version="1.0" encoding="utf-8"?>
<resources>
    <string-array name="items">
        <item>One</item>
        <item>Two</item>
        <item>Three<item></item></item>
    </string-array>
</resources>
```

（2）数据显示方式

Spinner 对象有两种数据显示方式，分别是当前项显示方式和下拉列表项显示方式。默认的显示方式定义在 API level 1，已经过时，而且显示时当前项悬空并与下划线距离较大。数据显示方式通过 layout 目录下的 xml 布局文件表示，这里当前项用 spinner_item.xml 表示，为了与下拉列表项对比，将字号设为 24sp。由于数据项都为 TextView 标签，因而，还可以设置背景和其他相关选项。

```xml
<TextView xmlns:android="http://schemas.android.com/apk/res/android"
    android:layout_width="wrap_content"
    android:layout_height="wrap_content"
    android:textSize="24sp"
    android:textColor="#000000" />
```

下拉列表项的显示用 spinner_dropdown_item.xml 表示，其中的 padding 属性指定显示的文本与其边框之间的距离,为了方便用户选择,此数据不宜太小,以免文本过于拥挤。

```
<TextView xmlns:android="http://schemas.android.com/apk/res/android"
    android:layout_width="wrap_content"
    android:layout_height="wrap_content"
    android:textSize="18sp"
    android:padding="10dp"
    android:textColor="#000000" />
```

(3)主要方法

setAdapter(SpinnerAdapter adapter) 方法指定适配器,初始化 Spinner 对象。
setSelection(int position) 方法显示指定位置处的选项,位置从 0 开始计数。
getItemAtPosition(int position) 方法获取指定位置处的 String 选项。

(4)事件处理

通过 Spinner 对象的 setOnItemClickListener(l) 方法注册侦听器。根据提示,l 是一个实例化的 OnItemClickListener 对象,因而可以写成如下形式,由系统自动提示实现 onItemSelected 和 onNothingSelected 接口。

```
setOnItemClickListener(new OnItemClickListener(){});
```

(5)基本原理与主要源代码

ArrayAdapter 是一个泛型类,可以存放指定类型的对象,由于下拉列表项是字符串数组,因而添加 <String> 标志,声明 adapter 对象。nIndex 是 Spinner 对象 sp 中当前项的索引,也即位置。

这里设置专门的 initSpinner 函数对 sp 进行初始化。字符串数组 strArray 可以直接声明并初始化,也可以从资源文件中读取;然后初始化 adapter 对象,需要三个参数,第一个为当前上下文环境,第二个就是 sp 当前项的布局样式,第三个为字符串数组;接着设置 adapter 的下拉列表项的布局样式;最后通过 sp 对象的 setAdapter 方法将处理好的 adapter 与 sp 相关联。

在 onItemSelected 事件处理函数中,主要需要获取当前项的字符串及其位置,position 直接表示位置,可以在全局变量 nIndex 中保存备用,id 表示当前行的位置,这里与 position 相等;parent 是选择发生时的父容器,getItemAtPosition(position) 方法获得当前 TextView 文本框对象,利用 toString() 方法转换为字符串;view 是被选择的 TextView 对象,可以通过 getText() 方法获取当前字符串。这里通过更简单的 adapter.getItem(nIndex) 方法获取当前字符串,并用 Toast 类的静态方法显示浮动消息。

sp 的 setSelection 方法用来设置当前项,但是只能传入整型位置值,而一般程序员更清楚列表项,因而设置自定义函数 getSpinnerItemIndex,将字符串形式的列表项转换为位置索引值。在 Java 源代码中,字符串的比较不能使用 "==",必须使用 String 对象的 equals 方法。主窗体的 MainActivity.java 的主要源代码如下所示。

```
package com.ch02.spinner;
import android.app.Activity;
import android.content.Context;
import android.os.Bundle;
```

```java
import android.view.Menu;
import android.view.MenuItem;
import android.view.View;
import android.widget.AdapterView;
import android.widget.AdapterView.OnItemSelectedListener;
import android.widget.ArrayAdapter;
import android.widget.Spinner;
import android.widget.Toast;
public class MainActivity extends Activity {
    private Spinner sp;
    ArrayAdapter<String> adapter;
    int nIndex;
    Context context;
    @Override
    protected void onCreate(Bundle savedInstanceState) {
        super.onCreate(savedInstanceState);
        context = this;
        setContentView(R.layout.activity_main);
        sp = (Spinner)findViewById(R.id.sp);
        initSpinner(sp);
    }
    private void initSpinner(Spinner sp){
        //String[] strArray = {"One", "Two", "Three"};
        String[] strArray = this.getResources().getStringArray(R.array.items);
        adapter = new ArrayAdapter<String>(this, R.layout.spinner_item, strArray);
        adapter.setDropDownViewResource(R.layout.spinner_dropdown_item);
        sp.setAdapter(adapter);
        sp.setOnItemSelectedListener(new OnItemSelectedListener(){
            @Override
            public void onItemSelected(AdapterView<?> parent, View view,
                    int position, long id) {
                // parent The AdapterView where the selection happened
                // view The view within the AdapterView that was clicked
                // position The position of the view in the adapter
                // id The row id of the item that is selected
                nIndex = position;
                Toast.makeText(context, adapter.getItem(nIndex) + " selected",
                        Toast.LENGTH_SHORT).show();
            }
            @Override
            public void onNothingSelected(AdapterView<?> parent) {
                //TODO
            }});
        sp.setSelection(getSpinnerItemIndex(sp, strArray[0]));
    }
    private int getSpinnerItemIndex(Spinner spinner, String strItem){
        int index = 0;
        for (int i=0;i<spinner.getCount();i++){
            if (spinner.getItemAtPosition(i).equals(strItem)){
                index = i;
            }
        }
        return index;
    }
}
```

（6）不使用适配器的方法

如果不使用适配器，可以直接使用 Spinner 的 entries 属性来定义列表项，列表项采用 values 目录下的 array.xml 文件表示。例如，将 entries 属性设置为"@array/items1"，也有相似的效果。

```xml
<?xml version="1.0" encoding="utf-8"?>
<resources>
    <string-array name="items1">
        <item>One</item>
        <item>Two</item>
        <item>Three<item></item></item>
    </string-array>
</resources>
```

（7）运行效果

Spinner 项目的运行效果如图 2-6 所示，当前项的字号比下拉列表项的字号大，如果选中"Two"，将用浮动消息显示"Two selected"。

2.8 ListView 组件

ListView 组件以垂直列表的形式列出所需要显示的列表项，应用非常广泛，如股票行情和工作方式选择等。标准的列表项比较简单，一般只有一个文本标签和一个选中标记，而实际项目中经常出现多个文本标签，如股市行情中就有证券名称、当前价、涨幅等。本节主要介绍其基本应用，在随后的实例中将介绍行内容更加丰富的列表项。

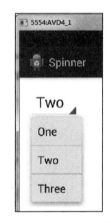

图 2-6　Spinner 运行效果

（1）主要属性与布局文件

ListView 的主要应用包括增加列表项、删除列表项和选中列表项。创建项目 ListView，使用线性布局 LinearLayout，垂直排列子线性布局和 ListView 组件。子线性布局采用水平排列，用 EditText 输入新的列表项，用一个 Button 增加列表项，另一个 Button 删除选中的列表项。由于列表项的数据可能较多，因而，ListView 一般位于屏幕的下部，layout_height 属性采用 match_parent 值，填满剩余的父空间。

drawSelectorOnTop 设置为 true 时，活动色会显示在上面（选中的列表项文字会被遮住），为 false 时则显示在下面，一般取 false 值。divider 属性用来设置列表项之间的分隔条，可采用颜色，也可采用图片，如果不需要分隔条，可以设置为"@null"，默认不需要设置，本项目设置为蓝色。dividerHeight 属性表示分隔条的高度，默认不设置，本项目设置为 1dp，如果设置为 0dp 将不显示分隔条。choiceMode 定义选中时的行为，这里为 singleChoice，即只允许选择一项，这样才能使得选中的列表项打钩，此时，layout_width 属性也设置为 match_parent，使得一个选项充满一行，方便用户选择。

```
<LinearLayout xmlns:android="http://schemas.android.com/apk/res/android"
    xmlns:tools="http://schemas.android.com/tools"
```

```xml
        android:layout_width="wrap_content"
        android:layout_height="match_parent"
        android:layout_marginBottom="@dimen/activity_horizontal_margin"
        android:layout_marginLeft="@dimen/activity_horizontal_margin"
        android:layout_marginRight="@dimen/activity_horizontal_margin"
        android:layout_marginTop="@dimen/activity_horizontal_margin"
        android:orientation="vertical"
        tools:context="com.ch02.listview.MainActivity" >
        <LinearLayout
            android:layout_width="match_parent"
            android:layout_height="wrap_content"
            android:gravity="center"
            android:orientation="horizontal" >
            <EditText
                android:id="@+id/edText"
                android:layout_width="wrap_content"
                android:layout_height="wrap_content"
                android:layout_marginRight="15dp"
                android:hint="Item"
                android:inputType="text"
                android:maxLength="6"
                android:text="@null" />
            <Button
                android:id="@+id/btAdd"
                android:layout_width="wrap_content"
                android:layout_height="wrap_content"
                android:layout_marginRight="15dp"
                android:onClick="onClick"
                android:text="Add"  />
            <Button
                android:id="@+id/btDelete"
                android:layout_width="wrap_content"
                android:layout_height="wrap_content"
                android:onClick="onClick"
                android:text="Delete" />
        </LinearLayout>
        <ListView
            android:id="@+id/list"
            android:layout_width="match_parent"
            android:layout_height="match_parent"
            android:choiceMode="singleChoice"
            android:divider="#ff1118ff"
            android:dividerHeight="1dp"
            android:drawSelectorOnTop="false" >
</LinearLayout>
```

(2) 数据处理

在 Spinner 组件中主要使用固定字符串数组来展示列表项，如果列表项固定，ListView 中也可使用这些方法。但是，ListView 一般需要处理更多、更灵活的列表项，因而使用动态数组更合适。ArrayList 类实现了 List 接口，使用类似数组方式来保存元素，元素按照索引位置依次存放（数据可重复），只需将元素新增或插入 ArrayList 对象，并不用事先声明大小，如同一个可自动调整大小的动态数组。声明一个 ArrayList<String> 类型的对象

listItems，就可以调用其 add 方法增加列表项，使用 remove 方法删除列表项，size 方法判断元素个数，isEmpty 方法查询是否为空。

（3）主要方法

ListView 对象 list 的 setItemChecked 方法可设置某选项是否打钩，第一个参数是该选项的位置，第二个参数是布尔值，true 表示打勾，false 显示灰色。当动态数组 listItems 中的数据发生变化时，需要调用适配器 ArrayAdapter<String> 对象 adapter 的 notifyDataSetChanged 方法来更新 list 中列表项的显示。

（4）事件处理

使用 list 对象的 setOnItemClickListener 注册侦听器，实现接口 OnItemClickListener 中的 onItemClick 方法，获取选中的列表项的位置或数据，标注是否选中（打勾）。

（5）屏幕旋转后数据丢失

屏幕旋转时会重新执行 onCreate 方法并实现初始化，list 中的数据将会丢失，因为 onCreate 方法中没有直接执行 adapter 对象的 notifyDataSetChanged 方法，从而通知重新显示列表项。可以在配置文件 AndroidManifest.xml 中将 Activity 标志中的 configChanges 属性设置为 "orientation|screenSize"，即监控屏幕方向和大小变化，然后在 MainActivity.java 的方法空白处右击，依次选择 "Source → Override/Implement Methods... → onConfigurationChanged"，重写该方法，添加 adapter.notifyDataSetChanged() 语句，这样即可重新显示列表项数据。

（6）基本原理与主要源代码

MainActivity.java 中的源代码实现以上全部功能的测试，这里使用了一个全局变量 nCurrent 表示当前选中的列表项的位置，初始值为 -1，表示没有选中任何列表项。第一次选中某列表项，nCurrent 中即为该列表项的位置值，在该列表项后打勾，如果再次点击该列表项，则去掉打勾，nCurrent 恢复为 -1，表示当前没有选中任何列表项。

当编辑框中有数据时，点击【Add】才能增加列表项时；当列表项中有数据并且选中该列表项时，点击【Delete】才能删除列表项。两个按钮都使用 Toast 浮动消息给出提示信息，而且，数据变化时都要调用 adapter 对象的 notifyDataSetChanged 方法，通知 list 对象更新列表项。

```
package com.ch02.listview;
import java.util.ArrayList;
import android.app.Activity;
import android.content.Context;
import android.content.res.Configuration;
import android.os.Bundle;
import android.text.TextUtils;
import android.view.Menu;
import android.view.MenuItem;
import android.view.View;
import android.widget.AdapterView;
import android.widget.ArrayAdapter;
```

```java
import android.widget.EditText;
import android.widget.ListView;
import android.widget.AdapterView.OnItemClickListener;
import android.widget.Toast;
public class MainActivity extends Activity {
    private EditText edText;
    private ListView list;
    private ArrayAdapter<String> adapter;
    private ArrayList<String> listItems = new ArrayList<String>();
    int nCurrent = -1;
    private Context context;
    @Override
    protected void onCreate(Bundle savedInstanceState) {
        super.onCreate(savedInstanceState);
        setContentView(R.layout.activity_main);
        context = this;
        edText = (EditText)findViewById(R.id.edText);
        list = (ListView)findViewById(R.id.list);
        adapter = new ArrayAdapter<String>(this,
                android.R.layout.simple_list_item_checked, listItems);
        list.setAdapter(adapter);
        list.setOnItemClickListener(new OnItemClickListener(){
            @Override
            public void onItemClick(AdapterView<?> parent, View view,
                    int position, long id) {
                // TODO Auto-generated method stub
                String strItem = parent.getItemAtPosition(position).toString();
                if(nCurrent != position){
                    nCurrent = position;
                    list.setItemChecked(position, true);
                    Toast.makeText(context, strItem + " selected",
                            Toast.LENGTH_SHORT).show();
                }
                else{
                    nCurrent = -1;
                    list.setItemChecked(position, false);
                    Toast.makeText(context, strItem + " unselected",
                            Toast.LENGTH_SHORT).show();
                }
            }});
    }
    public void onClick(View v){
        String strItem = edText.getText().toString();
        int id = v.getId();
        switch(id){
        case R.id.btAdd:
            if(TextUtils.isEmpty(strItem)){
                Toast.makeText(this, "Please input data.",
                        Toast.LENGTH_SHORT).show();
                return;
            }
            listItems.add(strItem);
            adapter.notifyDataSetChanged();
            break;
        case R.id.btDelete:
            if(listItems.isEmpty()) {
```

```
                Toast.makeText(this, "No any item.",
                        Toast.LENGTH_SHORT).show();
                return;
            }
            if(nCurrent == -1) {
                Toast.makeText(this, "Please selecct an item.",
                        Toast.LENGTH_SHORT).show();
                return;
            }
            list.setItemChecked(nCurrent, false);
            listItems.remove(nCurrent);
            adapter.notifyDataSetChanged();
            nCurrent = -1;
            break;
        }
    }
    @Override
    public void onConfigurationChanged(Configuration newConfig) {
        // TODO Auto-generated method stub
        super.onConfigurationChanged(newConfig);
        adapter.notifyDataSetChanged();
    }
}
```

（7）运行效果

运行程序，在编辑框中输入文本，点击【Add】可增加此列表项；选中某列表项，点击【Delete】能删除该列表项；选中或清除选中都有浮动消息提示。软件运行效果如图 2-7 所示。

图 2-7　ListView 运行效果

2.9　Switch 组件

Switch 组件是一个带有标识的可视化开关，通过移动滑块来选择一种状态。layout_width 一般设置为 match_parent，使得 Switch 组件可以占据一行，方便用户操作。checked 属性可以设置 Switch 组件的初始状态，true 或 false 可选，默认为 false。

（1）自定义 Switch 组件的实现

创建项目 SwitchApp，对比自定义 Switch 组件与系统组件的形式与应用。Switch 组件的滑块（Thumb）定义于 switch_thumb.xml 文件，采用白色椭圆形状、宽度和高度都是 30dp。

```xml
<?xml version="1.0" encoding="utf-8"?>
<shape xmlns:android="http://schemas.android.com/apk/res/android"
    android:shape="oval">
    <size
        android:width="30dp"
        android:height="30dp">
    </size>
    <solid
        android:color="@android:color/white">
    </solid>
</shape>
```

Switch 组件的主体（Track）采用不同的颜色来区分开关状态，亮色表示打开（true），灰色表示关闭（false），但形状都采用矩形加圆角的形式。打开的状态定义于 switch_track_on.xml 文件。

```xml
<?xml version="1.0" encoding="utf-8"?>
<shape xmlns:android="http://schemas.android.com/apk/res/android"
    android:shape="rectangle">
    <solid
        android:color="@android:color/holo_green_light">
    </solid>
    <corners
        android:radius="15dp">
    </corners>
</shape>
```

Switch 组件的关闭状态定义于 switch_track_off.xml 文件。

```xml
<?xml version="1.0" encoding="utf-8"?>
<shape xmlns:android="http://schemas.android.com/apk/res/android"
    android:shape="rectangle">
    <solid
        android:color="@android:color/darker_gray">
    </solid>
    <corners
        android:radius="15dp">
    </corners>
</shape>
```

打开和关闭如何使用可根据 switch_track.xml 文件的定义进行，当 state_checked 为 true 时，使用 switch_track_on；当 state_checked 为 false 时，使用 switch_track_off。

```xml
<?xml version="1.0" encoding="utf-8"?>
<selector xmlns:android="http://schemas.android.com/apk/res/android">
    <item
        android:state_checked="true"
        android:drawable="@drawable/switch_track_on"></item>
    <item
        android:state_checked="false"
        android:drawable="@drawable/switch_track_off"></item>
</selector>
```

（2）布局文件

布局文件 activity_main.xml 采用相对布局方式，上面为自定义 Switch 组件，下面为系统定义 Switch 组件，分别通过 checked 属性设置不同的初值。

```xml
<RelativeLayout xmlns:android="http://schemas.android.com/apk/res/android"
    xmlns:tools="http://schemas.android.com/tools"
    android:layout_width="match_parent"
    android:layout_height="match_parent"
    android:paddingBottom="@dimen/activity_vertical_margin"
    android:paddingLeft="@dimen/activity_horizontal_margin"
    android:paddingRight="@dimen/activity_horizontal_margin"
    android:paddingTop="@dimen/activity_vertical_margin"
    tools:context="com.ch02.switchapp.MainActivity" >
```

```xml
<Switch
    android:id="@+id/swItem1"
    android:layout_width="match_parent"
    android:layout_height="wrap_content"
    android:layout_alignParentLeft="true"
    android:checked="false"
    android:thumb="@drawable/switch_thumb"
    android:track="@drawable/switch_track"
    android:text="Switch1" />
<Switch
    android:id="@+id/swItem2"
    android:layout_width="match_parent"
    android:layout_height="wrap_content"
    android:layout_alignParentLeft="true"
    android:checked="true"
    android:layout_below="@+id/swItem1"
    android:layout_marginTop="10dp"
    android:text="Switch2" />
</RelativeLayout>
```

(3) 事件处理

Switch 组件通过 setOnCheckedChangeListener 方法来注册侦听器, 在其中实现 OnCheckedChangeListener 接口中的 onCheckedChanged 方法。

(4) 主要源代码

SwitchApp 的主要源代码 MainActivity.java 中用自定义类 SwitchListener 来实现侦听功能, 当 Switch 组件的状态发生变化时, 用 Toast 浮动消息显示 Switch 组件对象的名称及其状态。

```java
package com.ch02.switchapp;
import android.app.Activity;
import android.content.Context;
import android.os.Bundle;
import android.view.Menu;
import android.view.MenuItem;
import android.widget.CompoundButton;
import android.widget.Toast;
import android.widget.CompoundButton.OnCheckedChangeListener;
import android.widget.Switch;
public class MainActivity extends Activity {
    Switch swItem1, swItem2;
    Context context;
    @Override
    protected void onCreate(Bundle savedInstanceState) {
        super.onCreate(savedInstanceState);
        setContentView(R.layout.activity_main);
        context = this;
        SwitchListener listener = new SwitchListener();
        swItem1 = (Switch)findViewById(R.id.swItem1);
        swItem1.setOnCheckedChangeListener(listener);
        swItem2 = (Switch)findViewById(R.id.swItem2);
        swItem2.setOnCheckedChangeListener(listener);
    }
```

```
class SwitchListener implements OnCheckedChangeListener{
    @Override
    public void onCheckedChanged(CompoundButton buttonView,
            boolean isChecked) {
        // TODO Auto-generated method stub
        int id = buttonView.getId();
        String strId = "Switch";
        switch(id){
        case R.id.swItem1:
            strId += "1";
            break;
        case R.id.swItem2:
            strId += "2";
            break;
        }
        Toast.makeText(context, strId + " " +
                Boolean.toString(isChecked),
                Toast.LENGTH_SHORT).show();
    }
}
```

（5）运行效果

SwitchApp 的运行效果如图 2-8 所示，移动任一 Switch 的滑块都会在 Toast 浮动消息中显示 "Switch1/2 true/false" 信息。

2.10 DatePicker 组件

图 2-8 SwitchApp 运行效果

DatePicker 组件给用户提供一个选取日期的可视化界面。startYear 和 endYear 属性可以指定一个年份范围，calendarViewShown 为 true 时显示日历，为 false 时隐藏日历，其典型属性如下所示。

```
<DatePicker
    android:id="@+id/dtPicker"
    android:layout_width="wrap_content"
    android:layout_height="wrap_content"
    android:startYear="2015"
    android:endYear="2019"
    android:calendarViewShown="true"/>
```

（1）事件处理

DatePicker 组件的 init 方法完成日期的初始化并注册侦听器。可以分别输入年、月和日，然后实现 OnDateChangedListener 接口中的 onDateChanged 方法，在其中获取用户选择的日期。

```
init(int year, int monthOfYear, int dayOfMonth,
    OnDateChangedListener onDateChangedListener)
```

（2）主要源代码

创建项目 DatePicker，MainActivity.java 中的主要源代码如下所示。内部类 DtPicker-Listener 实现了 OnDateChangedListener 接口中的 onDateChanged 方法，获取用户选择的日期数据，然后用 Toast 浮动消息显示。由于 monthOfYear 用 0 表示 1 月，因而需要加 1 处理。

```java
package com.ch02.datepicker;
import android.app.Activity;
import android.content.Context;
import android.os.Bundle;
import android.view.Menu;
import android.view.MenuItem;
import android.widget.DatePicker;
import android.widget.DatePicker.OnDateChangedListener;
import android.widget.Toast;
public class MainActivity extends Activity{
    private DatePicker dtPicker;
    private Context context;
    @Override
    protected void onCreate(Bundle savedInstanceState) {
        super.onCreate(savedInstanceState);
        setContentView(R.layout.activity_main);
        context = this;
        dtPicker = (DatePicker)findViewById(R.id.dtPicker);
        DtPickerListener dtListener = new DtPickerListener();
        dtPicker.init(dtPicker.getYear(), dtPicker.getMonth(),
            dtPicker.getDayOfMonth(), dtListener);
    }
    class DtPickerListener implements OnDateChangedListener{
        @Override
        public void onDateChanged(DatePicker view, int year, int monthOfYear,
            int dayOfMonth) {
            Toast.makeText(context, monthOfYear+1 + "/" + dayOfMonth +
                "/" + year, Toast.LENGTH_SHORT).show();
        }
    }
}
```

（3）运行效果

DatePicker 项目的运行效果如图 2-9 所示，日期变化将通过 Toast 浮动消息显示相关信息。

2.11 AlertDialog 组件

在程序运行过程中，经常需要利用对话框与用户交互，根据用户选择做出响应，而 AlertDialog 组件正好能够完成多种对话框的定制。AlertDialog 是 Dialog 的一个子类，可以显示 1～3 个 Button，可方便地添加标题、消息和图标，定制对按钮的响应操作。如果需要显示用户自定义的对话框，则可以通过用户

图 2-9　DatePicker 运行效果

布局文件来实现。本节创建 AlertDialog 项目，通过不同的按钮显示不同的对话框。首先需要声明一个 AlertDialog.Builder 的构造器对象 builder，添加各种属性后创建 AlertDialog 对象来显示对话框。

（1）Simple Dialog

该对话框是最基本也是最常用的对话框，创建对象时所使用的 this 是上下文环境，setIcon 方法用于设置对话框的图标，setTitle 设置标题，setMessage 设置对话框内容，字符串在 strings.xml 文件中定义。

setPositiveButton 方法用于设置"确认"按钮并注册对应的侦听器，同时可以设置按钮的文本标识，当点击"确认"按钮时，通过 Toast 类的静态方法显示相关信息；setNegativeButton 方法则处理"取消"按钮相关事件，另外还可通过 setNeutralButton 方法设置一个中性按钮。

setCancelable 方法设置该对话框是否可以被取消，如果输入 false 参数，则该对话框将不可被取消，用户必须点击对话框的按钮进行处理；否则，用户点击对话框外面的屏幕，对话框即消失。设置好构造器的参数后，即可调用 create 方法生成 AlertDialog 对象 dialog，最后调用该对象的 show 方法即可显示对话框。

```
private void showSimpleDialog() {
    builder = new AlertDialog.Builder(this);
    builder.setIcon(R.drawable.ic_launcher);
    builder.setTitle(R.string.simple_dialog);
    builder.setMessage(R.string.dialog_message);
    builder.setPositiveButton(R.string.postive_button,
            new DialogInterface.OnClickListener() {
        @Override
        public void onClick(DialogInterface dialogInterface, int i) {
            Toast.makeText(getApplicationContext(),
                    R.string.toast_postive,
                    Toast.LENGTH_SHORT).show();
        }
    });
    builder.setNegativeButton(R.string.negative_button,
            new DialogInterface.OnClickListener() {
        @Override
        public void onClick(DialogInterface dialogInterface, int i) {
            Toast.makeText(getApplicationContext(),
                    R.string.toast_negative,
                    Toast.LENGTH_SHORT).show();
        }
    });
    builder.setCancelable(false);
    AlertDialog dialog = builder.create();
    dialog.show();
}
```

调用 showSimpleDialog 后的运行效果如图 2-10 所示，对话框中的文本和图标都在资源文件中定义，点击【Cancel】或【OK】按钮后才能关闭对话框，并显示相应的文本。

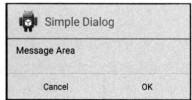

图 2-10　Simple Dialog

（2）Simple List Dialog

Simple List Dialog 是简单列表对话框，即用户从列表项中选择一项。items 是字符串数组，可以在资源文件 strings.xml 中定义，也可以在全局变量中定义。setItems 方法设置列表项并注册侦听器，onClick 函数中的 i 是列表项的序号，这里也通过浮动消息来显示用户选中的信息。

```java
private void showSimpleListDialog() {
    builder = new AlertDialog.Builder(this);
    builder.setIcon(R.drawable.ic_launcher);
    builder.setTitle(R.string.simple_list_dialog);
    final String[] items = {"Item One","Item Two","Item Three"};
    builder.setItems(items, new DialogInterface.OnClickListener() {
        @Override
        public void onClick(DialogInterface dialogInterface, int i) {
            Toast.makeText(getApplicationContext(),
                    items[i], Toast.LENGTH_SHORT).show();
        }
    });
    builder.setCancelable(false);
    AlertDialog dialog = builder.create();
    dialog.show();
}
```

调用 showSimpleListDialog 后的运行效果如图 2-11 所示，用户可以任选一项，此时对话框自动关闭，并显示用户选择的信息。

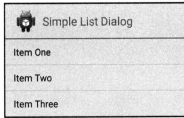

图 2-11　Simple List Dialog

（3）Single Choice Dialog

Single Choice Dialog 是一种附带单选按钮的列表对话框，与简单列表对话框有相似的视觉效果。通过 setSingleChoiceItems 方法添加列表项数据，该方法的第二个参数表示默认选中的列表项（从 0 开始计数）。

```java
private void showSingleChoiceDialog() {
    builder = new AlertDialog.Builder(this);
    builder.setIcon(R.drawable.ic_launcher);
    builder.setTitle(R.string.single_choice_dialog);
    final String[] items = {"Item One","Item Two","Item Three"};
    builder.setSingleChoiceItems(items, 2,
            new DialogInterface.OnClickListener() {
        @Override
        public void onClick(DialogInterface dialogInterface, int i) {
            Toast.makeText(getApplicationContext(), items[i],
                    Toast.LENGTH_SHORT).show();
        }
    });
    builder.setCancelable(true);
    AlertDialog dialog = builder.create();
    dialog.show();
}
```

调用 showSingleChoiceDialog 后的运行效果如图 2-12 所示，用户可以任选一项，侦听器中将显示用户选择的信息。如果给 setCancelable 方法传入 false，这时对话框不会自动关闭；也无法通过后退键关闭；可以传入 true，或者增加一个"确认"按钮。

（4）Multi Choice Dialog

Multi Choice Dialog 与 Single Choice Dialog 相似，但可选择多项，是一种多选对话框，通过 setMultiChoiceItems 方法添加列表数据，第二个参数是一个布尔值数组，表示该项是否被选中。

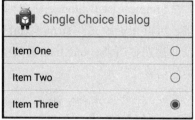

图 2-12　Single Choice Dialog

```
private void showMultiChoiceDialog() {
    builder = new AlertDialog.Builder(this);
    builder.setIcon(R.drawable.ic_launcher);
    builder.setTitle(R.string.multi_choice_dialog);
    final String[] items = {"Item One","Item Two","Item Three"};
    builder.setMultiChoiceItems(items, new boolean[]
        {false, false, false},
        new DialogInterface.OnMultiChoiceClickListener() {
        @Override
        public void onClick(DialogInterface dialogInterface,
            int i, boolean b) {
            Toast.makeText(getApplicationContext(),
                items[i] + " is "+ b,Toast.LENGTH_SHORT).show();
        }
    });
    builder.setCancelable(true);
    AlertDialog dialog = builder.create();
    dialog.show();
}
```

调用 showMultiChoiceDialog 后的运行效果如图 2-13 所示，每点击一项，将显示该项的布尔值。如果给 setCancelable 方法传入 false，该对话框将不可被关闭，可增加一个"确认"按钮。

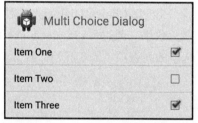

图 2-13　Multi Choice Dialog

（5）Custom Adapter Dialog

Custom Adapter Dialog 是用户自定义适配器对话框，与 Simple List Dialog 相似，只是内容由 custom_adapter.xml 文件定义，这里的列表项由两项组成，左边为一个图标 ImageView，右边是文本 TextView。color 在资源文件的 color.xml 中定义。

```xml
<?xml version="1.0" encoding="utf-8"?>
<LinearLayout xmlns:android="http://schemas.android.com/apk/res/android"
    android:layout_width="match_parent"
    android:layout_height="wrap_content"
    android:layout_marginBottom="10dp"
    android:layout_marginLeft="15dp"
    android:layout_marginRight="15dp"
    android:layout_marginTop="15dp"
    android:background="@color/light_grey"
```

```xml
        android:orientation="horizontal" >
    <ImageView
        android:id="@+id/id_image"
        android:layout_width="60dp"
        android:layout_height="60dp" />
    <TextView
        android:id="@+id/id_text"
        android:layout_width="wrap_content"
        android:layout_height="wrap_content"
        android:layout_gravity="center"
        android:layout_marginLeft="20dp"
        android:textColor="@color/colorPrimary" />
</LinearLayout>
```

在显示该对话框的函数 showCustomAdapterDialog 中，ArrayList 是一个泛型类，其中存放自定义类 ItemBean。

```java
private void showCustomAdapterDialog(){
    builder=new AlertDialog.Builder(this);
    builder.setIcon(R.drawable.ic_launcher);
    builder.setTitle(R.string.custom_adapter_dialog);
    final ArrayList<ItemBean> items = new ArrayList<ItemBean>();
    items.add(new ItemBean(R.drawable.ic_launcher,"Item One"));
    items.add(new ItemBean(R.drawable.ic_launcher, "Item Two"));
    CustomAdapter adapter = new CustomAdapter(this,
            R.layout.custom_adapter, items);
    builder.setAdapter(adapter, new DialogInterface.OnClickListener(){
        @Override
        public void onClick(DialogInterface dialogInterface, int i){
            String msg = items.get(i).getMessage();
            Toast.makeText(getApplicationContext(), msg,
                    Toast.LENGTH_SHORT).show();
        }
    });
    builder.setCancelable(false);
    AlertDialog dialog = builder.create();
    dialog.show();
}
```

ItemBean 类用来存放 custom_adapter.xml 文件中定义的图标和文本信息，包括一个构造函数、一个获得图标 ID 的函数以及一个获得文本信息的函数。

```java
private class ItemBean{
    private int imageId;
    private String message;
    public ItemBean(int imageId, String message) {
        this.imageId = imageId;
        this.message = message;
    }
    public String getMessage() {
        return message;
    }
    public int getImageId() {
        return imageId;
    }
}
```

CustomAdapter 继承自 ArrayAdapter，ArrayAdapter 也是一个泛型类，需要指明存放的数据类型。CustomAdapter 将文件 custom_adapter.xml 中定义的布局和 ItemBean 数据综合起来进行显示，构造函数由三个参数组成，分别是上下文环境、布局 ID 和数据。重载的 getView 函数用来装载需要显示的数据，通过 getLayoutInflater 方法获得 LayoutInflater 对象 inflater，由该对象填充 custom_adapter 布局，从而获得 ImageView 和 TextView 对象并填充数据，完成显示。

```
private class CustomAdapter extends ArrayAdapter<ItemBean> {
    private final Activity context;
    private int mLayout;
    private ArrayList<ItemBean> items;
    private ImageView image;
    private TextView text;
    public CustomAdapter(Activity context, int mLayout, ArrayList<ItemBean> items) {
        super(context, mLayout, items);
        this.context = context;
        this.mLayout = mLayout;
        this.items = items;
    }
    @Override
    public View getView(int position,View convertView,ViewGroup parent) {
        if(convertView == null){
            LayoutInflater inflater = context.getLayoutInflater();
            convertView = inflater.inflate(mLayout, parent, false);
            image = (ImageView) convertView.findViewById(R.id.id_image);
            text = (TextView) convertView.findViewById(R.id.id_text);
        }
        image.setImageResource(items.get(position).getImageId());
        text.setText(items.get(position).getMessage());
        return convertView;
    }
}
```

调用 showCustomAdapterDialog 后的运行效果如图 2-14 所示，用户可以任选一项，此时对话框自动关闭，并显示用户选择的文本信息。图标可以根据需要调整，这里采用了默认图标。

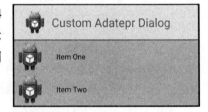

图 2-14　Custom Adapter Dialog

（6）Custom View Dialog

如果需要更具个性化的定制对话框，比如在程序运行过程中，需要弹出一个对话框来验证用户身份，则可以使用 Custom View Dialog。所谓 Custom View，即对话框的内容由用户来定义，这里用 custom_view.xml 文件来表达，提示用户输入用户名和密码，用按钮进行提交或取消输入。

```
<?xml version="1.0" encoding="utf-8"?>
<LinearLayout xmlns:android="http://schemas.android.com/apk/res/android"
    android:layout_width="match_parent"
    android:layout_height="match_parent"
    android:background="@color/blue"
    android:gravity="center"
```

```xml
    android:orientation="vertical" >
    <LinearLayout
        android:layout_width="match_parent"
        android:layout_height="wrap_content"
        android:layout_marginLeft="60dp"
        android:layout_marginRight="60dp"
        android:layout_marginTop="10dp"
        android:orientation="horizontal" >
        <TextView
            android:layout_width="wrap_content"
            android:layout_height="wrap_content"
            android:layout_marginRight="10dp"
            android:text="User"
            android:textColor="@color/white"
            android:textSize="20sp" />
        <EditText
            android:id="@+id/edUser"
            android:layout_width="0dp"
            android:layout_height="40dp"
            android:layout_weight="1"
            android:background="@color/white"
            android:inputType="text" />
    </LinearLayout>
    <LinearLayout
        android:layout_width="match_parent"
        android:layout_height="wrap_content"
        android:layout_marginLeft="60dp"
        android:layout_marginRight="60dp"
        android:layout_marginTop="10dp"
        android:orientation="horizontal" >
        <TextView
            android:layout_width="wrap_content"
            android:layout_height="wrap_content"
            android:layout_marginRight="10dp"
            android:text="Pass"
            android:textColor="@color/white"
            android:textSize="20sp" />
        <EditText
            android:id="@+id/edPass"
            android:layout_width="0dp"
            android:layout_height="40dp"
            android:layout_weight="1"
            android:background="@color/white"
            android:inputType="numberPassword" />
    </LinearLayout>
    <LinearLayout
        android:layout_width="match_parent"
        android:layout_height="wrap_content"
        android:layout_marginBottom="10dp"
        android:layout_marginLeft="60dp"
        android:layout_marginRight="60dp"
        android:layout_marginTop="10dp"
        android:orientation="horizontal" >
        <Button
            android:id="@+id/btCancel"
            android:layout_width="0dp"
```

```xml
            android:layout_height="wrap_content"
            android:layout_marginRight="10dp"
            android:layout_weight="1"
            android:background="@color/light_grey"
            android:text="Cancel"
            android:textColor="@color/blue" />
        <Button
            android:id="@+id/btLogin"
            android:layout_width="0dp"
            android:layout_height="wrap_content"
            android:layout_weight="1"
            android:background="@color/light_grey"
            android:text="Login"
            android:textColor="@color/blue" />
    </LinearLayout>
</LinearLayout>
```

Custom View Dialog 同样通过 getLayoutInflater 函数获得 LayoutInflater 对象，并填充 custom_view 布局得到 loginDialog 视图，从而进一步获得 EditText 和 Button 对象。这里只处理用户名输入，没有处理密码输入。调用 setView 方法设置对话框视图 loginDialog。无论 setCancelable 方法中传入 true 或 false，点击命令按钮，对话框都不能消失。这时，可以调用 AlertDialog 对象的 dismiss 方法使得对话框从屏幕消失。

对于在 onClick 事件处理程序中调用的 showCustomViewDialog 函数，其中声明的局部变量都用 final 进行了修饰。如果用 final 修饰类，则该类不能被继承；如果用 final 修饰方法，则该方法不能被重写；如果用 final 修饰全局变量，则该全局变量不能被修改，相当于声明了一个常量。但是，如果用 final 来修饰局部变量，则该局部变量可以在内联方法里作为变量使用，这也是系统要求。

```java
private void showCustomViewDialog(){
    builder = new AlertDialog.Builder(this);
    builder.setIcon(R.drawable.ic_launcher);
    builder.setTitle(R.string.custom_view_dialog);
    LinearLayout loginDialog = (LinearLayout)
            getLayoutInflater().inflate(R.layout.custom_view,null);
    final EditText edUser = (EditText)
            loginDialog.findViewById(R.id.edUser);
    builder.setView(loginDialog);
    builder.setCancelable(false);
    final AlertDialog dialog=builder.create();
    dialog.show();
    Button btLogin = (Button)loginDialog.findViewById(R.id.btLogin);
    btLogin.setOnClickListener(new OnClickListener(){
        @Override
        public void onClick(View v) {
            String strUser = edUser.getText().toString();
            Toast.makeText(getApplicationContext(),
                    "Login: " + strUser, Toast.LENGTH_SHORT).show();
            dialog.dismiss();
        }});
    Button btCancel = (Button)loginDialog.findViewById(R.id.btCancel);
    btCancel.setOnClickListener(new OnClickListener(){
```

```
@Override
public void onClick(View v) {
    Toast.makeText(getApplicationContext(),
            "Cancel", Toast.LENGTH_SHORT).show();
    dialog.dismiss();
}});
}
```

调用 showCustomViewDialog 函数后的运行效果如图 2-15 所示，如果在 User 文本框中输入"Ma"，点击【Login】，将显示"Login: Ma"，同时对话框消失。

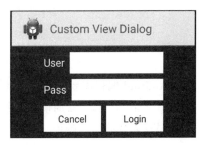

图 2-15 Custom View Dialog

2.12 本章小结

本章详细介绍了 Android 编程常用组件，包括主要属性、方法和事件处理。不同于 PC 端的快速可视化编程，Android 中的组件对用户响应需要事先注册侦听器。实现事件处理主要有三种方法：①在组件的相关属性中指定处理方法；②直接调用组件的相关方法来注册侦听器，实现相关接口；③通过内嵌类实现相关接口。Android 组件还有很多，随着版本的升级，相关属性、方法与事件都会有一定的变化，但是只要掌握基本的方法，就可以很快掌握新组件的使用。组件只是编程的基本元素，需要与下一章的常用技术搭配，才能编写一些基本的实用程序。

第3章 常用技术

Android 编程需要与后台技术相结合才能完成更多的功能。本章主要在上一章的基础之上介绍 Activity 的生命周期，介绍绘图技术的同时引进创建新类的方法。在设置手机情景模式的同时创建静态函数库，这样可以在其他项目中方便地引用而无须另外编写代码。随后介绍消息机制、多线程技术和数据存储技术，最后介绍百度地图项目的基本研发过程。

3.1 进一步了解 Activity

Activity 是 Android 系统提供的一个可视化用户交互接口，相当于 Windows 下的应用程序窗体，只是在一般情况下，Activity 使用全屏模式。Activity 也可以被理解为一个容器，上一章中的各个组件都可以在 Activity 的界面显示。

（1）Activity 中的方法

Activity 从创建到销毁会经历多种方法，需要在合适的方法中完成初始化工作、保存数据或恢复数据。因而，需要掌握各种常用方法的含义以及执行顺序。创建 ActivityLife 项目，同时利用调试信息和 Toast 浮动信息输出对应方法的提示，以观察 Activity 方法中的执行顺序。

- onCreate 方法

该方法由系统自动生成，用来创建 Activity 对象，在其中需要完成各个组件对象的初始化工作及数据的初始化工作。

```java
@Override
protected void onCreate(Bundle savedInstanceState) {
    super.onCreate(savedInstanceState);
    setContentView(R.layout.activity_main);
    Log.d("DEBUG", "onCreate");
    Toast.makeText(this, "onCreate", Toast.LENGTH_SHORT).show();
}
```

- onResume 方法

该方法在 Activity 处于前台可与用户交互时执行，可以在该方法中对界面进行辅助性初始化工作，实现该方法的代码如下所示。

```java
@Override
protected void onResume() {
    Log.d("DEBUG", "onResume");
    Toast.makeText(this, "onResume", Toast.LENGTH_SHORT).show();
    super.onResume();
}
```

- onConfigurationChanged 方法

为了避免一种或多种配置改变时重新启动 Activity，可以在配置文件 AndroidManifest.xml 中设置 Activity 标记的 configChanges 属性，如设置为 orientation|screenSize，这样当屏幕转动时，系统将自动调用 onConfigurationChanged 方法，而不会执行 onCreate 方法重新启动 Activity。

```
@Override
public void onConfigurationChanged(Configuration newConfig) {
    Log.d("DEBUG", "onConfigurationChanged");
    Toast.makeText(this, "onConfigurationChanged",
            Toast.LENGTH_SHORT).show();
    super.onConfigurationChanged(newConfig);
}
```

- onSaveInstanceState 方法

在旋转屏幕时，该方法可用来保存临时信息，临时信息可以在 onCreate 方法中得到应用。如果监控 onConfigurationChanged 方法有效，就不会执行该方法。

```
@Override
protected void onSaveInstanceState(Bundle outState) {
    Log.d("DEBUG", "onSaveInstanceState");
    Toast.makeText(this, "onSaveInstanceState",
            Toast.LENGTH_SHORT).show();
    super.onSaveInstanceState(outState);
}
```

- onDestroy 方法

该方法在销毁 Activity 时执行，可以在该方法中释放资源。

```
@Override
protected void onDestroy() {
    Log.d("DEBUG", "onDestroy");
    Toast.makeText(this, "onDestroy", Toast.LENGTH_SHORT).show();
    super.onDestroy();
}
```

（2）测试结论

屏蔽 Activity 标记的 configChanges 属性，竖屏启动程序，依次显示 onCreate→onResume，点击手机后退键将显示 onDestroy。运行状态竖屏后转横屏，依次显示 onSaveInstanceState→onDestroy→onCreate→onResume，由于竖屏转横屏需要重新执行 onCreate 方法，对组件对象进行初始化，因而，可以在 onSaveInstanceState 方法中保存数据，然后再到 onCreate 方法中恢复数据。

恢复 Activity 标记的 configChanges 属性，竖屏启动程序，依次显示 onCreate→onResume，转动屏幕，只显示 onConfigurationChanged。对于不少商业应用程序，竖屏和横屏采用不同的布局，这时，可以在 onConfigurationChanged 方法中重新布局和执行初始化工作。

3.2 绘图

在游戏应用中，图形图像技术不可或缺。随着物联网的普及，手机应用于实时控制也越来越普遍，因而，将实时数据通过图形展示在手机屏幕也得到了广泛的应用。最简单的绘图只需要两个类，即画笔 Paint 类和画布 Canvas 类。

（1）Paint 类

Paint 类代表画笔，可以用来画线，也可以用来画点。用 new 初始化以后，就可以调用画笔对象的方法来设置线宽、颜色和透明度等效果。Paint 类的常用方法如表 3-1 所示。

表 3-1　Paint 类的常用方法

方　　法	说　　明
setARGB(int a, int r, int g, int b)	用于设置颜色，各参数值在 0～255 之间，分别表示透明度、红色、绿色和蓝色
setColor(int color)	用于设置颜色，参数可以使用 Color 类提供的颜色常量，也可通过 Color.rgb 方法指定
setAlpha(int a)	单独设置透明度，参数值在 0～255 之间
setAntiAlias(boolean aa)	指定是否使用抗锯齿功能，如果使用会使得绘图速度变慢
setStyle(Style style)	设置填充风格，有 Style.STROKE、Style.FILL 和 Style.FILL_AND_STROKE 三种
setStrokeWidth(float width)	设置画笔的宽度

以下代码完成对画笔的初始化，并设置了填充风格、是否抗锯齿、线宽以及颜色。如有需要，还可以设置其他参数。

```
Paint linePaint = new Paint();              //生成Paint对象
linePaint.setStyle(Style.STROKE);           //设置STROKE填充风格
linePaint.setAntiAlias(false);              //不使用抗锯齿功能
linePaint.setStrokeWidth(3);                //设置线宽为3
linePaint.setColor(Color.RED);              //设置颜色为红色
```

（2）Canvas 类

画笔只是一个工具，具体绘图工作是由 Canvas 类的对象来完成的。Canvas 类提供了丰富的绘制几何图形的方法，画笔一般是这些方法中的一个参数。常用的绘图方法如表 3-2 所示。

表 3-2　Canvas 类的常用绘图方法

方　　法	说　　明
drawPoint(float x, float y, Paint paint)	绘制一个点，参数表示点的坐标及画笔
drawLine(float startX, float startY, float stopX, float stopY, Paint paint)	绘制一条直线，参数依次表示开始点的坐标、结束点的坐标及画笔
drawCircle(float cx, float cy, float radius, Paint paint)	绘制圆，参数依次表示圆心坐标、半径及画笔
drawRect(float left, float top, float right, float bottom, Paint paint)	绘制矩形，参数分别表示左上角和右下角的坐标
drawLines(float[] pts, Paint paint)	绘制多条直线，第一个参数为点坐标的数组，第二个参数为画笔

（3）主窗体布局文件 activity_main.xml

创建项目 Paint，画圆心和对应的圆。主窗体采用相对布局，设计一个自定义视图组件 DrawView，在其中完成图形绘制。DrawView 下面设置两个按钮，一个用于清除图形，另一个完成图形绘制。自定义视图组件用完整的包名作为标记。

```xml
<RelativeLayout xmlns:android="http://schemas.android.com/apk/res/android"
    xmlns:tools="http://schemas.android.com/tools"
    android:layout_width="match_parent"
    android:layout_height="match_parent"
    android:paddingBottom="@dimen/activity_vertical_margin"
    android:paddingLeft="@dimen/activity_horizontal_margin"
    android:paddingRight="@dimen/activity_horizontal_margin"
    android:paddingTop="@dimen/activity_vertical_margin"
    tools:context="com.ch03.paint.MainActivity" >
    <com.ch03.paint.DrawView
        android:id="@+id/vRealTime"
        android:layout_width="160dp"
        android:layout_height="160dp" />
    <Button
        android:id="@+id/btClear"
        android:layout_width="wrap_content"
        android:layout_height="wrap_content"
        android:layout_below="@+id/vRealTime"
        android:onClick="onClick"
        android:text="Clear" />
    <Button
        android:id="@+id/btDraw"
        android:layout_width="wrap_content"
        android:layout_height="wrap_content"
        android:layout_below="@+id/vRealTime"
        android:layout_toRightOf="@+id/btClear"
        android:onClick="onClick"
        android:text="Draw" />
</RelativeLayout>
```

（4）DrawView 源代码

右击项目源代码包名 com.ch03.paint，依次选择 New→Class，打开如图 3-1 所示的窗口，在 Name 文本框中输入类名 DrawView，在 Superclass 文本框中输入 android.view.View，点击【Finish】完成类的基本创建。

图 3-1　创建新的视图类 DrawView

接着重载构造函数,完成画圆和圆心的画笔的初始化。还需要重载 onDraw 函数,计算组件的高度和宽度,调用 drawBackGround 函数绘制背景,调用 drawCircle 函数绘制图形。在 onDraw 中是否绘制图形由 boolean 变量 bClear 决定,如果 bClear 为 true,即需要清除图形,则不再绘制图形。setClear 方法用来设置是否清除图形。

```java
package com.ch03.paint;
import android.content.Context;
import android.graphics.Canvas;
import android.graphics.Color;
import android.graphics.Paint;
import android.graphics.Paint.Style;
import android.util.AttributeSet;
import android.view.View;
public class DrawView extends View {
    private Paint linePaint;                            //画线用的画笔
    private Paint ptPaint;
    private float fHeight;                              //组件的高度
    private float fWidth;                               //组件的宽度
    private boolean bClear = false;
    final int WIDTH_LINE = 5;
    public DrawView(Context context, AttributeSet attrs) {
        super(context, attrs);
        linePaint = new Paint();                        //画线
        linePaint.setStyle(Style.STROKE);               //设置填充风格
        linePaint.setAntiAlias(false);                  //不使用抗锯齿功能
        linePaint.setStrokeWidth(WIDTH_LINE);           //线宽
        linePaint.setColor(Color.RED);
        ptPaint= new Paint();                           //画圆心
        ptPaint.setStrokeWidth(WIDTH_LINE*2);
        ptPaint.setColor(Color.BLUE);
    }
    @Override
    protected void onDraw(Canvas canvas) {
        fHeight = this.getHeight();
        fWidth = this.getWidth();
        drawBackGround(canvas);
        if(!bClear) drawCircle(canvas, linePaint);
        super.onDraw(canvas);
    }
    private void drawBackGround(Canvas canvas){
        Paint bkPaint=new Paint();
        bkPaint.setColor(Color.WHITE);
        canvas.drawRect(0, 0, fWidth, fHeight, bkPaint);
    }
    private void drawCircle(Canvas canvas, Paint paint){
        float cx = fWidth/2;
        float cy = cx;
        float radius = cx - 10;
        canvas.drawPoint(cx, cy, ptPaint);
        canvas.drawCircle(cx, cy, radius, linePaint);
    }
    public void setClear(boolean bClear){
        this.bClear = bClear;
    }
}
```

（5）主窗体源代码

由于两个按钮使用了 onClick 属性，因而，在 onCreate 方法中只需要对 DrawView 对象 vRealTime 进行初始化，并读取保存的 Bundle 对象，获取 boolean 变量 bClear 的值以控制是否清除图形。在 onClick 事件处理函数中设置 bClear 值，并调用 vRealTime 对象的 invalidate 方法立即刷新视图。重载 onSaveInstanceState 方法，当发生转屏时保存 bClear 值，以便在 onCreate 方法中读取使用。

```
package com.ch03.paint;
import android.app.Activity;
import android.os.Bundle;
import android.view.View;
public class MainActivity extends Activity {
    private DrawView vRealTime;
    boolean bClear = false;
    @Override
    protected void onCreate(Bundle savedInstanceState) {
        super.onCreate(savedInstanceState);
        setContentView(R.layout.activity_main);
        vRealTime = (DrawView)findViewById(R.id.vRealTime);
        if(savedInstanceState!=null)
            bClear = savedInstanceState.getBoolean("Status");
        vRealTime.setClear(bClear);
    }
    public void onClick(View v){
        int id = v.getId();
        switch(id){
        case R.id.btClear:
            bClear = true;
            break;
        case R.id.btDraw:
            bClear = false;
            break;
        }
        vRealTime.setClear(bClear);
        vRealTime.invalidate();
    }
    @Override
    protected void onSaveInstanceState(Bundle outState) {
        // TODO Auto-generated method stub
        outState.putBoolean("Status", bClear);
        super.onSaveInstanceState(outState);
    }
}
```

（6）运行效果

程序的运行效果如图 3-2 所示，点击【Clear】可以立即清除图形，点击【Draw】则立即完成图形的绘制。

3.3 用静态库函数设置手机情景模式和音量

设置手机情景模式和音量是常态工作，使用静态库函数实现，供

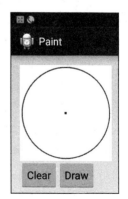

图 3-2 Paint 运行效果

多个项目共享代码，这样可以较好地提高编程效率，增加应用程序的可靠性。本节将创建两个项目：一个是静态库函数项目，以后的可复用代码都将放入该项目中；另一个项目将调用前者的静态库函数。

（1）创建 Android Library

创建项目 Library，包名设置为 com.walkerma.library，点击【Next】。如图 3-3 所示，不要选择第一和第二项（icon 和 activity），但选择第三项以将该项目标识为 Library，点击【Finish】完成创建。

图 3-3　创建 Android Library

上一节创建了新类 DrawView，这里也创建一个新类 Sound，在图 3-1 的 Package 文本框中输入 com.walkerma.library，因为这是第一个源代码类，需要指定包名。在 Name 文本框中输入 Sound，点击【Finish】完成 Sound 类的创建工作。

Sound 类没有构造函数，仅提供共享函数，因而可供外部调用的函数都是 public 类型且用 static 修饰，这样不需要生成对象即可使用"类名.函数名"的形式来调用。声音管理程序一般需要在上下文中调用 getSystemService 方法并指明 AUDIO_SERVICE 来获得 AudioManager 对象 am，调用 am 的 setRingerMode 方法即可设置情景模式，整型常量 RINGER_MODE_NORMAL 为正常模式，RINGER_MODE_SILENT 为静音模式，RINGER_MODE_VIBRATE 为振动模式。

AudioManager 对象 am 的 getRingerMode 方法读取当前情景模式，返回三个整型常量之一，getSoundMode 函数调用了 am 的 getRingerMode 方法。

Sound 类的 setMaxInCallVolume 函数设置免提最大音量（减 1）。首先通过 am 对象的 setMode 方法设置为免提模式（MODE_IN_CALL），打开扬声器，然后通过 getStreamMaxVolume 方法获得最大音量，setStreamVolume 方法设置最大音量（减 1）。

```
package com.walkerma.library;
import android.content.Context;
import android.media.AudioManager;
```

```
public class Sound {
    public static void setSoundMode(Context context, int ringerMode){
        AudioManager am = (AudioManager)context.
                getSystemService(Context.AUDIO_SERVICE);
        if(am == null) return;
        am.setRingerMode(ringerMode);
    }
    public static int getSoundMode(Context context){
        AudioManager am = (AudioManager)context.
                getSystemService(Context.AUDIO_SERVICE);
        if(am == null) return AudioManager.RINGER_MODE_NORMAL;
        return am.getRingerMode();
    }
    public static void setMaxInCallVolume(Context context){
        AudioManager am = (AudioManager)context.
                getSystemService(Context.AUDIO_SERVICE);
        am.setMode(AudioManager.MODE_IN_CALL);
        am.setStreamVolume(AudioManager.STREAM_RING,
                am.getStreamMaxVolume(AudioManager.STREAM_RING) - 1, 0);
    }
}
```

（2）设置手机情景模式和音量

新建项目 RingerMode，设置手机情景模式和音量需要权限 WRITE_SETTINGS、WRITE_SECURE_SETTINGS 和 MODIFY_AUDIO_SETTINGS，可以打开 AndroidManifest 配置文件，如图 3-4 所示，点击底部的 Permissions 标签，在右侧的 Name 下拉框（未显示）中选择合适的权限，然后点击【Add】按钮添加。

图 3-4　设置应用程序权限

右击 RingerMode 项目，选择属性 Properties，如图 3-5 所示，在左侧选择"Android"，在右下方点击【Add】按钮，从弹出的对话框中选择刚刚创建的 Library，完成准备工作。

RingerMode 项目的布局文件比较简单，设置了四个按钮，分别完成三种情景模式的切换及设置音量。在 MainActivity 源代码中，直接通过 Sound 类名调用其中的公用函数/方法，而无须对 Sound 类初始化。系统自动引导添加对 Sound 类的引用，见源代码中的黑体所示。

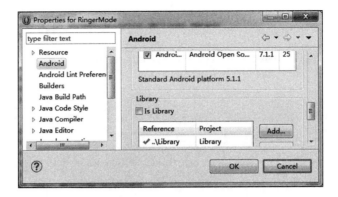

图 3-5　添加 Android Library 的引用

```java
package com.ch03.ringermode;
import com.walkerma.library.Sound;
import android.app.Activity;
import android.media.AudioManager;
import android.os.Bundle;
import android.view.View;
import android.view.View.OnClickListener;
import android.widget.Button;
public class MainActivity extends Activity implements OnClickListener{
    private Button btNormal, btSilent, btVibrate, btMaxVolume;
    @Override
    protected void onCreate(Bundle savedInstanceState) {
        super.onCreate(savedInstanceState);
        setContentView(R.layout.activity_main);
        btNormal = (Button)findViewById(R.id.btNormal);
        btNormal.setOnClickListener(this);
        btSilent = (Button)findViewById(R.id.btSilent);
        btSilent.setOnClickListener(this);
        btVibrate = (Button)findViewById(R.id.btVibrate);
        btVibrate.setOnClickListener(this);
        btMaxVolume = (Button)findViewById(R.id.btMaxVolume);
        btMaxVolume.setOnClickListener(this);
    }
    @Override
    public void onClick(View v) {
        // TODO Auto-generated method stub
        int id = v.getId();
        switch(id){
        case R.id.btNormal:
            Sound.setSoundMode(this, AudioManager.RINGER_MODE_NORMAL);
            break;
        case R.id.btSilent:
            Sound.setSoundMode(this, AudioManager.RINGER_MODE_SILENT);
            break;
        case R.id.btVibrate:
            Sound.setSoundMode(this, AudioManager.RINGER_MODE_VIBRATE);
            break;
        case R.id.btMaxVolume:
            Sound.setMaxInCallVolume(this);
            break;
```

                }
            }
        }

RingerMode 程序运行后，可以方便地设置三种情景模式和音量，但并无任何核心代码，因为核心代码都放在静态函数库中。

3.4 播放音频

Android 可以支持多种音频格式，来电、短信与微信及监控信号的变化等都需要播放音频。MediaPlayer 类可以播放音频，只需要创建该类的对象，并指定需要播放的音频文件，然后调用它的 start 方法即可。

（1）创建项目

创建项目 MediaPlayer，在该项目中仅设置一个按钮来播放指定的音频 notify.wav。在资源目录 res 下创建子目录 raw，并将需要播放的音频文件复制到这里。为了达到较好的播放效果，需要调节音量，因而需要在配置文件中设置 MODIFY_AUDIO_SETTINGS 权限。

（2）源代码分析

本项目通过免提最大音量方式播放音频，当程序退出时需要恢复以前的声音参数设置。在主窗体的 onCreate 方法中调用 MediaPlayer 类的静态方法 create 指定播放文件，得到初始化对象 mPlayer，然后调用自定义函数 setSoundMode 保存当前的声音设置，包括模式、音量和扬声器状态等，然后设置成免提最大音量模式。

在命令按钮的 onClick 事件处理函数中播放音频。首先调用 mPlayer 对象的 getCurrentPosition 方法，获取当前播放位置，如果该位置大于 0，那么需要调用 seekTo(0) 方法将位置恢复为 0，然后才能调用 start 方法播放音频。

按手机的后退键，程序将退出，在主窗体的 onDestroy 方法中调用自定义函数 restoreSoundMode 来恢复原来的声音参数设置。

```
package com.ch03.mediaplayer;
import android.app.Activity;
import android.content.Context;
import android.media.AudioManager;
import android.media.MediaPlayer;
import android.os.Bundle;
import android.view.View;
public class MainActivity extends Activity {
    private AudioManager am;
    private int nMode;
    private int nCurrentVolume;
    private boolean bSpeakerStatus;
    private MediaPlayer mPlayer;
    @Override
    protected void onCreate(Bundle savedInstanceState) {
        super.onCreate(savedInstanceState);
```

```java
        setContentView(R.layout.activity_main);
        mPlayer = MediaPlayer.create(this, R.raw.notify);
        setSoundMode();
    }
    public void onClick(View v){
        try{
            if(mPlayer.getCurrentPosition()>0)
                mPlayer.seekTo(0);
            mPlayer.start();
        } catch (IllegalStateException e) {
            e.printStackTrace();
        }
    }
    @Override
    protected void onDestroy() {
        restoreSoundMode();
        super.onDestroy();
    }
    private void setSoundMode(){
        am = (AudioManager)getSystemService(Context.AUDIO_SERVICE);
        if(am == null) return;
        nMode = am.getMode();
        am.setMode(AudioManager.MODE_IN_CALL);
        nCurrentVolume = am.getStreamVolume(AudioManager.STREAM_MUSIC);
        am.setStreamVolume(AudioManager.STREAM_MUSIC,
                am.getStreamMaxVolume(AudioManager.STREAM_MUSIC ),0);
        bSpeakerStatus = am.isSpeakerphoneOn();
        bSpeakerStatus = am.isSpeakerphoneOn();
        am.setSpeakerphoneOn(true);
    }
    private void restoreSoundMode(){
        if(am == null) return;
        am.setMode(nMode);
        am.setStreamVolume(AudioManager.STREAM_MUSIC, nCurrentVolume, 0);
        am.setSpeakerphoneOn(bSpeakerStatus);
    }
}
```

（3）运行效果

程序运行后，哪怕是在静音模式下，每次点击命令按钮都能从头开始以最大音量播放音频。程序关闭后，将恢复为原来的情景模式。

3.5 利用消息机制处理后退键

一个标准的 Android 设备包含了多个能够触发击键事件的按键。智能手机中的常用物理按键及相关事件与说明如表 3-3 所示，在应用程序中需要处理的最常用的按键为后退键，因为该键会导致用户意外退出程序。在 Android 中处理物理按键事件时，提供的回调方法有 onKeyUp 和 onKeyDown 等。为了避免用户意外退出应用程序，可使用 onKeyDown 方法监控按键，同时结合消息机制。

表 3-3 Android 设备常用物理按键

按键	KeyEvent	说　　明
后退键	KEYCODE_BACK	返回到前一个界面
菜单键	KEYCODE_MENU	显示当前应用的可用菜单
HOME 键	KEYCODE_HOME	返回到 HOME 界面
相机键	KEYCODE_CAMERA	启动相机
电源键	KEYCODE_POWER	启动或唤醒设备
音量键	KEYCODE_VOLUME_UP KEYCODE_VOLUME_DOWN	控制当前上下文音量，如手机的通话音量、音乐播放器的音量等

（1）消息处理类 Handler

消息处理类 Handler 允许发送 Message 或 Runnable 对象到其所在线程的消息队列中并进行处理，每个 Handler 类的实例都与一个线程及其消息队列相联系，当创建一个 Handler 实例时，这个实例就绑定到创建该实例的线程和消息队列中。

Handler 类采用 post 方法将 Runnable 对象发送到所在线程的消息队列中排队执行，可以立即发送，也可以延迟发送，延迟的时间单位为毫秒。Handler 类通过 sendMessage 方法将消息发送到所在线程，可以立即发送（空）消息，也可以延迟发送（空）消息，消息中含有需要传递到主线程中的数据以及消息识别代码。Handler 的 handleMessage 是一个自动回调方法，需要根据实际功能重写。Handler 类提供的常用方法如表 3-4 所示。

表 3-4 Handler 类提供的常用方法

方　　法	说　　明
handleMessage(Message msg)	处理消息的方法，自动回调，需要重写
post(Runnable r)	立即发送 Runnable 对象
postDelayed(Runnable r, long delayMillis)	延迟 delayMillis 毫秒发送 Runnable 对象
sendMessage(Message msg)	立即发送消息
sendMessageDelayed(Message msg, long delayMillis)	延迟 delayMillis 毫秒发送消息
sendEmptyMessage(int what)	立即发送空消息
sendEmptyMessageDelayed(int what,long delayMillis)	延迟 delayMillis 毫秒发送空消息

（2）消息类 Message

尽管 Message 类有默认的构造方法，但通常采用 Handler 对象的 obtainMessage 方法来获得并初始化一个 Message 对象，指定表 3-5 所示的参数，其中 what 用来标识消息，如果需要向主线程传递整型变量，直接设置 arg1 和 arg2 的值，如果不需要，默认设置为 0 即可。obj 是 Object 类型，也就是通用类型，可以是任何类型的数据。关于 obtainMessage 方法的具体应用和 Message 的进一步说明详见下一节。

表 3-5 Message 类的常用属性

属性	类型	说　　明
what	int	用户自定义的消息标识，一般为整型常量
arg1	int	用来存放发送给接收器的整型数据
arg2	int	用来存放发送给接收器的整型数据
obj	Object	用来存放发送给接收器的 Object 类型的数据

（3）创建项目与源代码分析

创建项目 Back，采用默认的布局文件，设置 boolean 变量 bCanExit，初始值为 false，表示不退出，MSG_DELAY 为整型消息标识，mHandler 为 Handler 对象。新建一个内部类 IncomingHandlerCallback，实现 Handler 的回调接口，重写 handleMessage 方法，当有消息时调用 processMessage 函数，其中采用 switch 语句，可以处理多种消息。在 onCreate 方法中对 mHandler 进行初始化，确保有消息到达时能够最终调用 processMessage 函数。

重写 onKeyDown 方法，其中也采用 switch 语句，以利于程序的扩展。当按下后退键时，bCanExit 为 false，if 条件满足，设置 bCanExit 为 true，显示"Press again to quit."（再次按下退出），发送延迟 1000 毫秒的空消息，最后返回 true，屏蔽后退键。如果在 1000 毫秒之内再次按下后退键，这时 bCanExit 为 true，将执行后退键的功能，即退出本程序（只有一个界面）。

如果在发送空消息 1000 毫秒之内没有再次按下后退键，将执行消息处理函数 processMessage 中的内容，将 bCanExit 恢复为 false，这时再按下后退键，程序还是不能退出。

程序退出时，在 onDestroy 方法中调用 mHandler 的 removeCallbacksAndMessages 方法，并传入 null 作为参数，确保移除所有回调和消息，避免内存泄漏。

```
package com.ch03.back;
import android.app.Activity;
import android.os.Bundle;
import android.os.Handler;
import android.os.Message;
import android.view.KeyEvent;
import android.widget.Toast;
public class MainActivity extends Activity {
    private Handler mHandler;
    boolean bCanExit = false;
    final int MSG_DELAY = 10001;
    class IncomingHandlerCallback implements Handler.Callback{
        @Override
        public boolean handleMessage(Message msg) {
            processMessage(msg);
            return true;
        }
    }
    private void processMessage(Message msg){
        switch(msg.what){
            case MSG_DELAY:
```

```
            bCanExit = false;
            break;
        }
    }
    @Override
    protected void onCreate(Bundle savedInstanceState) {
        super.onCreate(savedInstanceState);
        setContentView(R.layout.activity_main);
        mHandler = new Handler(new IncomingHandlerCallback());
    }
    @Override
    protected void onDestroy() {
        mHandler.removeCallbacksAndMessages(null);
        super.onDestroy();
    }
    @Override
    public boolean onKeyDown(int keyCode, KeyEvent event) {
        switch(keyCode){
        case KeyEvent.KEYCODE_BACK:
            if(!bCanExit){
                bCanExit = true;
                Toast.makeText(getApplicationContext(),
                        "Press again to quit.", Toast.LENGTH_SHORT).show();
                mHandler.sendEmptyMessageDelayed(MSG_DELAY, 1000);
                return true;                         //屏蔽后退键
            }
            else break;
        }
        return super.onKeyDown(keyCode, event);     //执行后退键功能
    }
}
```

（4）运行效果

程序运行后，按一次后退键，将显示"Press again to quit."；如果在 1 秒之内连续按两次后退键，程序将退出，否则保持原界面不变。

3.6 利用多线程和消息机制获取 IP 地址

在 Android 系统中，所有的操作在默认情况下都是在主线程中进行的，这个主线程负责管理与用户界面（User Interface，UI）相关的事件。对于一些比较耗时的操作或者需要并行完成的操作，通常使用一个独立的线程来执行，否则容易引起程序崩溃。在主线程之外的子线程中，不能直接更新 UI 组件，但可以通过上一节介绍的消息传递机制来完成。获取 IP 地址是一个耗时的工作，需要使用多线程。

（1）多线程类 Thread

多线程类 Thread 的常用构造函数如表 3-6 所示。Java 线程是 Thread 类的对象，可以以两种方式创建多线程的 Java 应用程序：实现 Runnable 接口或者直接继承 Thread 类。

表 3-6　Thread 类的常用构造函数

构造函数	说　　明
Thread()	无参数
Thread(Runnable runnable)	含 Runnable 对象
Thread(String threadName)	含线程名称
Thread(Runnable runnable, String threadName)	含 Runnable 对象和线程名称

Thread 类的常用方法如表 3-7 所示，其中的静态方法可以直接以类名调用。Thread 类的 stop 方法不安全，且容易使得应用程序或者虚拟机处于不可预测的状态，已经不再使用。

表 3-7　Thread 类的常用方法

常用方法	说　　明
static int　activeCount()	返回目前具有多少个执行中的线程
static Thread　currentThread()	返回当前的线程对象
static boolean　interrupted()	返回当前线程是否存在一个待处理的中断请求
static void sleep(long time)	线程休眠 time 毫秒
void start()	启动线程
final boolean　isAlive()	返回线程是否在执行中
void　interrupt()	发出中断线程的请求

一个简洁的 thread 对象可以如此创建，该线程对象可以用 thread.start() 方法启动，也可以方便地用 thread. interrupt() 方法中断。调用 Thread 类的非静态方法，必须先检查所在对象是否为 null，如果为 null 则将抛出异常。

```
private Thread thread = new Thread(new Runnable(){
    @Override
    public void run() {
        // TODO Auto-generated method stub
        while(!Thread.currentThread().isInterrupted()){
            //你的操作
        }
}});
```

（2）创建项目与源代码分析

创建项目 ThreadMsgIP，在配置文件中添加 ACCESS_WIFI_STATE 权限。因为获取 IP 地址是一种耗时操作，因而需要借助多线程来实现；在多线程中又不能更新 UI 组件的内容，因而需要采用消息机制。该项目在默认布局的基础之上添加一个按钮，用来启动新线程。

在 onCreate 方法中完成初始化工作，在 onClick 方法中启动多线程，进而获取 IP 地址。获取 IP 地址的多线程 ThreadGetIP 直接继承自 Thread 类，重写了 run 方法。首先获取当前 WifiManager 对象 wm，判断 WiFi 是否可用，如果可用就调用 getLocalWifiIP 函数获取字符串形式的 IP 地址。

在 getLocalWifiIP 函数中通过 wm 对象的 getConnectionInfo 方法获取 WifiInfo 对象 wfInfo，

从而进一步获得整型格式的 IP 地址，该 IP 地址需要格式化成文本字符串，然后返回。

Handler 对象 mHandler 的 obtainMessage 方法获得消息对象，再调用消息对象的 sendToTarget 方法将数据发送到主线程，由主线程的 processMessage 函数来更新文本框组件中的内容。

```java
package com.ch03.threadmsgip;
import java.util.Locale;
import android.app.Activity;
import android.content.Context;
import android.net.wifi.WifiInfo;
import android.net.wifi.WifiManager;
import android.os.Bundle;
import android.os.Handler;
import android.os.Message;
import android.view.View;
import android.widget.TextView;
public class MainActivity extends Activity {
    private TextView txtIP;
    private Handler mHandler;
    final int MSG_IP = 10001;
    class ThreadGetIP extends Thread{
        public void run(){
            String strIP;//
            WifiManager wm = (WifiManager)
                    getSystemService(Context.WIFI_SERVICE);
            if (wm.isWifiEnabled())
                strIP = getLocalWifiIP();
            else
                strIP = "NULL";
            mHandler.obtainMessage(MSG_IP, 0, 0, strIP).sendToTarget();
        }
    }
    class IncomingHandlerCallback implements Handler.Callback{
        @Override
        public boolean handleMessage(Message msg) {
            processMessage(msg);
            return true;
        }
    }
    private void processMessage(Message msg){
        switch(msg.what){
        case MSG_IP:
            txtIP.setText("WIFI " + (String)msg.obj);
            break;
        }
    }
    private String getLocalWifiIP(){
        WifiManager wm = (WifiManager) getSystemService(WIFI_SERVICE);
        WifiInfo wfInfo = wm.getConnectionInfo();
        int ipAddress = wfInfo.getIpAddress();
        return String.format(Locale.ENGLISH, "%d.%d.%d.%d",
                (ipAddress & 0xff),
                (ipAddress >> 8 & 0xff),
                (ipAddress >> 16 & 0xff),
```

```
                (ipAddress >> 24 & 0xff));
    }
    @Override
    protected void onCreate(Bundle savedInstanceState) {
        super.onCreate(savedInstanceState);
        setContentView(R.layout.activity_main);
        txtIP = (TextView)findViewById(R.id.txtIP);
        mHandler = new Handler(new IncomingHandlerCallback());
    }
    @Override
    protected void onDestroy() {
        // TODO Auto-generated method stub
        mHandler.removeCallbacksAndMessages(null);
        super.onDestroy();
    }
    public void onClick(View v){
        if(v.getId() != R.id.btGetIP) return;
        ThreadGetIP thGetIP = new ThreadGetIP();
        thGetIP.start();
    }
}
```

在多线程中，Handler 对象 mHandler 的 obtainMessage 相关方法可以由如下语句完成，显然没有 obtainMessage 方法简单高效。

```
Message msg = new Message();          //生成消息对象
msg.what=MSG_IP;                       //填写消息标识
msg.obj=strIP;                         //填写需要发送到主线程中的数据
msg.setTarget(mHandler);               //设置sendToTarget方法的发送目标
msg.sendToTarget();                    //调用sendToTarget方法向mHandler发送消息
```

（3）运行效果

运行程序，点击命令按钮，将生成多线程对象，并启动多线程，在多线程中将获得的 IP 地址字符串通过消息的形式发送到主线程，从而在主线程中完成文本框内容的更新。运行效果如图 3-6 所示。如果打开了 WiFi 开关而没有建立 Internet 连接，得到的 IP 地址为 "0.0.0.0"。

图 3-6　利用多线程与消息机制获取 IP 地址

3.7　定时功能的实现

以移动智能终端为控制中枢的多屏互动、智能家居等应用不断发展，如将 Android 手机作为遥控器控制照明灯、洗碗机、落地灯等家用电器。手机控制家用电器需要定时查询家用电器的状态，以便进行显示或控制决策。本节介绍基本定时功能的实现，该技术可用于对家用电器状态的定时查询，并在第 15 章用该技术查询设备的温度。

（1）创建项目与源代码分析

创建项目 Timer，利用默认的布局文件，再加一个按钮，用于启动定时，从而显示时间。整型常量 DELAY_MS 为间隔时间 1000 毫秒，MSG_TIME 为消息标识。runnable 对象发送消息后调用 Handler 对象 mHandler 的 postDelayed 方法延迟 1000 毫秒后再次发送消息。

最终的消息处理函数 processMessage 调用 getCurrentTime 函数获取当前时间，并在 txtTime 文本框中显示。在 getCurrentTime 函数中首先调用 Calendar 的静态方法 getInstance 以获得 Calendar 对象 calendar，再调用 calendar 对象的 getTime 方法以得到 Date 对象 date，最后通过 SimpleDateFormat 类的对象得到格式化的时间字符串。

在 onCreate 方法中对 mHandler 对象进行初始化，在 onClick 方法中通过调用 mHandler 对象的 postDelayed 方法发送 runnable 消息，启动定时操作。程序关闭时需要在 onDestroy 方法中移除所有回调和消息。

```
package com.ch03.timer;
import java.text.SimpleDateFormat;
import java.util.Calendar;
import java.util.Date;
import java.util.Locale;
import android.app.Activity;
import android.os.Bundle;
import android.os.Handler;
import android.os.Message;
import android.view.View;
import android.widget.TextView;
public class MainActivity extends Activity {
    private TextView txtTime;
    private Handler mHandler;
    final int DELAY_MS = 1000;
    final int MSG_TIME = 10001;
    Runnable runnable = new Runnable() {
        @Override
        public void run() {
            mHandler.obtainMessage(MSG_TIME, 0, 0, null).sendToTarget();
            mHandler.postDelayed(this, DELAY_MS);
        }
    };
    class IncomingHandlerCallback implements Handler.Callback{
        @Override
        public boolean handleMessage(Message msg) {
            processMessage(msg);
            return true;
        }
    }
    private void processMessage(Message msg){
        switch(msg.what){
        case MSG_TIME:
            txtTime.setText(getCurrentTime());
            break;
        }
    }
    @Override
    protected void onCreate(Bundle savedInstanceState) {
        super.onCreate(savedInstanceState);
        setContentView(R.layout.activity_main);
        txtTime = (TextView)findViewById(R.id.txtTime);
        mHandler = new Handler(new IncomingHandlerCallback());
    }
```

```
public void onClick(View v){
    mHandler.postDelayed(runnable, DELAY_MS);
}
@Override
protected void onDestroy() {
    // TODO Auto-generated method stub
    mHandler.removeCallbacksAndMessages(null);
    super.onDestroy();
}
private String getCurrentTime(){
    Calendar calendar = Calendar.getInstance();
    SimpleDateFormat simpleDateFormat = new SimpleDateFormat(
            "HH:mm:ss", Locale.getDefault());
    Date date = calendar.getTime();
    return simpleDateFormat.format(date);
}
```

（2）运行效果

程序运行后，每隔 1 秒自动更新文本框中的时间，效果如图 3-7 所示。

3.8 SQLite 与自定义 ListView

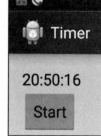

图 3-7 定时器

SQLite 是一款轻型数据库，是遵守 ACID 的关系型数据库管理系统。SQLite 的设计目标是嵌入式的，占用的资源非常少，广泛应用于嵌入式系统设备中。与常见的客户机/服务器工作方式不同，SQLite 引擎直接连接到程序中并成为它的一个主要部分，因而，主要通信协议通过在编程语言内直接调用 API 来实现。本节利用 SQLite 数据库给 ListView 添加数据项，展示 SQLite 数据库的操作技术。

（1）SQLiteOpenHelper 类

SQLiteOpenHelper 类用于数据库的创建及版本管理，可以重写其 onCreate (SQLite-Database)、onUpgrade(SQLiteDatabase, int, int) 和 onOpen(SQLiteDatabase) 方法，如果存在该数据库就打开它，如果不存在该数据库则创建它。SQLiteOpenHelper 类的构造函数的参数说明见表 3-8，第三项参数 factory 一般取 null 值。

表 3-8 SQLiteOpenHelper 类的构造函数参数

参　　数	说　　明
Context context	数据库的上下文环境
String name	数据库的名字（与数据表的名字不同）
SQLiteDatabase.CursorFactory factory	用于在执行查询时返回 Cursor 的一个子类
int version	数据库的版本号

SQLiteOpenHelper 类的常用方法如表 3-9 所示。onCreate 为抽象方法，需要重写，在数据库第一次创建时自动执行，调用数据库对象 db 的 execSQL 方法执行 SQL 语句的

CREATE TABLE 指令来创建数据库。

表 3-9　SQLiteOpenHelper 类的常用方法

方　　法	说　　明
abstract void　onCreate(SQLiteDatabase db)	创建数据库
SQLiteDatabase　getReadableDatabase()	创建或打开一个数据库实例
SQLiteDatabase　getWritableDatabase()	创建或打开一个用于读写的数据库实例
abstract void　onUpgrade(SQLiteDatabase db, int old Version, int newVersion)	更新数据库
synchronized void　close()	关闭数据库

getReadableDatabase 和 getWritableDatabase 方法都返回一个数据库对象，但是当磁盘满时，前者返回一个只读数据库，后者会抛出 SQLiteException 异常。

当数据库是旧版（参照表 3-8 中的版本号）时调用 onUpgrade 方法来修改数据库，先执行 SQL 语句的 DROP TABLE 指令来删除数据表，再调用 onCreate 方法重新建立数据表。onUpgrade 方法也是抽象方法，需要重写。

数据库不再使用时，需要调用 close 方法关闭。

（2）其他相关类

ContentValues 类采用（key, value）的形式存储数据，key 是 String 类型的数据的名称，value 是具体数据，只限于基本类型。存储数据采用 put 方法，读取数据采用 get 方法。在 SQLite 数据库系统中，该类主要用于存储字段名和数值对。

SQLiteDatabase 是数据库类，insert 方法用于插入一条记录，第一个参数为数据表的名称，一个数据库中可能包含多张表，所以插入记录时要指定表的名称，第二个参数一般为 null，第三个参数就是 ContentValues 对象。delete 方法用于删除一条记录，update 方法用于更新一条记录，参数 whereClause 是删除或更新的条件，whereArgs 是条件的进一步说明，以下两种方法等价：如果 whereClause 直接指定了字段的值，whereArgs 就不需要了，填写 null 即可；如果 whereClause 中用 "?" 来代替字段值，那么 whereArgs 就要用字符串数组来表示字段值。

```
"fieldName = '" + strValue + "'", null
"fieldName = ?", new String[]{"value"}
```

SQLiteDatabase 类的 rawQuery 方法的第一个参数为一个 Select 查询语句，第二个参数的用法与 whereArgs 类似，该方法返回一个 Cursor 对象，即记录的集合。

Cursor 类的常用方法如表 3-10 所示。获得 Cursor 对象后，需要逐条处理记录，因而，涉及记录指针的前后移动、是否到达首尾等方法。获取某条记录的某个字段的数据可采用 getInt、getFloat 和 getString 方法，需要传入字段的索引号。在编程实践中，一般字段的索引号不方便记忆，而字段的名称比较清楚，因而为了读取字段数据，可以通过 Cursor 类的 getColumnIndex 方法由字段名得到字段的索引号，从而通过 getInt 等方法读取数据。

表 3-10　Cursor 类的常用方法

方　法	说　明
boolean moveToFirst()	将指针移到第一条记录处，如果集合为空则返回 false，否则返回 true
boolean moveToNext ()	将指针移到下一条记录处，如果超过最后一条记录则返回 false，否则返回 true
boolean moveToPrevious()	将指针移到上一条记录处，如果超过第一条记录则返回 false，否则返回 true
boolean moveToLast()	将指针移到最后一条记录处，如果集合为空则返回 false，否则返回 true
boolean isFirst()	指针是否位于第一条记录
boolean isLast()	指针是否位于最后一条记录
boolean isAfterLast()	指针是否在最后一条记录之后
boolean isBeforeFirst()	指针是否在第一条记录之前
int getCount()	返回记录数
int getColumnCount()	返回字段数
int getColumnIndex(String columnName)	返回字段名的索引号
String getColumnName(int columnIndex)	返回索引号对应的字段名
float getFloat(int columnIndex)	获取指定索引号字段的浮点数据
int getInt(int columnIndex)	获取指定索引号字段的整数
String getString(int columnIndex)	获取指定索引号字段的字符串
boolean isNull(int columnIndex)	返回指定索引号字段是否为空值

在操作 Cursor 对象的时候，一般要检查是否为空值或记录为空的情况。

（3）常用数据类型及应用

SQLite 中常用的数据类型有 TEXT（文本）、INTEGER（整型）和 REAL（实型）。创建表格的 SQL 语句为：

```
CREATE TABLE tableName (
    _id INTEGER PRIMARY KEY,
    name TEXT NOT NULL,
    course TEXT NULL,
    grade REAL DEFAULT 0)
```

对于多字段名作为关键字的情况，只需在 CREATE 语句的最后使用"PRIMARY KEY (field1, field2, …)"来指定关键字。

（4）通用 SQLite 数据库类

在 3.3 节的 com.walkerma.library 中新建 DatabaseHelper 数据库类，它继承自 SQLite-OpenHelper 类，其中，字符串变量 strTableName 是表的名称，strCreateDb 是创建数据库的 CREATE 语句，db 是 SQLiteDatabase 数据库对象，rowRecords 是 Cursor 对象，表示这是一个记录集合。

第一个构造函数直接调用父类的构造函数。第二个构造函数另外传入表的名称与字段列表，调用第一个构造函数后，生成 CREATE 语句并存放在变量 strTableName 中，以便在 onCreate 方法中使用。

DatabaseHelper 类初始化后，调用 refreshReadDB 并传入查询字符串（SELECT 语句），获得 db 和 rowRecords，即可进行插入、删除与更新操作。getCount 方法获得当前 rowRecords 中的记录数。数据库使用完毕，必须调用 close 方法关闭。

```java
package com.walkerma.library;
import android.content.ContentValues;
import android.content.Context;
import android.database.Cursor;
import android.database.sqlite.SQLiteDatabase;
import android.database.sqlite.SQLiteDatabase.CursorFactory;
import android.database.sqlite.SQLiteOpenHelper;
public class DatabaseHelper extends SQLiteOpenHelper {
    private String strTableName;
    private String strCreateDb;
    private SQLiteDatabase db=null;
    private Cursor rowRecords = null;
    public DatabaseHelper(Context context, String strDbName,
            CursorFactory factory, int nVersion) {
        super(context, strDbName, factory, nVersion);
    }
    public DatabaseHelper(Context context, String strDbName,
            int nVersion, String strTableName, String strFields)  {
        this(context, strDbName, null, nVersion);
        this.strTableName = strTableName;
        strCreateDb = "CREATE TABLE " + strTableName + "(" + strFields + ")";
    }
    @Override
    public void onCreate(SQLiteDatabase db) {
        db.execSQL(strCreateDb);
    }
    @Override
    public void onUpgrade(SQLiteDatabase db, int oldVersion, int newVersion) {
        db.execSQL("DROP TABLE IF EXISTS " + strTableName); //EXISTS
        onCreate(db);
    }
    @Override
    public synchronized void close() {
        if(rowRecords != null) rowRecords.close();
        if(db != null) db.close();
        super.close();
    }
    public void updateRecords(ContentValues values,
            String whereClause, String[] whereArgs){
        db.update(strTableName, values, whereClause, whereArgs);
    }
    public void deleteRecords(String whereClause, String[] whereArgs){
        db.delete(strTableName, whereClause, whereArgs);
    }
    public void insertRecords(ContentValues values){
        db.insert(strTableName, null, values);
    }
    public void refreshReadDB(String strQuery, String[] selecttionArgs){
```

```
        db = getWritableDatabase();
        rowRecords = db.rawQuery(strQuery, selecttionArgs);
    }
    public Cursor getRecords(){
        return rowRecords;
    }
    public int getCount(){
        if(rowRecords!=null)
            return rowRecords.getCount();
        else
            return 0;
    }
```

（5）适配器的处理

创建项目 SQLiteListView，引用 Library 类库，并在配置文件中添加 VIBRATE 权限。ListView 组件的行元素来自 SQLite 数据库，且行元素是自定义的，包括 TextView 文本内容（水果名称）、是否选中的 ImageView 标志和上下移动的 ImageView 箭头，其布局文件为 list_combined.xml。

```xml
<?xml version="1.0" encoding="utf-8"?>
<RelativeLayout xmlns:android="http://schemas.android.com/apk/res/android"
    android:layout_width="match_parent"
    android:layout_height="match_parent"
    android:orientation="horizontal" >
    <TextView
        android:id="@+id/txtItem"
        android:layout_width="wrap_content"
        android:layout_height="wrap_content"
        android:layout_marginLeft="10dp"
        android:layout_marginRight="10dp"
        android:layout_alignParentLeft="true"
        android:layout_centerVertical="true"
        android:lines="1"
        android:textAppearance="?android:attr/textAppearanceMedium"
        android:text="Fruit" />
    <ImageView
        android:id="@+id/check_mark"
        android:layout_width="40dp"
        android:layout_height="40dp"
        android:layout_centerVertical="true"
        android:layout_toRightOf="@id/txtItem"
        android:src="@drawable/check_mark_pure"
        android:visibility="invisible"
        android:contentDescription="@null"/>
    <ImageView
        android:id="@+id/circled_up"
        android:layout_width="40dp"
        android:layout_height="40dp"
        android:layout_alignParentRight="true"
        android:layout_centerVertical="true"
        android:layout_marginRight="10dp"
        android:layout_marginLeft="20dp"
        android:contentDescription="@null"
        android:src="@drawable/circled_up"
```

```
            android:visibility="invisible" />
    <ImageView
        android:id="@+id/circled_down"
        android:layout_width="40dp"
        android:layout_height="40dp"
        android:layout_toLeftOf="@id/circled_up"
        android:layout_centerVertical="true"
        android:contentDescription="@null"
        android:src="@drawable/circled_down"
        android:visibility="invisible" />
</RelativeLayout>
```

在适配器源代码 getView 方法中，首先完成布局文件 list_combined.xml 到 rowItem 视图的映射，然后进一步初始化行元素中的每个组件，注册上下箭头的 OnClick 侦听器。nItemPos 变量存放当前选中行的位置，位置初始为 –1，表示未选中选项，选中的选项要显示对勾图片和上下箭头图片（最上选项没有向上箭头，最下选项没有向下箭头）。

upMoveItem 函数将本选项向上移动一个位置，首先删除本选项，然后在本选项的上一位置加上该选项；downMoveItem 函数将本选项向下移动一个位置，原理与前面相似。

```java
package com.ch03.sqlitelistview;
import java.util.ArrayList;
import android.app.Activity;
import android.content.Context;
import android.os.Vibrator;
import android.view.LayoutInflater;
import android.view.View;
import android.view.ViewGroup;
import android.widget.ArrayAdapter;
import android.widget.ImageView;
import android.widget.TextView;
public class CustomListAdapter extends ArrayAdapter<String> {
    private int nItemPos = -1;
    private boolean bCanVibrate = false;
    private final int nVibrateTime = 30;
    private final Activity context;
    private ArrayList<String> listItems;
    private int mLayout;
    public CustomListAdapter(Activity context, int mLayout,
            ArrayList<String> listItems){
        super(context, mLayout, listItems);
        this.context = context;
        this.mLayout = mLayout;    // row xml
        this.listItems = listItems;
    }
    public View getView(int position,View convertView,ViewGroup parent) {
        View rowItem = convertView;
        final ViewHolder holder;
        if(rowItem == null){
            LayoutInflater inflater = context.getLayoutInflater();
            rowItem = inflater.inflate(mLayout, parent, false);
            holder = new ViewHolder();
            holder.txtItem = (TextView)rowItem.
                    findViewById(R.id.txtItem);
            holder.imgCheckMark = (ImageView)rowItem.
```

```java
                    findViewById(R.id.check_mark);
            holder.imgCircledDown = (ImageView)rowItem.
                    findViewById(R.id.circled_down);
            holder.imgCircledUp = (ImageView)rowItem.
                    findViewById(R.id.circled_up);
            rowItem.setTag(holder);
        }
        else holder = (ViewHolder) rowItem.getTag();
        holder.imgCircledDown.setOnClickListener(new View.OnClickListener() {
            @Override
            public void onClick(View view) {downMoveItem(nItemPos); }});
        holder.imgCircledUp.setOnClickListener(new View.OnClickListener() {
            @Override
            public void onClick(View view) {upMoveItem(nItemPos); }});
        String strData = listItems.get(position);
        holder.txtItem.setText(strData);
        if(nItemPos>=0 && nItemPos<listItems.size()){
            if(strData.equals(listItems.get(nItemPos))){
                holder.imgCheckMark.setVisibility(View.VISIBLE);
                if(nItemPos < listItems.size() - 1)
                    holder.imgCircledDown.setVisibility(View.VISIBLE);
                if(nItemPos > 0)
                    holder.imgCircledUp.setVisibility(View.VISIBLE);
            }
            else{
                holder.imgCheckMark.setVisibility(View.INVISIBLE);
                holder.imgCircledDown.setVisibility(View.INVISIBLE);
                holder.imgCircledUp.setVisibility(View.INVISIBLE);
            }
        }
        if(nItemPos == -1){
            holder.imgCheckMark.setVisibility(View.INVISIBLE);
            holder.imgCircledDown.setVisibility(View.INVISIBLE);
            holder.imgCircledUp.setVisibility(View.INVISIBLE);
        }
        return rowItem;
    }
    public void setSelectedPos(int nPos){
        nItemPos = nPos;
    }
    public void setVibrateStatus(boolean bCanVibrate){
        this.bCanVibrate = bCanVibrate;
    }
    public void vibrateCell(){
        //long pattern[] = {50,100,100,250,150,350};
        Vibrator v = (Vibrator)context.getSystemService(
                Context.VIBRATOR_SERVICE);
        //v.vibrate(pattern,3);
        v.vibrate(nVibrateTime);
    }
    public int getSelectedPos(){
        return nItemPos;
    }
    private void upMoveItem(int which){
        if(which <= 0 || which >= listItems.size()) return;
        String strTmp = listItems.get(which);
```

```
            listItems.remove(which);
            listItems.add(which - 1, strTmp);
            nItemPos = which - 1;    // adjust current position
            notifyDataSetChanged();
            if(bCanVibrate) vibrateCell();
        }
        private void downMoveItem(int which){
            if(which < 0 || which >= listItems.size() -1 ) return;
            String strTmp = listItems.get(which);
            listItems.remove(which);
            listItems.add(which+1, strTmp);
            nItemPos = which + 1;    // adjust current position
            notifyDataSetChanged();
            if(bCanVibrate) vibrateCell();
        }
        static class ViewHolder {
            TextView txtItem;
            ImageView imgCheckMark;
            ImageView imgCircledDown;
            ImageView imgCircledUp;
        }
    }
```

（6）主窗体布局与源代码

在主窗体布局文件 activity_main.xml 中，用 TextView 标记显示水果价格，ListView 的选项中只有水果名称。

```
<RelativeLayout xmlns:android="http://schemas.android.com/apk/res/android"
    xmlns:tools="http://schemas.android.com/tools"
    android:layout_width="match_parent"
    android:layout_height="match_parent"
    android:paddingBottom="@dimen/activity_vertical_margin"
    android:paddingLeft="@dimen/activity_horizontal_margin"
    android:paddingRight="@dimen/activity_horizontal_margin"
    android:paddingTop="@dimen/activity_vertical_margin"
    tools:context="com.ch03.sqlitelistview.MainActivity" >
    <TextView
        android:id="@+id/txtPrice"
        android:layout_width="wrap_content"
        android:layout_height="wrap_content"
        android:textAppearance="?android:attr/textAppearanceMedium"
        android:layout_alignParentTop="true"
        android:layout_centerInParent="true"
        android:text="@string/price" />
    <ListView
        android:id="@+id/list"
        android:layout_width="match_parent"
        android:layout_height="wrap_content"
        android:layout_alignParentLeft="true"
        android:layout_below="@+id/txtPrice"
        android:layout_marginTop="20dp" >
    </ListView>
</RelativeLayout>
```

在主窗体源代码的 onCreate 方法中，主要完成 CustomListAdapter 类的对象 adapter 的初始化工作，通过 setVibrateStatus 函数设置振动有效，同时将 adapter 绑定到 ListView 对象 list 中，并注册侦听器。

在 onItemClick 事件处理函数中，首次点击时，由于 adapter 对象的 getSelectedPos 函数读取的有效位置为 –1，与 position 不相等，因而调用 setSelectedPos 函数设置 position 为当前有效位置，获取水果名称，并根据水果名称获取其价格。如果水果名称被二次重复点击，则取消该选项，水果价格设置为 0。

接着生成自定义数据库类对象 db，数据库的名称为 dbFruit，其中数据表的名称为 fruit，该表只有两个字段 _fruit 和 _price，前者为关键字。最后调用 refreshItems 函数刷新数据库并给 list 添加数据项。

在 refreshItems 函数中，如果调用 refreshReadDB 函数得到的记录数为 0，那么调用 addDataToDB 函数添加三条记录。然后通过 getRecords 函数获取 Cursor 对象 rowRecords，通过循环将水果名称添加到 ArrayList 对象 listItems 中，最后调用 adapter 对象的 notifyDataSetChanged 方法通知数据改变，刷新 ListView 的内容。

```java
package com.ch03.sqlitelistview;
import java.util.ArrayList;
import com.walkerma.library.DatabaseHelper;
import android.app.Activity;
import android.content.ContentValues;
import android.database.Cursor;
import android.os.Bundle;
import android.view.View;
import android.widget.AdapterView;
import android.widget.ListView;
import android.widget.TextView;
import android.widget.AdapterView.OnItemClickListener;
public class MainActivity extends Activity {
    private TextView txtPrice;
    private ListView list;
    private DatabaseHelper db;
    private final String strQuery = "SELECT * FROM fruit";
    Cursor rowRecords;
    private ArrayList<String> listItems = new ArrayList<String>();
    private CustomListAdapter adapter;
    @Override
    protected void onCreate(Bundle savedInstanceState) {
        super.onCreate(savedInstanceState);
        setContentView(R.layout.activity_main);
        txtPrice = (TextView)findViewById(R.id.txtPrice);
        list = (ListView)findViewById(R.id.list);
        adapter = new CustomListAdapter(this, R.layout.list_combined, listItems);
        adapter.setVibrateStatus(true);
        list.setAdapter(adapter);
        list.setOnItemClickListener(new OnItemClickListener(){
            @Override
            public void onItemClick(AdapterView<?> parent, View view,
                    int position, long id) {
                if(adapter.getSelectedPos() != position){
```

```java
                    adapter.setSelectedPos(position);    //setItemChecked
                    String strItem = parent.getItemAtPosition(position).toString();
                    txtPrice.setText(Float.toString(getPrice(strItem)));
                }
                else{
                    adapter.setSelectedPos(-1);    //setItemUnChecked
                    txtPrice.setText("0");
                }
                adapter.notifyDataSetChanged();    // call getView automatically
        }});
        db = new DatabaseHelper(this, "dbFruit", 1, "fruit",
                "_fruit TEXT PRIMARY KEY, " +
                "_price REAL DEFAULT 0");
        refreshItems();
    }
    private void refreshItems(){
        db.refreshReadDB(strQuery, null);
        if(db.getCount() == 0) addDataToDB();
        db.refreshReadDB(strQuery, null);
        rowRecords = db.getRecords();
        rowRecords.moveToFirst();
        for(int i=0; i<db.getCount(); i++){
            listItems.add(rowRecords.getString(0));
            rowRecords.moveToNext();
        }
        adapter.notifyDataSetChanged();
    }
    private void addDataToDB(){
        ContentValues cv = new ContentValues();
        cv.put("_fruit", "Apple");
        cv.put("_price", 5.62);
        db.insertRecords(cv);
        cv.put("_fruit", "Banana");
        cv.put("_price", 4.5);
        db.insertRecords(cv);
        cv.put("_fruit", "Pear");
        cv.put("_price", 2.8);
        db.insertRecords(cv);
    }
    private float getPrice(String name){
        db.refreshReadDB("SELECT * FROM fruit WHERE _fruit ='" +
                name + "'", null);
        if(db.getCount()==0) return 0;
        rowRecords = db.getRecords();
        rowRecords.moveToFirst();
        return rowRecords.getFloat(1);
    }
    @Override
    protected void onDestroy() {
        db.close();
        super.onDestroy();
    }
}
```

（7）运行效果

SQLiteListView 项目的运行效果如图 3-8 所示，第一次运行时数据库为空，调用 add-DataToDB 函数添加三种水果及其价格到数据库中，然后再通过 ListView 显示，以后就直接从数据库中读取并显示。点击箭头，选项可以移动，同时手机振动。

3.9 查询联系人

手机联系人信息是一种通过内容提供程序（Content Provider）来实现的可分享的 SQLite 数据库的数据。

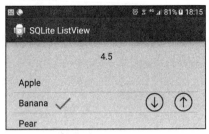

图 3-8　SQLiteListView 的运行效果

内容提供程序的主要目的是将保存数据和实际使用它的应用程序隔离开来，以便增加整个 Android 操作系统的弹性。内容提供程序是一组封装数据的容器，提供了不同应用程序之间分享数据的接口。

（1）内容提供程序的基础知识

Android 操作系统内置多个内容提供程序来存储共享信息，如通讯录、浏览器书签、通话记录等。内容提供程序中的数据通过统一资源定位器（Universal Resource Indicator，URI）来区分，其基本格式为：

`content://<内容提供程序名称>/<数据路径>/<记录编号>`

"内容提供程序名称"可以是系统定义的名称，也可以是自定义或系统定义的完整的包名，"数据路径"就是数据集合下面的分支，"记录编号"确定具体的记录。例如"联系人电话"的 URI 为：

`content://com.android.contacts/data/phones`

其中内容提供程序名称为具体的包名 com.android.contacts，数据路径为 data 下面的 phones 数据集。

URI 字符串不能直接使用，需要解析为 Uri 对象：

`Uri uri = Uri.parse("content://com.android.contacts/data/phones");`

为了使用方便，Android 定义了如下静态常量来表示该 Uri 对象：

ContactsContract.CommonDataKinds.Phone.CONTENT_URI

这时需要导入如下包名，ContactsContract 部分是重复的：

`import android.provider.`**`ContactsContract`**

为了在编程中撰写方便，直接使用 **Phone**.CONTENT_URI，可以增加导入包名的深度为：

`import android.provider.ContactsContract.CommonDataKinds.`**`Phone`**

为了读取内容提供程序中的数据，首先需要调用 getContentResolver 方法取得 Content Resolver 对象，然后调用该对象的 query 方法得到 Cursor 对象的记录集合。query 方法有 5 个参数，如表 3-11 所示。

表 3-11 查询参数说明

参　　数	说　　明
Uri uri	查询内容提供程序的 Uri 对象，字符串资源需要解析
String[] projection	字符串数组表示的字段列表，null 表示返回全部字段
String selection	筛选条件，SQL 语句中的 WHERE 子句，null 返回所有记录
String[] selectionArgs	如果筛选条件中有 "?"，就在此用字符串数组提供其值，否则为 null
String sortOrder	查询结果排序，null 表示默认排序

（2）适配器的布局说明

创建项目 ContactsQuery，在配置文件中添加 READ_CONTACTS 权限。该项目查找姓名开头字符相同的所有联系人的姓名和电话号码，需要定制适配器，其布局文件为 list_combined.xml，采用相对布局，姓名 TextView 靠左显示，电话号码 TextView 靠右显示。

```xml
<?xml version="1.0" encoding="utf-8"?>
<RelativeLayout xmlns:android="http://schemas.android.com/apk/res/android"
    android:layout_width="match_parent"
    android:layout_height="match_parent"
    android:orientation="horizontal" >
    <TextView
        android:id="@+id/txtName"
        android:layout_width="wrap_content"
        android:layout_height="wrap_content"
        android:layout_alignBaseline="@+id/txtPhone"
        android:layout_alignParentLeft="true"
        android:layout_centerVertical="true"
        android:layout_marginLeft="10dp"
        android:text="Name"
        android:textAppearance="?android:attr/textAppearanceMedium" />
    <TextView
        android:id="@+id/txtPhone"
        android:layout_width="wrap_content"
        android:layout_height="wrap_content"
        android:layout_alignParentRight="true"
        android:layout_centerVertical="true"
        android:layout_marginRight="10dp"
        android:text="Phone"
        android:textAppearance="?android:attr/textAppearanceMedium" />
</RelativeLayout>
```

（3）适配器源代码

本适配器的内置静态类 ViewHolder 当作结构使用，包括用于显示姓名和电话的两个 TextView。在适配器源代码 getView 方法中，当刷新的位置 position 与保存的有效位置 nItemPos 相等时，就设置数据行的背景色，否则清除背景色（为透明）。

```java
package com.ch03.contactsquery;
import java.util.ArrayList;
import android.app.Activity;
import android.graphics.Color;
import android.view.LayoutInflater;
```

```java
import android.view.View;
import android.view.ViewGroup;
import android.widget.ArrayAdapter;
import android.widget.TextView;
public class ContactListAdapter extends ArrayAdapter<RowHolder> {
    private int nItemPos = -1;
    private final Activity context;
    private ArrayList<RowHolder> listItems;
    private int mLayout;
    public ContactListAdapter(Activity context,
            int mLayout, ArrayList<RowHolder> listItems) {
        super(context, mLayout, listItems);
        this.context = context;
        this.mLayout = mLayout;     // row xml
        this.listItems = listItems;
    }
    public View getView(int position,View convertView,ViewGroup parent) {
        View rowItem = convertView;
        final ViewHolder holder;
        if(rowItem == null){
            LayoutInflater inflater = context.getLayoutInflater();
            rowItem = inflater.inflate(mLayout, parent, false);
            holder = new ViewHolder();
            holder.txtName = (TextView) rowItem.findViewById(R.id.txtName);
            holder.txtPhone = (TextView) rowItem.findViewById(R.id.txtPhone);
            rowItem.setTag(holder);
        }
        else holder = (ViewHolder) rowItem.getTag();
        if(nItemPos == position)
            rowItem.setBackgroundColor(Color.CYAN);
        else
            rowItem.setBackgroundColor(Color.TRANSPARENT);
        RowHolder rowData= new RowHolder();
        rowData = listItems.get(position);
        holder.txtName.setText(rowData.Name);
        holder.txtPhone.setText(rowData.Phone);
        return rowItem;
    }
    public void setSelectedPos(int nPos){
        nItemPos = nPos;
    }
    static class ViewHolder {
        TextView txtName;
        TextView txtPhone;
    }
}
```

(4) 主窗体布局文件

主窗体布局文件中用 EditText 来输入姓名的开头字符，用一个 Button 查询以这些开头字符开始的姓名和电话，并以升序排列，另一个 Button 完成相似的功能，但以降序排列。ListView 占用剩余的空间显示姓名和电话的列表。

```xml
<LinearLayout xmlns:android="http://schemas.android.com/apk/res/android"
    xmlns:tools="http://schemas.android.com/tools"
```

```xml
        android:layout_width="wrap_content"
        android:layout_height="match_parent"
        android:layout_marginBottom="@dimen/activity_horizontal_margin"
        android:layout_marginLeft="@dimen/activity_horizontal_margin"
        android:layout_marginRight="@dimen/activity_horizontal_margin"
        android:layout_marginTop="@dimen/activity_horizontal_margin"
        android:orientation="vertical"
        tools:context="com.ch03.contactsquery.MainActivity">
        <LinearLayout
            android:layout_width="match_parent"
            android:layout_height="wrap_content"
            android:gravity="center"
            android:orientation="horizontal" >
            <EditText
                android:id="@+id/edText"
                android:layout_width="wrap_content"
                android:layout_height="wrap_content"
                android:layout_marginRight="15dp"
                android:hint="Name"
                android:inputType="text"
                android:maxLength="6"
                android:text="@null" />
            <Button
                android:id="@+id/btUp"
                android:layout_width="wrap_content"
                android:layout_height="wrap_content"
                android:layout_marginRight="15dp"
                android:onClick="onClick"
                android:text="Up" />
            <Button
                android:id="@+id/btDown"
                android:layout_width="wrap_content"
                android:layout_height="wrap_content"
                android:onClick="onClick"
                android:text="Down" />
        </LinearLayout>
        <ListView
            android:id="@+id/list"
            android:layout_width="match_parent"
            android:layout_height="match_parent"
            android:divider="#ff1118ff"
            android:dividerHeight="1dp"
            android:drawSelectorOnTop="false" >
        </ListView>
</LinearLayout>
```

（5）主窗体源代码

主窗体源代码中的 onItemClick 事件显示行数据是否被选中。initAdapter 初始化适配器，第一个参数是姓名字符串，第二个参数为排序标识，其中调用 getData 函数获取数据行，并装载到 ArrayList 对象 listItems 中。

getData 函数使用"LIKE"关键字进行模糊查询，Phone.CONTENT_URI 是联系人资源的静态常量，projection 是映射字符串数组，包括姓名和电话，由于 selection 中没有"?"，因而其后的参数为"null",最后以电话号码字段排序输出。

```java
package com.ch03.contactsquery;
import java.util.ArrayList;
import android.app.Activity;
import android.content.Context;
import android.content.res.Configuration;
import android.database.Cursor;
import android.net.Uri;
import android.os.Bundle;
import android.provider.ContactsContract.Contacts;
import android.provider.ContactsContract.CommonDataKinds.Phone;
import android.text.TextUtils;
import android.view.View;
import android.widget.AdapterView;
import android.widget.AdapterView.OnItemClickListener;
import android.widget.EditText;
import android.widget.ListView;
import android.widget.Toast;
public class MainActivity extends Activity {
    private EditText edText;
    private ListView list;
    private ArrayList<RowHolder> listItems = 
            new ArrayList<RowHolder>();
    private ContactListAdapter adapter;
    int nCurrent = -1;
    private Context context;
    String[] projection = new String[] {Phone.DISPLAY_NAME, Phone.NUMBER};
    @Override
    protected void onCreate(Bundle savedInstanceState) {
        super.onCreate(savedInstanceState);
        setContentView(R.layout.activity_main);
        context = this;
        edText = (EditText)findViewById(R.id.edText);
        list = (ListView)findViewById(R.id.list);
        adapter = new ContactListAdapter(this,
                R.layout.list_combined, listItems);
        list.setAdapter(adapter);
        list.setOnItemClickListener(new OnItemClickListener(){
            String strName;
            @Override
            public void onItemClick(AdapterView<?> parent, View view,
                    int position, long id) {
                // TODO Auto-generated method stub
                RowHolder row = (RowHolder)parent.getItemAtPosition(position);
                strName = row.Name;
                if(nCurrent != position){
                    nCurrent = position;
                    Toast.makeText(context, strName + " selected",
                            Toast.LENGTH_SHORT).show();
                }
                else{
                    nCurrent = -1;
                    list.setItemChecked(position, false);
                    Toast.makeText(context, strName + " unselected",
                            Toast.LENGTH_SHORT).show();
```

```
            }
            adapter.setSelectedPos(nCurrent);
            adapter.notifyDataSetChanged();
        }});
}
private void initAdapter(String condition, String direction){
    listItems.clear();
    Cursor rowRecords = getData(condition, direction);
    if(rowRecords == null) {
        adapter.notifyDataSetChanged();
        return;
    }
    int nLen = rowRecords.getCount();
    if(nLen == 0) {
        adapter.notifyDataSetChanged();
        return;
    }
    rowRecords.moveToFirst();
    for(int i=0; i<nLen; i++){
        RowHolder rowData = new RowHolder(); //create every time
        rowData.Name = rowRecords.getString(0);
        rowData.Phone = rowRecords.getString(1);
        listItems.add(rowData);
        rowRecords.moveToNext();
    }
    adapter.notifyDataSetChanged();
}
private Cursor getData(String condition, String direction){
    String selection = Contacts.DISPLAY_NAME + " LIKE '" +
            condition + "%'";
    Cursor cr = getContentResolver().query(
            Phone.CONTENT_URI,
            projection, selection, null,
            Phone.NUMBER + " " + direction);
    return cr;
}
public void onClick(View v){
    String condition = edText.getText().toString();
    if(TextUtils.isEmpty(condition)){
        Toast.makeText(this, "Please input data.",
                Toast.LENGTH_SHORT).show();
        return;
    }
    int id = v.getId();
    switch(id){
    case R.id.btUp:
        initAdapter(condition, "ASC");
        break;
    case R.id.btDown:
        initAdapter(condition, "DESC");
        break;
    }
}
@Override
public void onConfigurationChanged(Configuration newConfig) {
    super.onConfigurationChanged(newConfig);
```

```
        adapter.notifyDataSetChanged();
    }
}
```

（6）运行效果

程序运行效果如图 3-9 所示，专门设置了三个电话号码，输入"安卓"，由于采用模糊查询，"安卓"开头的电话全部找到。这里按照电话号码排序，点击【Up】采用升序排序，点击【Down】采用降序排序，图中为降序排序的情况。

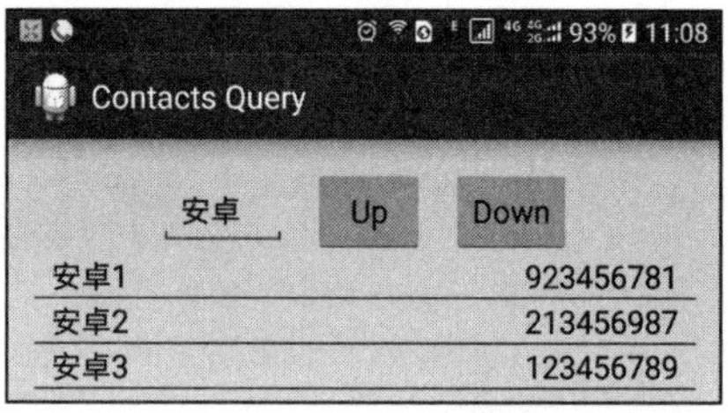

图 3-9 ContactsQuery 的运行效果

（7）查询的注意事项与查询的完整形式

ContentResolver 对象的 query 方法将 SELECT 语句分解成多个参数，以下面的 SELECT 语句和 getData 函数为例进行对照，SELECT 选择的字段名列表通过字符串数组 projection 给出，table1 相当于 Uri 对象（这里通过字符串进行解析），WHERE 后面的字段条件相当于 selection 参数，ORDER BY 后面的参数则是 query 方法的最后一个参数。如果条件 selection 中含有 "?"，则随后的参数通过字符串数组给出其值。

在使用 query 方法查询数据的过程中，参数中不包含 SELECT、FROM、WHERE 与 ORDER BY 等关键字。

```
SELECT field1, field2 FROM table1 WHERE field1="value" ORDER BY field1 ASC
private Cursor getData(String condition, String direction){
    String selection= Contacts.DISPLAY_NAME + "=?";
    Cursor cr = getContentResolver().query(
            Uri.parse("content://com.android.contacts/data/phones"),
            projection, selection, new String[]{condition},
            Phone.NUMBER + " " + direction);
    return cr;
}
```

3.10 使用 SharedPreferences 对象存储数据

SharedPreferences 类供开发人员保存和获取基本数据类型的键值对，主要用于基本类型，如 boolean、int、long、float 和 String 等，所存储的数据在程序退出后仍旧会保存。有两种方法可以获得 SharedPreferences 对象：① getSharedPreferences 方法，它有两个参数，第一个参数是保存数据的文件名，第二个参数为操作模式，一般默认用 0 或者 MODE_PRIVATE，这种方法可以保存以文件名作为分组的多组数据。② getPreferences 方法，它只有一个操作模式参数，只能保存一组数据，而且键值对只能用于本 Activity 对象。

（1）主窗体布局文件

创建项目 SharedListView，用 SharedPreferences 对象存储用户数据。EditText 组件用来输入数据，一个 Button 组件用来将数据加入 ListView，另一个 Button 组件将数据从 ListView 中删除。ListView 组件位于屏幕下方。

```xml
<LinearLayout xmlns:android="http://schemas.android.com/apk/res/android"
    xmlns:tools="http://schemas.android.com/tools"
    android:layout_width="wrap_content"
    android:layout_height="match_parent"
    android:layout_marginBottom="@dimen/activity_horizontal_margin"
    android:layout_marginLeft="@dimen/activity_horizontal_margin"
    android:layout_marginRight="@dimen/activity_horizontal_margin"
    android:layout_marginTop="@dimen/activity_horizontal_margin"
    android:orientation="vertical"
    tools:context="com.ch03.sharedlistview.MainActivity" >
    <LinearLayout
        android:layout_width="match_parent"
        android:layout_height="wrap_content"
        android:gravity="center"
        android:orientation="horizontal" >
        <EditText
            android:id="@+id/edText"
            android:layout_width="wrap_content"
            android:layout_height="wrap_content"
            android:layout_marginRight="15dp"
            android:hint="Item"
            android:inputType="text"
            android:maxLength="6"
            android:text="@null" />
        <Button
            android:id="@+id/btAdd"
            android:layout_width="wrap_content"
            android:layout_height="wrap_content"
            android:layout_marginRight="15dp"
            android:onClick="onClick"
            android:text="Add" />
        <Button
            android:id="@+id/btDelete"
            android:layout_width="wrap_content"
            android:layout_height="wrap_content"
            android:onClick="onClick"
            android:text="Delete" />
    </LinearLayout>
    <ListView
        android:id="@+id/list"
        android:layout_width="match_parent"
        android:layout_height="match_parent"
        android:choiceMode="singleChoice"
        android:divider="#ff1118ff"
        android:dividerHeight="1dp"
        android:drawSelectorOnTop="false" >
    </ListView>
</LinearLayout>
```

（2）主窗体源代码

getSharedString 函数将 ArrayList 对象 listItems 中的列表选项保存到一个字符串中，中间用"/"分隔。writeSharedData 函数使用上述第一种方法获取 SharedPreferences 对象 sp，以应用程序名字作为文件名，调用 sp 的 edit 方法来获取 Editor 对象，然后将列表字符串通过 putString 方法保存,最后通过 commit 方法提交。

readStorage 函数通过 SharedPreferences 对象的 getString 方法获取保存的字符串数据,然后通过 readSharedData 函数对字符串进行处理,采用字符串对象的 split 方法将由 "/" 分隔的字符串分解为字符串数组,然后逐个加入 listItems 中。

在删除一个选项时,必须调用 ListView 的 setItemChecked 方法将该选项设置为 false,然后再利用 listItems 对象的 remove 方法删除该选项,否则直接删除该选项,该位置的选项还是选中状态。

```java
package com.ch03.sharedlistview;
import java.util.ArrayList;
import java.util.HashSet;
import android.app.Activity;
import android.content.Context;
import android.content.SharedPreferences;
import android.content.SharedPreferences.Editor;
import android.content.res.Configuration;
import android.os.Bundle;
import android.text.TextUtils;
import android.view.View;
import android.widget.AdapterView;
import android.widget.ArrayAdapter;
import android.widget.EditText;
import android.widget.ListView;
import android.widget.Toast;
import android.widget.AdapterView.OnItemClickListener;
public class MainActivity extends Activity {
    private EditText edText;
    private ListView list;
    private ArrayAdapter<String> adapter;
    private ArrayList<String> listItems = new ArrayList<String>();
    private int nCurrent = -1;
    private Context context;
    private boolean bChanged = false;
    @Override
    protected void onCreate(Bundle savedInstanceState) {
        super.onCreate(savedInstanceState);
        setContentView(R.layout.activity_main);
        context = this;
        edText = (EditText)findViewById(R.id.edText);
        list = (ListView)findViewById(R.id.list);
        adapter = new ArrayAdapter<String>(this,
                android.R.layout.simple_list_item_checked, listItems);
        list.setAdapter(adapter);
        list.setOnItemClickListener(new OnItemClickListener(){
            @Override
            public void onItemClick(AdapterView<?> parent, View view,
                    int position, long id) {
                // TODO Auto-generated method stub
                String strItem = parent.getItemAtPosition(position).toString();
                if(nCurrent != position){
                    nCurrent = position;
                    list.setItemChecked(position, true);
                    Toast.makeText(context, strItem + " selected",
```

```
                    Toast.LENGTH_SHORT).show();
            }
            else{
                nCurrent = -1;
                list.setItemChecked(position, false);
                Toast.makeText(context, strItem + " unselected",
                        Toast.LENGTH_SHORT).show();
            }
        }});
        readSharedData();
    }
    private void readSharedData(){
        String strDisk = readStorage();
        if(strDisk.isEmpty()) return;
        String[] strArray = strDisk.split("/");
        listItems.clear();
        for(int i=0; i<strArray.length; i++){
            listItems.add(strArray[i]);
        }
    }
    private String readStorage(){
        String strAppName = getString(R.string.app_name);
        SharedPreferences sp = getSharedPreferences(strAppName, MODE_PRIVATE);
        return sp.getString("SharedItems", "");
    }
    public void onClick(View v){
        String strItem = edText.getText().toString();
        int id = v.getId();
        switch(id){
        case R.id.btAdd:
            if(TextUtils.isEmpty(strItem)){
                Toast.makeText(this, "Please input data.",
                        Toast.LENGTH_SHORT).show();
                return;
            }
            listItems.add(strItem);
            adapter.notifyDataSetChanged();
            bChanged = true;
            break;
        case R.id.btDelete:
            if(listItems.isEmpty()) {
                Toast.makeText(this, "No any item.",
                        Toast.LENGTH_SHORT).show();
                return;
            }
            if(nCurrent == -1) {
                Toast.makeText(this, "Please selecct an item.",
                        Toast.LENGTH_SHORT).show();
                return;
            }
            list.setItemChecked(nCurrent, false);
            listItems.remove(nCurrent);
            adapter.notifyDataSetChanged();
            nCurrent = -1;
            bChanged = true;
            break;
```

```
            }
        }
        @Override
        protected void onDestroy() {
            // TODO Auto-generated method stub
            if(bChanged){
                writeSharedData();
            }
            super.onDestroy();
        }
        private void writeSharedData(){
            String strAppName = getString(R.string.app_name);
            SharedPreferences sp = getSharedPreferences(strAppName, MODE_PRIVATE);
            Editor editor = sp.edit();
            if(listItems.isEmpty()) return;
            String strInput = getSharedString();
            editor.putString("SharedItems", strInput);
            editor.commit();
        }
        private String getSharedString(){
            String strTmp = "";
            for(int i=0; i<listItems.size(); i++){
                strTmp += listItems.get(i);
                if(i!=listItems.size()-1) strTmp += "/";
            }
            return strTmp;
        }
        @Override
        public void onConfigurationChanged(Configuration newConfig) {
            // TODO Auto-generated method stub
            super.onConfigurationChanged(newConfig);
            adapter.notifyDataSetChanged();
        }
    }
```

（3）运行效果

程序运行效果如图 3-10 所示，输入字符串并点击【Add】将加入 ListView 中，选中列表项，点击【Delete】，该列表项将被删除。退出程序后重新打开，列表项将恢复退出时的状态。

（4）HashSet 的应用

HashSet 类使用散列表（Hash Table）算法改进添加、删除和访问集合对象元素的执行效率，集合中的元素不重复且无序。SharedPreferences 对象也可用来保存 HashSet 对象，采用 Editor 对象的 putStringSet 方法即可。

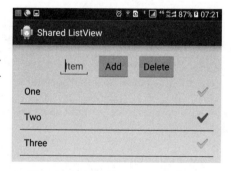

图 3-10 SharedListView 项目运行效果

```
HashSet<String> hash = new HashSet<String>();
hash.add("张三");
hash.add("李四");
editor.putStringSet("HashSet", hash);
```

读取 HashSet 对象时采用 SharedPreferences 对象的 getStringSet 方法，但需要对结果进行强制类型转换。如果 HashSet 对象不为空，还可以将其转换为数组。

```
HashSet<String> hash = new HashSet<String>();
hash = (HashSet<String>) sp.getStringSet("HashSet", null);
if(hash!=null){
    String[] strArray = new String[hash.size()];
    hash.toArray(strArray);
}
```

3.11 内部文本文件存取

使用 Java 提供的 I/O 流体系可以很方便地对本地存储的数据进行读写操作，其中 FileOutputStream 类的 openFileOutput 方法用来打开相应的输出流；FileInputStream 类的 openFileInput 方法用来打开相应的输入流。默认情况下，使用 I/O 流保存的文件仅对当前应用程序可见，对于其他应用程序（包括用户）是不可见的（即不能访问其中的数据）。如果用户卸载了该应用程序，则保存数据的文件也会被一起删除。由于不需使用外部资源，这些读写操作都不需要申请任何权限。

（1）核心代码解析

新建项目 FilesListView，该项目与上一节的项目类似，采用相同的布局方式，只是核心代码有差异。在 Library 项目中创建新类 FileProcess，在其中创建内部读写函数，以方便其他程序共享。FilesListView 需要引用 Library 类库，导入 FileProcess 类，这样才能使用读写文件的静态函数。

静态函数 writeInternalData 完成将文本数据写入文件的操作，第一个参数为上下文环境，第二个参数是文件名（可包含路径），第三个参数为需要保存的字符串。首先通过 openFileOutput 方法获得 FileOutputStream 对象 fos，然后调用 fos 的 write 方法写入数据。由于写入的数据是字节数组，因而需要调用字符串对象的 getBytes 方法对字符串进行转换。如果写入数据成功，则返回 true，否则返回 false。

```
public static boolean writeInternalData(Context context,
        String strPath, String strTxt){
    FileOutputStream fos = null;
    try {
        fos = context.openFileOutput(strPath, 0);
        try {
            fos.write(strTxt.getBytes());
            return true;
        } catch (IOException e) {
            e.printStackTrace();
            return false;
        }
    } catch (FileNotFoundException e) {
        e.printStackTrace();
        return false;
    }
}
```

静态函数 readInternalData 读取存储的字符串，只需要传入上下文环境和文件名（可包含路径）即可。首先通过 openFileInput 方法获得 FileInputStream 对象 fis，然后调用 fis 的 read 方法读取数据。由于读取的数据是字节数组，因而需要调用 String 类的构造方法将字节数组转换为字符串。如果读取数据成功，则返回读取的字符串；否则返回 null。

```java
public static String readInternalData(Context context, String strPath){
    FileInputStream fis = null;
    byte[] buffer = null;
    try {
        fis = context.openFileInput(strPath);
        try {
            buffer = new byte[fis.available()];
            fis.read(buffer);
            return new String(buffer);
        } catch (IOException e) {
            e.printStackTrace();
            return null;
        }
    } catch (FileNotFoundException e) {
        e.printStackTrace();
        return null;
    }
}
```

（2）运行效果

本节的运行效果跟上一节的一样，只是本节使用内部文件存取信息，上一节使用 SharedPreferences 对象存取信息。

3.12 百度地图

百度地图 Android SDK 是一套基于 Android 4.0 及以上版本设备的应用程序接口。可以使用该套 SDK 开发适用于 Android 系统移动设备的地图应用，通过调用地图 SDK 接口，可以轻松访问百度地图服务和数据，构建功能丰富、交互性强的地图类应用程序。该套地图 SDK 免费对外开放，接口使用无次数限制。百度地图 Android 资源的网址为：

http://lbsyun.baidu.com/index.php?title=androidsdk

（1）开发步骤

申请百度账号并申请成为百度地图开发者，最好进行认证，这样可以得到更好的服务。然后在百度地图控制台申请 SDK 开发密钥，其网址为：

http://lbsyun.baidu.com/apiconsole/key

点击"创建应用"，在图 3-11 的"应用名称"中输入应用名称，BaiduMapTest 就是需要创建的 Android 项目名称，"应用类型"处选择 Android 端，"启用服务"选项中不需要的可以不选，"发布版 SHA1"是一个开发工具指纹信息，通过 Eclipse 菜单 Window→Preferences→Android-→Build→SHA1 fingerprint 获取，"包名"即开发项目的完整包名。输入以上信息后，点击【提交】，即可得到应用 AK(API Key)，这个应用 AK 将用于项目配置文件。

读者可以输入自己的开发工具指纹信息，在配置文件中填上自己的应用 AK。

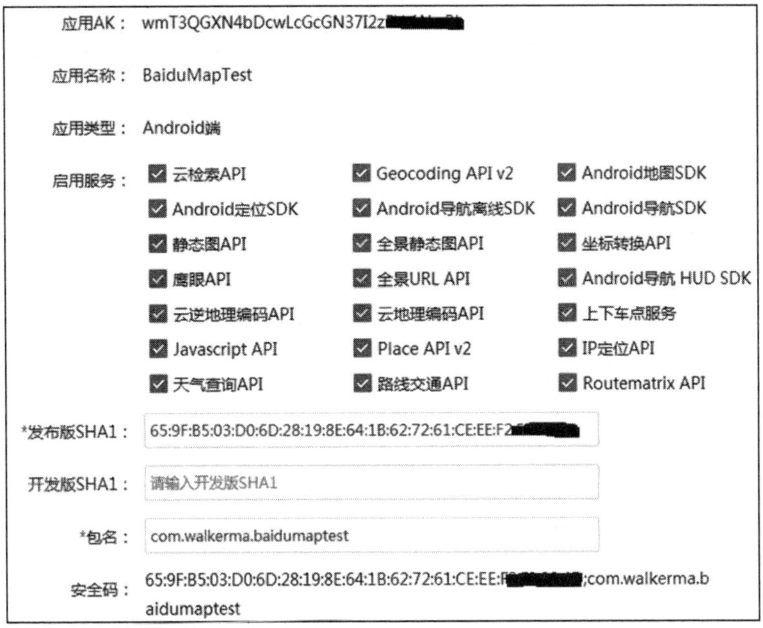

图 3-11 申请应用 AK 的方法

从下面的网址下载开发包，包括基础定位、离线定位、检索功能、各种导航以及计算工具等各种开发包，以及相应的示例代码和类参考。

http://lbsyun.baidu.com/index.php?title=sdk/download&action

下载完相关开发包后，需要导入项目中。如图 3-12 所示，通过项目的 Properties→Java Build Path->Libraries 标签选择【Add External JARs...】，选定 baidumapapi_ X.jar，确定后返回。

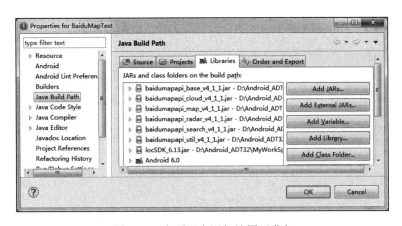

图 3-12　向项目中添加地图开发包

（2）布局文件和配置文件

完成以上基本工作后，下面需要实现 BaiduMapTest 项目。布局文件中的组件名称使用全称，且宽度和高度属性都使用"match_parent"，使得地图充满整个屏幕空间。

```xml
<?xml version="1.0" encoding="utf-8"?>
<RelativeLayout xmlns:android="http://schemas.android.com/apk/res/android"
    android:layout_width="match_parent"
    android:layout_height="match_parent" >
    <com.baidu.mapapi.map.MapView
        android:id="@+id/bmapView"
        android:layout_width="match_parent"
        android:layout_height="match_parent"
        android:clickable="true" />
</RelativeLayout>
```

配置文件与一般项目不同的地方主要体现在三个方面：第一，需要使用与地图相关的权限，以获取定位信息；第二，需要增加 meta-data 标记，并在其中填写申请的 AK；第三，声明需要调用百度地图服务。

```xml
<?xml version="1.0" encoding="utf-8"?>
<manifest xmlns:android="http://schemas.android.com/apk/res/android"
    package="com.walkerma.baidumaptest"
    android:versionCode="1"
    android:versionName="1.0" >
    <uses-sdk
        android:minSdkVersion="14"
        android:targetSdkVersion="22" />
    <!--访问网络，网络定位需要上网-->
    <uses-permission android:name="android.permission.INTERNET" />
    <!--网络定位-->
    <uses-permission android:name=
        "android.permission.ACCESS_COARSE_LOCATION" />
    <!--访问GPS定位-->
    <uses-permission android:name=
        "android.permission.ACCESS_FINE_LOCATION" />
    <!--获取运营商信息，用于支持提供运营商信息相关的接口-->
    <uses-permission android:name=
        "android.permission.ACCESS_NETWORK_STATE" />
    <!--用于访问WiFi网络信息，WiFi信息用于进行网络定位-->
    <uses-permission android:name=
        "android.permission.ACCESS_WIFI_STATE" />
    <!--写入扩展存储，向扩展卡写入数据，用于写入离线定位数据-->
    <uses-permission android:name=
        "android.permission.WRITE_EXTERNAL_STORAGE" />
    <application
        android:allowBackup="true"
        android:icon="@drawable/ic_launcher"
        android:label="@string/app_name"
        android:theme="@style/AppTheme" >
        <meta-data
            android:name="com.baidu.lbsapi.API_KEY"
            **android:value="填写你申请的AK"** />
        <activity
            android:name=".MainActivity"
            android:configChanges="orientation|keyboardHidden"
            android:label="@string/app_name"
            android:screenOrientation="portrait" >
            <intent-filter>
                <action android:name="android.intent.action.MAIN" />
```

```xml
            <category android:name="android.intent.category.LAUNCHER" />
        </intent-filter>
    </activity>
    <service
        android:name="com.baidu.location.f"
        android:enabled="true"
        android:process=":remote" >
    </service>
</application>
</manifest>
```

(3) 主窗体源代码

主窗体源代码主要包括 onCreate 方法和自定义地址侦听类 MyLocationListenner 的实现。onCreate 方法完成基本的初始化工作，LocationClientOption 是定位相关类的选项设置类，这里设置为每隔 1000 毫秒查询一次地址，并要求返回地址信息。setCoorType 设置坐标类型，取值有三种，"gcj02"返回国家测绘地理信息局经纬度坐标系，"bd09"返回百度墨卡托坐标系，"bd09ll"返回百度经纬度坐标系。

MyLocationListenner 实现了 BDLocationListener 接口，在 onReceiveLocation 方法中，首先构建生成定位数据对象 locData，然后设置定位精度、方向及经纬度数据，最后在地图上显示。如果是首次显示地图，还需要设置地图中心点和缩放级别，然后以动画形式更新地图。

```java
package com.walkerma.baidumaptest;
import com.baidu.location.BDLocation;
import com.baidu.location.BDLocationListener;
import com.baidu.location.LocationClient;
import com.baidu.location.LocationClientOption;
import com.baidu.mapapi.SDKInitializer;
import com.baidu.mapapi.map.BaiduMap;
import com.baidu.mapapi.map.MapStatus;
import com.baidu.mapapi.map.MapStatusUpdateFactory;
import com.baidu.mapapi.map.MapView;
import com.baidu.mapapi.map.MyLocationData;
import com.baidu.mapapi.model.LatLng;
import android.app.Activity;
import android.os.Bundle;
public class MainActivity extends Activity {
    MapView mMapView = null;
    //定位相关
    LocationClient mLocClient;
    // UI相关
    boolean isFirstLoc = true; //是否首次定位
    BaiduMap mBaiduMap;
    public MyLocationListenner myListener = new MyLocationListenner();
    @Override
    protected void onCreate(Bundle savedInstanceState) {
        super.onCreate(savedInstanceState);
        //在使用SDK各组件之前初始化context信息,传入ApplicationContext
        //注意该方法要在setContentView方法之前实现
        SDKInitializer.initialize(getApplicationContext());
        setContentView(R.layout.activity_main);
```

```java
        //获取地图控件引用
        mMapView = (MapView) findViewById(R.id.bmapView);
        mBaiduMap = mMapView.getMap();
        //开启定位图层
        mBaiduMap.setMyLocationEnabled(true);
        //定位初始化
        mLocClient = new LocationClient(this);
        mLocClient.registerLocationListener(myListener);
        LocationClientOption option = new LocationClientOption();
        option.setOpenGps(true);            //是否打开GPS进行定位
        option.setCoorType("bd09ll");       //设置坐标类型
        option.setAddrType("all");          //默认值为all时,表示返回地址信息
        option.setScanSpan(1000);           //扫描时间为1000毫秒
        mLocClient.setLocOption(option);
        mLocClient.start();
    }
    @Override
    protected void onDestroy() {
        super.onDestroy();
        //实现地图生命周期管理
        if(mLocClient!=null)
            mLocClient.unRegisterLocationListener(myListener);
        mMapView.onDestroy();
    }
    @Override
    protected void onResume() {
        super.onResume();
        //实现地图生命周期管理
        mMapView.onResume();
    }
    @Override
    protected void onPause() {
        super.onPause();
        //实现地图生命周期管理
        mMapView.onPause();
    }
    public class MyLocationListenner implements BDLocationListener {
        @Override
        public void onReceiveLocation(BDLocation location) {
            // map view销毁后不再处理新接收的位置
            if (location == null || mMapView == null)
                return;
            //构建生成定位数据对象
            MyLocationData locData = new MyLocationData.Builder()
            //定位数据,accuracy单位为米
                .accuracy(location.getRadius())
            //此处设置开发者获取到的方向信息,顺时针0°~360°
                .direction(location.getDirection())
                .latitude(location.getLatitude())
                .longitude(location.getLongitude()).build();
            mBaiduMap.setMyLocationData(locData);
            if (isFirstLoc) {
                isFirstLoc = false;
                LatLng ll = new LatLng(location.getLatitude(),
                        location.getLongitude());
                //创建地图状态对象
```

```
                MapStatus.Builder builder = new MapStatus.Builder();
                //设置地图中心点和缩放级别
                builder.target(ll).zoom(18.0f);
                //动画形式更新地图
                mBaiduMap.animateMapStatus(MapStatusUpdateFactory.
                    newMapStatus(builder.build()));
            }
        }
        public void onReceivePoi(BDLocation poiLocation) {
        }
    }
}
```

（4）运行效果

程序运行效果如图 3-13 所示，图中显示了当前位置，如果移动手机的位置，地图会随着更新，当前位置也会一起变化。

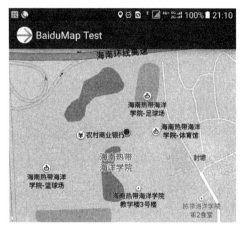

图 3-13　BaiduMapTest 运行效果

3.13　本章小结

掌握 Activity 的生命周期可以在程序设计过程中较好地对项目进行初始化、解决横屏和竖屏的切换问题及资源回收工作等。绘图部分介绍了自定义组件的设计方法和使用步骤，其中 2D 绘图技术还可以用来显示远程实时数据，第 15 章将采用该技术来显示远程温度变化的实时趋势线。手机的情景模式设置和音量调整及播放音频是一种常规工作，本章使用静态函数实现，以便于代码共享，随后的 SQLite 数据库类与文件存取等工作都实现了静态函数的功能。

消息机制可以实现后台线程向前台 UI 界面传递消息和数据，实现 UI 界面实时显示后台数据的功能。多线程的启用可以解决耗时任务的执行而不影响前台 UI 界面的操作，实现 Runnable 接口的定时功能可以用于物联网实时数据查询工作，第 15 章的实时温度查询就是采用这一技术。

SQLite 数据库部分实现了一个自定义数据库类，可以方便地进行数据库的创建，以及记录的查询、更新和删除工作。将该数据库用于 ListView，可以有效保存数据。本章在此基础之上介绍了内容共享程序，详细介绍了联系人查询方法及各个参数的详细使用，并提供了切实可行的例程；随后介绍了内部文件存取方法，提供了两个通用的内部文件存取的静态函数。

地图应用在现代生活中越来越重要，本章介绍了开发地图应用程序的基本步骤，如果需要进一步开发其他应用程序，可以下载相应的开发包和例程，重新设计即可形成符合要求的新的应用程序。但需要注意的是，若更换了不同的开发环境，百度地图的 AK 码需要根据开发环境的指纹 SHA1 重新生成，否则无法显示百度地图。

第 4 章　Intent 的综合应用

在 Android 应用程序中，三大核心组件 Activity、BroadcastReceiver 和 Service 都是通过 Intent 消息来激活的。Intent 消息是同一或不同应用程序中组件之间延迟运行时绑定的机制，Intent 本身是一个对象，其中包含对 Intent 有兴趣的组件的信息（如需要执行的动作和需要传递的数据），以及 Android 系统感兴趣的信息（如处理 Intent 组件的分类信息和如何启动目标活动的指令等）。

4.1　Intent 的基础知识

Intent 在激活三大组件的过程中，需要指定对象名称和动作名称，添加数据、种类并补充附加信息，有时还需要设置标志信息。本节介绍 Intent 的相关基础知识，并通过简单实例加以说明。

（1）利用 setClass 方法启动 Activity

Intent 的 setClass 方法可以用来设置需要启动的 Activity，其语法格式为：

public Intent setClass (Context packageContext, Class<?> cls)

packageContext 为当前 Activity 的 this 对象，cls 是需要启动的 Activity 的 class 对象，本方法返回 Intent 对象。如下代码即可从 firstActivity 启动 nextActivity。

```
Intent intent = new Intent();
intent.setClass(firstActivity.this, nextActivity.class);
startActivity(intent);
```

（2）动作和数据

Intent 的动作（ACTION）与作用对象相关。Intent 对象的动作通过 setAction 方法设置，该方法需要提供字符串形式的动作名称，并返回 Intent 对象，而获取动作字符串的方法 getAction 不需要提供参数。常用的 Intent 类的动作常量如表 4-1 所示。

表 4-1　Intent 类的常用动作常量举例

动作常量	作用对象	说　　明
ACTION_CALL	Activity	直接拨打电话
ACTION_DIAL	Activity	打开拨打电话界面
ACTION_MAIN	Activity	设置默认启动的 Activity
ACTION_SENDTO	Activity	发送信息
Settings.ACTION_WIFI_SETTINGS	Activity	打开 WiFi 设置界面

(续)

动作常量	作用对象	说 明
ACTION_BOOT_COMPLETED	BroadcastReceiver	系统完成启动
ACTION_BATTERY_LOW	BroadcastReceiver	显示电量低的警告信息

有的动作需要携带数据，这可以通过 setData 方法进行设置，其参数为 Uri 对象。以下代码先设置拨打电话的动作，然后设置 Uri 对象，实现直接拨打电话。拨打电话需要设置 CALL_PHONE 权限。

```
Intent intent = new Intent();
intent.setAction(Intent.ACTION_CALL);
intent.setData(Uri.parse("tel:"+"1234"));
startActivity(intent);
```

通常情况下，设置的数据类型能够从 URI 中推测，特别是"content:URIs"，表示数据位于设备上且被内容提供程序管理。当然，还可以显式地通过 setType 方法设置数据类型。

（3）种类

Intent 中还可以包含组件类型信息（作为被执行动作的附加信息），并且在一个 Intent 对象中可以指定任意数量的种类信息。下面的代码将返回系统桌面，如果没有种类信息 CATEGORY_HOME，将返回另外一个界面。

```
Intent intent = new Intent();
intent.setAction(Intent.ACTION_MAIN);
intent.addCategory(Intent.CATEGORY_HOME);
startActivity(intent);
```

相应地，removeCategory 方法从 Intent 对象中删除指定的种类信息，getCategory 方法用来获取所有与 Intent 对象相关的种类信息。

（4）标志

标志主要用来指示应用程序如何启动一个 Activity 和启动之后如何处理。所有标志都定义在 Intent 类中，常用的标志常量（整型）如表 4-2 所示。可以使用 setFlags 或 addFlags 方法添加一个标志到 Intent 对象中，使用 getFlags 方法获取 Intent 对象中的所有标志。

表 4-2　Intent 类的常用标志常量举例

动作常量	说 明
FLAG_ACTIVITY_NEW_TASK	系统将检查当前所有已创建的 Task 中是否有需要启动的 Activity 的 Task。如果有，则在该 Task 上创建 Activity；如果没有，则新建具有该 Activity 属性的 Task，并在该新建的 Task 上创建 Activity
FLAG_ACTIVITY_CLEAR_TOP	如果在当前 Task 中有要启动的 Activity，那么将该 Activity 之前的所有 Activity 都关掉，并将该 Activity 置前以避免创建 Activity 的实例
FLAG_ACTIVITY_EXCLUDE_FROM_RECENTS	使得新的 Activity 不会在最近启动的 Activity 列表中保存
FLAG_ACTIVITY_NO_HISTORY	使得新的 Activity 不在历史 Stack 中保留，用户一旦离开 Activity，这个 Activity 将自动关闭

设置标志信息的代码为：

```
Intent intent = new Intent();
intent.setFlags(Intent. FLAG_ACTIVITY_NEW_TASK);
```

（5）过滤器

Intent 过滤器一般在配置文件中用 <intent-filter> 标记表示，每个过滤器表示对应的组件能够接收的一组 Intent（含有动作、数据和种类）。过滤器要检查 Intent 的所有三个条件，任何一个条件不匹配，<intent-filter> 标记所对应的组件都不会响应。

如下代码在配置文件中设置了一个广播接收器，BootReceiver 是一个自定义类，用来处理系统启动以后需要执行的工作，BootReceiver 类被触发执行的条件是 Intent 对象所携带的动作信息必须是"android.intent.action.BOOT_COMPLETED"（与表 4-1 中的 ACTION_BOOT_COMPLETED 常量相等），种类信息也必须相等。此外，Intent 对象还可携带数据信息。

```
<receiver
    android:name=".BootReceiver"
    android:enabled="true">
    <intent-filter>
        <action android:name="android.intent.action.BOOT_COMPLETED" />
        <category android:name="android.intent.category.DEFAULT" />
    </intent-filter>
</receiver>
```

Intent 过滤器是 IntentFilter 类的实例，也可以在 Java 源代码中通过 registerReceiver 方法进行动态注册，具体实现将在 4.5 节中介绍。

（6）附加信息

在传递 Intent 对象的过程中，可以调用 putExtra 方法以键值对的方式补充附加信息，其定义如下，其中第二个参数不限于字符串类型，可以是 boolean、byte 或 double 等基本数据类型，也可以是数组信息。附加信息的获取在下一节介绍。

```
public Intent putExtra(String name, String value);
```

4.2 在 Activity 之间传递数据

在 Activity 之间传递数据可以分为两种情况，第一个 Activity 将数据传递到第二个 Activity，然后第二个 Activity 在关闭后将数据返回给第一个 Activity。

（1）创建新的 Activity

创建项目 MultiActivity，采用默认设置。通过菜单 New→Other 打开如图 4-1 所示的对话框，点击【Next】，创建第二个 Activity，名称填写为 SecondActivity。

（2）布局文件

主窗体布局文件为 activity_main.xml，TextView 中的 text 属性为"Wait for the second"，Button 按钮用于启动第二个窗体。

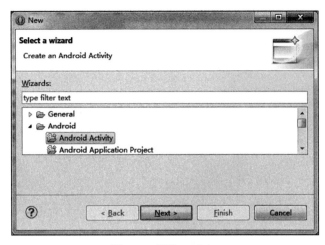

图 4-1 创建 Activity

```
<RelativeLayout xmlns:android="http://schemas.android.com/apk/res/android"
    xmlns:tools="http://schemas.android.com/tools"
    android:layout_width="match_parent"
    android:layout_height="match_parent"
    android:paddingBottom="@dimen/activity_vertical_margin"
    android:paddingLeft="@dimen/activity_horizontal_margin"
    android:paddingRight="@dimen/activity_horizontal_margin"
    android:paddingTop="@dimen/activity_vertical_margin"
    tools:context="com.ch03.multiactivity.MainActivity" >
    <TextView
        android:id="@+id/txtData"
        android:layout_width="wrap_content"
        android:layout_height="wrap_content"
        android:textSize="18sp"
        android:text="Wait for the second" />
    <Button
        android:id="@+id/btMain"
        android:layout_width="wrap_content"
        android:layout_height="wrap_content"
        android:onClick="onClick"
        android:layout_toRightOf="@+id/txtData"
        android:layout_alignBaseline="@+id/txtData"
        android:text="New Activity" />
</RelativeLayout>
```

第二个窗体布局文件为 activity_second.xml，用 EditText 标记输入文本，通过 Button 关闭窗体并将输入的文本传递到主窗体的 TextView 标记中显示。

```
<RelativeLayout xmlns:android="http://schemas.android.com/apk/res/android"
    xmlns:tools="http://schemas.android.com/tools"
    android:layout_width="match_parent"
    android:layout_height="match_parent"
    android:paddingBottom="@dimen/activity_vertical_margin"
    android:paddingLeft="@dimen/activity_horizontal_margin"
    android:paddingRight="@dimen/activity_horizontal_margin"
    android:paddingTop="@dimen/activity_vertical_margin"
    tools:context="com.ch03.multiactivity.SecondActivity" >
    <EditText
```

```xml
            android:id="@+id/edText"
            android:layout_width="wrap_content"
            android:layout_height="wrap_content"
            android:hint="Input data here" />
    <Button
            android:id="@+id/btRet"
            android:layout_width="wrap_content"
            android:layout_height="wrap_content"
            android:onClick="onClick"
            android:layout_alignParentTop="true"
            android:layout_toRightOf="@+id/edText"
            android:text="Return" />
</RelativeLayout>
```

（3）源代码

在主窗体源代码 MainActivity.java 的 onClick 方法中，通过 Intent 对象 it 的 setClass 方法设置需要启动的窗体 SecondActivity，然后调用 putExtra 方法设置键值对，调用 setFlags 方法设置标志，最后调用 startActivityForResult 方法启动下一个窗体并要求返回结果，其中的 REQUEST_CODE 是一个自定义整型常量，作为请求标识。

```java
package com.ch03.multiactivity;
import android.app.Activity;
import android.content.Intent;
import android.os.Bundle;
import android.view.View;
import android.widget.TextView;
public class MainActivity extends Activity {
    private TextView txtData;
    final int REQUEST_CODE = 10011;
    @Override
    protected void onCreate(Bundle savedInstanceState) {
        super.onCreate(savedInstanceState);
        setContentView(R.layout.activity_main);
        txtData = (TextView)findViewById(R.id.txtData);
    }
    public void onClick(View v){
        Intent it = new Intent();
        it.setClass(MainActivity.this, SecondActivity.class);
        it.putExtra("TextData", "From the first");
        it.setFlags(Intent.FLAG_ACTIVITY_CLEAR_TOP);
        startActivityForResult(it, REQUEST_CODE);
    }
    @Override
    protected void onActivityResult(int requestCode, int resultCode, Intent data) {
        // TODO Auto-generated method stub
        if (requestCode == REQUEST_CODE){
            if (resultCode == SecondActivity.RESULT_OK){
                Bundle bundle = new Bundle();
                bundle = data.getExtras();
                txtData.setText(bundle.getString("EditData"));
            }
        }
        super.onActivityResult(requestCode, resultCode, data);
    }
}
```

在第二个窗体源代码 SecondActivity.java 的 onCreate 方法中，首先获得 Intent 对象，然后得到绑定键值对的 Bundle 对象，最后从中取出数据并在 EditText 对象中显示。在编程实践中，这几条语句一般合并如下：

```
edText.setText(getIntent().getExtras().getString("TextData"));
```

在 EditText 对象中输入新的字符串后，通过 onClick 方法向第一个窗体返回数据。首先利用 Bundle 对象的 putString 方法存入键值对，然后通过 Intent 对象的 putExtras 方法存入 Bundle 对象，最后调用 setResult 返回数据，其中 RESULT_OK 是自定义静态公用整型常量，作为返回数据的标识。SecondActivity 中的工作完成后，调用 finish 方法关闭窗体，并返回到主窗体的 onActivityResult 方法中。

在主窗体的 onActivityResult 方法中，首先判断 requestCode 与当初的请求标识是否相同，然后再判断 resultCode 与返回数据的标识是否一致，如果两者一致，则取出数据并在 TextView 对象中显示。

```
package com.ch03.multiactivity;
import android.app.Activity;
import android.content.Intent;
import android.os.Bundle;
import android.view.View;
import android.widget.EditText;
public class SecondActivity extends Activity {
    private EditText edText;
    public final static int RESULT_OK = 1;
    @Override
    protected void onCreate(Bundle savedInstanceState) {
        super.onCreate(savedInstanceState);
        setContentView(R.layout.activity_second);
        edText = (EditText)findViewById(R.id.edText);
        Intent intent = getIntent();
        Bundle bundle = intent.getExtras();
        String strData = bundle.getString("TextData");
        edText.setText(strData);
    }
    public void onClick(View v){
        Bundle bundle = new Bundle();
        bundle.putString("EditData", edText.getText().toString());
        Intent it = new Intent();
        it.addFlags(Intent.FLAG_ACTIVITY_NO_HISTORY);
        it.putExtras(bundle);
        setResult(RESULT_OK, it);
        finish();
    }
}
```

（4）运行效果

程序运行后如图 4-2 所示，上图 TextView 文本框中首先显示"From the first"，点击【New Activity】，显示 SecondActivity，并且 EditText 编辑框中显示"From the first"。在下图的编辑框中输入"From the second"，点击【Return】返回上图，上图将显示"From the second"。

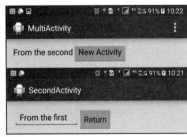

图 4-2　MultiActivity 运行效果

4.3 状态栏通知

Toast 类可用于告知用户一些提示信息，而另一种告知用户信息的常用方法是使用状态栏（Status Bar）。状态栏是移动设备最上方的一条横向的长条区域，可显示提醒信息的图标、标题文字和时间，往下拖动，可以展开通知的项目列表，比如未接来电的电话号码，如果点击该通知项目则可以回拨电话。下面通过实例来展示状态栏的显示及随后的操作。

（1）主窗体源代码

新建项目 Notify，布局文件中使用两个 Button，其中 btNotify 用来显示通知图标，btCancel 用来删除通知图标。NOTIFY_ID 是自定义整型常量，用来标识图标，显示和删除图标时都需要用到该标识常量。点击命令按钮时，首先调用 getSystemService 方法以获取图标通知管理类 NotificationManager 的对象 manager，然后调用 manager 的 notify 方法显示图标，调用 cancel 方法删除图标。

notify 方法的第二个参数是一个 Notification 对象，由 buildNotification 函数产生，在其中依次调用构造器 Builder 的方法：setContentTitle 方法用于设置图标的标题；setContentText 方法用于设置图标的文本内容；setSmallIcon 方法用于设置小图标的资源；setDefaults 方法用于设置默认的提醒方式，DEFAULT_SOUND 为声音提醒，VIBRATE 为振动提醒，LIGHTS 为灯光提醒，ALL 则集成了这三种提醒；setContentIntent 方法采用 PendingIntent 对象作为参数，表示点击图标以后将会执行的动作。

PendingIntent 用于处理即将发生的事情，例如在通知 Notification 中用于跳转页面（查看短信或未接来电等），但不是马上跳转。Intent 是即时启动，Intent 对象随所在的 Activity 对象消失而消失。PendingIntent 可以看作对 Intent 的包装，一般通过 getActivity、getBroadcast 和 getService 方法来得到 PendingIntent 的实例，但是，当前 Activity 对象并不马上启动它所包含的 Intent 对象，而是在外部执行 PendingIntent 对象时再调用 Intent 对象。

在调用 PendingIntent 的静态方法 getActivity 来获取 PendingIntent 的实例时，最后一个参数是一个整型常量标志，主要有以下几种情况：FLAG_CANCEL_CURRENT 表示如果当前系统中已经存在一个相同的 PendingIntent 对象，那么先将已有的 PendingIntent 对象取消，再重新生成一个 PendingIntent 对象；FLAG_NO_CREATE 表示如果当前系统中不存在相同的 PendingIntent 对象，系统将不会创建该 PendingIntent 对象而是直接返回 null；FLAG_ONE_SHOT 表示该 PendingIntent 只作用一次；FLAG_UPDATE_CURRENT 表示如果系统中已存在该 PendingIntent 对象，那么系统将保留该 PendingIntent 对象，但是会使用新的 Intent 对象来更新之前 PendingIntent 对象中 Intent 对象的数据。

```
package com.ch03.notify;
import android.app.Activity;
import android.app.Notification;
import android.app.NotificationManager;
import android.app.PendingIntent;
import android.content.Intent;
import android.os.Bundle;
import android.view.View;
public class MainActivity extends Activity {
```

```java
    final int NOTIFY_ID = 10123;
    @Override
    protected void onCreate(Bundle savedInstanceState) {
        super.onCreate(savedInstanceState);
        setContentView(R.layout.activity_main);
    }
    public void onClick(View v){
        int id = v.getId();
        NotificationManager manager = (NotificationManager)
                getSystemService(NOTIFICATION_SERVICE);
        switch(id){
        case R.id.btNotify:
            manager.notify(NOTIFY_ID, buildNotification("dialing..."));
            break;
        case R.id.btCancel:
            manager.cancel(NOTIFY_ID);
            break;
        }
    }
    private Notification buildNotification(String strContent) {
        Intent intent = new Intent(Intent.ACTION_DIAL);
        intent.setFlags(Intent.FLAG_ACTIVITY_NEW_TASK);
        PendingIntent pi = PendingIntent.getActivity(this, 0, intent,
                PendingIntent.FLAG_UPDATE_CURRENT);
        Notification notify = new Notification.Builder(this)
        .setContentTitle("Notify Example")
        .setContentText(strContent)
        .setSmallIcon(R.drawable.ic_launcher)
        .setContentIntent(pi)
        .setDefaults(Notification.DEFAULT_SOUND|
                Notification.FLAG_SHOW_LIGHTS))
        .build();
        return notify;      // API level 16
    }
}
```

（2）运行效果

Notify 项目的运行效果如图 4-3 所示，点击通知图标，将显示拨打电话页面。

这里用于构建 Notification 对象的方法为 API level 16，因而配置文件中的 minSdkVersion 属性应做相应的调整。另外，Notification 对象还有其他多种属性，如还可通过构造器的 setPriority 方法设置权限的等级，相关内容可以参阅 Android 的官方文档，根据需要进行设置。

图 4-3　Notify 项目的运行效果

4.4　广播接收器与开机自动启动

在接到来电、收到短信或系统开机等情况下，Android 系统都会发出广播，任何注册了相应接收器的应用程序都可以收到广播并执行相应的处理。应用程序本身也可以自行送出广播。大部分系统广播接收器需要申请系统权限，广播接收器的核心代码是实现 Broadcast-Receiver 类的 onReceive 抽象方法。本节新建项目 BootStart，接收系统启动广播，并在其中启动本应用程序，显示提示信息。

（1）配置文件

配置文件中需要填写接收系统启动的权限，接收器标记 receiver 中的 name 属性表示处理系统广播的类，这里为 BootReceiver，该类前面的圆点是一种省略写法，这样不需要写完整的包名，在 Intent 过滤器中填写需要接收的 Intent 对象的动作。

```xml
<?xml version="1.0" encoding="utf-8"?>
<manifest xmlns:android="http://schemas.android.com/apk/res/android"
    package="com.walkerma.bootstart"
    android:versionCode="1"
    android:versionName="1.0" >
    <uses-sdk
        android:minSdkVersion="14"
        android:targetSdkVersion="23" />
    <uses-permission android:name=
        "android.permission.RECEIVE_BOOT_COMPLETED"/>
    <application
        android:allowBackup="true"
        android:icon="@drawable/ic_launcher"
        android:label="@string/app_name"
        android:theme="@style/AppTheme" >
        <activity
            android:name=".MainActivity"
            android:label="@string/app_name" >
            <intent-filter>
                <action android:name="android.intent.action.MAIN" />
                <category android:name="android.intent.category.LAUNCHER" />
            </intent-filter>
        </activity>
        <receiver
            android:name=".BootReceiver"
            android:enabled="true" >
            <intent-filter>
                <action android:name="android.intent.action.BOOT_COMPLETED" />
            </intent-filter>
        </receiver>
    </application>
</manifest>
```

（2）广播接收器源代码

新建 BootReceiver 类，继承自 BroadcastReceiver，实现 onReceive 方法，在其中启动主窗体。

```java
package com.walkerma.bootstart;
import android.content.BroadcastReceiver;
import android.content.Context;
import android.content.Intent;
public class BootReceiver extends BroadcastReceiver {
```

```
    @Override
    public void onReceive(Context context, Intent intent) {
        // TODO Auto-generated method stub
        Intent i = new Intent(context, MainActivity.class);
        i.addFlags(Intent.FLAG_ACTIVITY_NEW_TASK);
        context.startActivity(i);
    }
}
```

（3）程序运行效果

采用模拟器进行验证，首先在模拟器上安装 BootStart，然后关闭模拟器，重新运行任何一个应用程序，BootStart 随后都将自动启动，运行效果如图 4-4 所示。从 Android 6.0（API level 23）开始，为了安全起见，系统对权限执行更严格的管理，对于系统认为危险的权限，每次执行程序都需要用户亲自授权。为了便于使用含有这种权限的应用程序，一般在配置文件中将 targetSdkVersion 属性设置为 23 以下。

图 4-4　BootStart 运行效果

4.5　显示来电和接收短信

显示来电和接收短信是手机最基本的功能。上一节通过在配置文件中使用 receiver 标记注册广播接收器，并用继承于 BroadcastReceiver 的派生类实现了对系统启动的监控，本节将利用代码注册广播接收器，并且使用一个继承于 BroadcastReceiver 的内置派生类来实现广播接收器。

（1）布局文件

新建项目 MsgCaller，使用 txtPhone 显示电话号码，txtMsg 显示短信的文本内容。

```
<RelativeLayout xmlns:android="http://schemas.android.com/apk/res/android"
    xmlns:tools="http://schemas.android.com/tools"
    android:layout_width="match_parent"
    android:layout_height="match_parent"
    android:paddingBottom="@dimen/activity_vertical_margin"
    android:paddingLeft="@dimen/activity_horizontal_margin"
    android:paddingRight="@dimen/activity_horizontal_margin"
    android:paddingTop="@dimen/activity_vertical_margin"
    tools:context="com.ch03.msgcaller.MainActivity" >
    <TextView
        android:id="@+id/txtPhone"
        android:layout_width="wrap_content"
        android:layout_height="wrap_content"
        android:layout_alignParentTop="true"
        android:layout_centerHorizontal="true"
        android:textAppearance="?android:attr/textAppearanceMedium"
        android:text="####" />
    <TextView
        android:id="@+id/txtMsg"
        android:layout_width="wrap_content"
```

```
            android:layout_height="wrap_content"
            android:layout_below="@+id/txtPhone"
            android:layout_centerHorizontal="true"
            android:text="Message content here..."
            android:textAppearance="?android:attr/textAppearanceMedium"
            android:maxLines="10" />
</RelativeLayout>
```

(2) 配置文件

要显示来电，需要设置读取电话状态的权限，要显示短信内容，则需要设置接收和读取短信的权限。这里没有使用 receiver 广播接收标记，而是在源代码中进行动态注册。

```
<?xml version="1.0" encoding="utf-8"?>
<manifest xmlns:android="http://schemas.android.com/apk/res/android"
    package="com.ch03.msgcaller"
    android:versionCode="1"
    android:versionName="1.0" >
    <uses-sdk
        android:minSdkVersion="19"
        android:targetSdkVersion="21" />
    <uses-permission android:name="android.permission.READ_PHONE_STATE" />
    <uses-permission android:name="android.permission.READ_SMS" />
    <uses-permission android:name="android.permission.RECEIVE_SMS" />
    <application
        android:allowBackup="true"
        android:icon="@drawable/ic_launcher"
        android:label="@string/app_name"
        android:theme="@style/AppTheme" >
        <activity
            android:name=".MainActivity"
            android:label="@string/app_name" >
            <intent-filter>
                <action android:name="android.intent.action.MAIN" />
                <category android:name="android.intent.category.LAUNCHER" />
            </intent-filter>
        </activity>
    </application>
</manifest>
```

(3) 显示短信的广播接收器源代码

由于通过源代码动态注册侦听短信的广播信息，因而可以方便地定义构造函数传递参数，这里为两个文本框，用于显示电话号码与短信。采用 Telephony 的静态方法 getMessagesFromIntent 获取短信数组，依次得到电话号码与短信内容。由于电话号码可能是空值（即没有开通来电显示或对方隐藏电话号码），因而应做检测，如果为空值，就显示"Private"。getMessagesFromIntent 为 API level 19 中的函数，因而，配置文件中的 minSdkVersion 不得低于 19。

```
package com.ch03.msgcaller;
import android.content.BroadcastReceiver;
import android.content.Context;
import android.content.Intent;
import android.provider.Telephony .Sms.Intents;
```

```java
import android.telephony.SmsMessage;
import android.text.TextUtils;
import android.widget.TextView;
public class MsgReceiver extends BroadcastReceiver {
    private TextView txtPhone, txtMsg;
    public MsgReceiver(TextView txtPhone, TextView txtMsg){
        this.txtPhone = txtPhone;
        this.txtMsg = txtMsg;
    }
    @Override
    public void onReceive(Context context, Intent intent) {
        // TODO Auto-generated method stub
        SmsMessage[] inMsgs = Telephony.
                getMessagesFromIntent(intent);     //    API level 19
        String strPhone = inMsgs[0].getOriginatingAddress().trim();
        String strBody = inMsgs[0].getMessageBody();
        if(TextUtils.isEmpty(strPhone))
            txtPhone.setText("Private");
        else
            txtPhone.setText(strPhone);
        txtMsg.setText(strBody);
    }
}
```

为了能够在低于 API level 19 的手机上接收短信，可以对手机的版本号进行判断：如果手机的版本号低于 API level 19，则使用旧函数接收短信，否则使用新函数接收短信，其源代码如下所示。

```java
@SuppressLint("NewApi") @SuppressWarnings("deprecation")
// KITKAT, Android 4.4, API 19
public static SmsMessage[] getIncomingMessage(Intent intentSMS) {
    Bundle bundle = intentSMS.getExtras();
    if (Build.VERSION.SDK_INT >= Build.VERSION_CODES.KITKAT) {
        return Intents.getMessagesFromIntent(intentSMS);
    } else {
        Object pdus[] = (Object[]) bundle.get("pdus");
        int nLen = pdus.length;
        SmsMessage[] msgs = new SmsMessage[nLen];
        for(int i=0; i<nLen; i++)
            msgs[i] = SmsMessage.createFromPdu((byte[])pdus[i]);
        return msgs;
    }
}
```

（4）主窗体源代码

动态注册广播接收器主要通过调用 registerReceiver 方法来实现，该方法需要传入两个参数，第一个为侦听广播信息类的对象，第二个为 IntentFilter 对象。显示来电的广播接收器类 CallerReceiver 是一个嵌入类，可以方便地调用主类中的组件对象，也就是直接更新主类用于显示电话号码的 TextView 对象。

在程序退出时，需要调用 unregisterReceiver 方法注销广播接收器。

```java
package com.ch03.msgcaller;
import android.app.Activity;
```

```java
import android.content.BroadcastReceiver;
import android.content.Context;
import android.content.Intent;
import android.content.IntentFilter;
import android.os.Bundle;
import android.telephony.TelephonyManager;
import android.text.TextUtils;
import android.widget.TextView;
public class MainActivity extends Activity {
    private TextView txtPhone, txtMsg;
    private CallerReceiver caller;
    private MsgReceiver msg;
    @Override
    protected void onCreate(Bundle savedInstanceState) {
        super.onCreate(savedInstanceState);
        setContentView(R.layout.activity_main);
        txtPhone = (TextView)findViewById(R.id.txtPhone);
        txtMsg = (TextView)findViewById(R.id.txtMsg);
        caller = new CallerReceiver();
        registerReceiver(caller, new IntentFilter(
                "android.intent.action.PHONE_STATE"));
        msg = new MsgReceiver(txtPhone, txtMsg);
        registerReceiver(msg, new IntentFilter(
                "android.provider.Telephony.SMS_RECEIVED"));
    }
    public class CallerReceiver extends BroadcastReceiver {
        @Override
        public void onReceive(Context context, Intent intent) {
            // TODO Auto-generated method stub
            String state = intent.getStringExtra(
                    TelephonyManager.EXTRA_STATE);
            if (state.equals(TelephonyManager.EXTRA_STATE_RINGING)){
                String strPhone = intent.getStringExtra(
                        TelephonyManager.EXTRA_INCOMING_NUMBER);
                if(TextUtils.isEmpty(strPhone))
                    txtPhone.setText("Private");
                else
                    txtPhone.setText(strPhone);
            }
            return;
        }
    }
    @Override
    protected void onDestroy() {
        // TODO Auto-generated method stub
        unregisterReceiver(caller);
        unregisterReceiver(msg);
        super.onDestroy();
    }
}
```

（5）运行效果

MsgCaller 项目的运行效果如图 4-5 所示，上面的文本框显示电话号码，下面的文本框显示短信内容。

图 4-5　MsgCaller 运行效果

4.6　带回执的短信发送

在发送手机短信时，短信有没有发送出去？对方有没有收到？本节使用 PendingIntent 对象来捕获短信发送情况。

（1）布局文件

新建项目 SmsReceipt，布局文件中的 EditText 标记用来输入手机号码，Button 标记用来发送短信，短信内容直接在源代码中给出。

```
<RelativeLayout xmlns:android="http://schemas.android.com/apk/res/android"
    xmlns:tools="http://schemas.android.com/tools"
    android:layout_width="match_parent"
    android:layout_height="match_parent"
    android:paddingBottom="@dimen/activity_vertical_margin"
    android:paddingLeft="@dimen/activity_horizontal_margin"
    android:paddingRight="@dimen/activity_horizontal_margin"
    android:paddingTop="@dimen/activity_vertical_margin"
    tools:context="com.ch04.smsreceipt.MainActivity" >
    <EditText
        android:id="@+id/edPhone"
        android:layout_width="wrap_content"
        android:layout_height="wrap_content"
        android:inputType="number"
        android:hint="###########"
        android:layout_alignParentLeft="true"
        android:layout_alignParentTop="true">
        <requestFocus />
    </EditText>
    <Button
        android:id="@+id/btSend"
        android:layout_width="wrap_content"
        android:layout_height="wrap_content"
        android:onClick="onClick"
        android:layout_alignLeft="@+id/edPhone"
        android:layout_below="@+id/edPhone"
        android:text="Send" />
</RelativeLayout>
```

（2）短信发送完成广播接收器

在短信发送完成广播接收器 SmsSentBroadcastReceiver 中，通过 getResultCode 方法获取整型结果常量，再利用 switch 语句转换为字符串，最后采用 Toast 类进行显示。

```
package com.ch04.smsreceipt;
import android.app.Activity;
import android.content.BroadcastReceiver;
import android.content.Context;
import android.content.Intent;
import android.telephony.SmsManager;
import android.widget.Toast;
public class SmsSentBroadcastReceiver extends BroadcastReceiver {
    @Override
    public void onReceive(Context context, Intent intent) {
        // TODO Auto-generated method stub
```

```
            String strResult = "";
            switch (getResultCode()) {
            case Activity.RESULT_OK:
                strResult = "Alarm sent OK";
                break;
            case SmsManager.RESULT_ERROR_GENERIC_FAILURE:
                strResult = "Transmission failed";
                break;
            case SmsManager.RESULT_ERROR_RADIO_OFF:
                strResult = "Radio off";
                break;
            case SmsManager.RESULT_ERROR_NULL_PDU:
                strResult = "No PDU defined";
                break;
            case SmsManager.RESULT_ERROR_NO_SERVICE:
                strResult = "No service";
            }
            Toast.makeText(context.getApplicationContext(), strResult,
                Toast.LENGTH_SHORT).show();
        }
    }
```

（3）短信送达广播接收器

短信送达广播接收器类 SmsDeliveredBroadcastReceiver 也采用 getResultCode 方法获取整型结果常量，再进行转换和显示。

```
    package com.ch04.smsreceipt;
    import android.app.Activity;
    import android.content.BroadcastReceiver;
    import android.content.Context;
    import android.content.Intent;
    import android.widget.Toast;
    public class SmsDeliveredBroadcastReceiver extends BroadcastReceiver {
        @Override
        public void onReceive(Context context, Intent intent) {
            // TODO Auto-generated method stub
            String strResult = "";
            switch (getResultCode()) {
            case Activity.RESULT_OK:
                strResult = "Alarm delivered";
                break;
            case Activity.RESULT_CANCELED:
                strResult = "Alarm not delivered";
            }
            Toast.makeText(context.getApplicationContext(),
                    strResult, Toast.LENGTH_SHORT).show();
        }
    }
```

（4）主窗体源代码

onClick 方法通过短信管理器类 SmsManager 的静态方法 getDefault 获取 SmsManager 对象 sm，然后调用 sendTextMessage 方法发送文本内容的短信。sendTextMessage 方法有 5 个参数，第 1 个为目标手机号码（由 EditText 对象提供），第 2 个为服务商提供的短信中心号码（一般为 null，自动获取），第 3 个为短信内容，第 4 个为 PendingIntent 对象，当短信被成功发送或发送失败时将广播此对象，第 5 个参数也为 PendingIntent 对象，当短信发送到接收者或者发送失败时将广播此对象。

  PendingIntent 对象的初始化和广播接收器的注册在 initSMS 函数中实现。首先生成 Intent 对象 sentIntent，动作为自定义字符串，通过 SENT 常量给出；然后通过 PendingIntent 类的静态方法 getBroadcast 获得 PendingIntent 类的实例 sentPI，其中的 FLAG_UPDATE_CURRENT 标志表示如果 PendingIntent 实例已经存在，那么继续保留，但是使用当前 Intent 对象提供的附加数据；最后生成广播接收器对象，调用 registerReceiver 方法完成注册。同理，对于短信送达也采用相似的方法。这里生成的 PendingIntent 实例也在 sendTextMessage 方法中使用。

```java
package com.ch04.smsreceipt;
import android.app.Activity;
import android.app.PendingIntent;
import android.content.Intent;
import android.content.IntentFilter;
import android.os.Bundle;
import android.telephony.SmsManager;
import android.view.View;
import android.widget.EditText;
public class MainActivity extends Activity {
    private EditText edPhone;
    private SmsSentBroadcastReceiver sentReceiver;
    private SmsDeliveredBroadcastReceiver deliveredReceiver;
    private Intent sentIntent;
    private PendingIntent sentPI;
    private Intent deliveryIntent;
    private PendingIntent deliveredPI;
    private final String SENT = "sent";
    private final String DELIVERED = "delivered";
    @Override
    protected void onCreate(Bundle savedInstanceState) {
        super.onCreate(savedInstanceState);
        setContentView(R.layout.activity_main);
        edPhone = (EditText)findViewById(R.id.edPhone);
        initSMS();
    }
    private void initSMS(){
        sentIntent = new Intent(SENT);
        sentPI = PendingIntent.getBroadcast(
                getApplicationContext(), 0, sentIntent,
                PendingIntent.FLAG_UPDATE_CURRENT);
        sentReceiver = new SmsSentBroadcastReceiver();
        registerReceiver(sentReceiver, new IntentFilter(SENT));
        deliveryIntent = new Intent(DELIVERED);
        deliveredPI = PendingIntent.getBroadcast(
                getApplicationContext(), 0, deliveryIntent,
                PendingIntent.FLAG_UPDATE_CURRENT);
        deliveredReceiver = new SmsDeliveredBroadcastReceiver();
        registerReceiver(deliveredReceiver, new IntentFilter(DELIVERED));
    }
    public void onClick(View v){
```

```
        SmsManager sm = SmsManager.getDefault();
        sm.sendTextMessage(edPhone.getText().toString(), null,
            "OK", sentPI, deliveredPI);
    }
}
```

(5)运行效果

程序运行效果如图 4-6 所示,在 EditText 中输入本机号码(这里为作者单位的短号),点击【Send】,在屏幕下方将依次显示"SMS sent OK"和"SMS delivered",接着手机收到文本内容为"OK"的短信。

图 4-6　SmsReceipt 项目运行效果

4.7 服务的基础知识

Service(服务)是一个应用程序组件,它能够在后台执行一些耗时较长的操作,并且不提供用户界面。服务能被其他应用程序的组件启动,即使用户切换到另外的应用时仍能保持在后台运行。此外,应用程序组件还能与服务绑定,并与服务进行交互,甚至能进行进程间通信(Inter-Process Communication,IPC)。比如,服务可以处理网络传输、音乐播放、执行文件 I/O 或者与 Content Provider 进行交互,所有这些都是在后台进行的。

(1)Service 概述

服务有两种基本类型,即 Started(启动)和 Bound(绑定)服务。Started 服务是应用程序组件通过调用 startService 来启动的,一旦被启动,该服务就处于"started"状态,服务就能在后台一直运行下去,即使启动它的组件已经被销毁。通常,Started 服务执行单一的操作并且不会向调用者返回结果。例如,它可以通过网络下载或上传文件。当操作完成后,服务应该自行终止。

Bound 服务是应用程序组件通过调用 bindService 来绑定到服务上的,绑定后服务处于"bound"状态。Bound 服务提供了一个客户端/服务器接口,允许组件与服务进行交互、发送请求或获取结果,甚至可以利用 IPC 跨进程执行这些操作。绑定服务的生存期与被绑定的应用程序组件一致。多个组件可以同时与一个服务绑定,不过在所有组件解除绑定后,服务也随之被销毁。

服务也可以同时以两种方式工作,但需要实现两个回调方法,即实现 onStartCommand 方法以允许组件启动服务,实现 onBind 方法以允许组件绑定服务。

服务运行于宿主进程的主线程中,它不创建自己的线程并且不运行在单独的进程中(除非明确指定)。因而,如果服务要执行一些很耗 CPU 的工作或者阻塞的操作(比如播放 MP3 或网络操作),则应该在服务中创建一个新的线程来执行这些操作。利用单独的线程将减少 Activity 发生应用程序停止响应(Application Not Responding,ANR)错误的风险。

(2)Service 中的回调方法

为了创建一个服务,必须新建一个 Service 的子类(或一个已有 Service 的子类)。在实现代码中,可按需重写一些回调方法,用于对服务生命周期中的关键节点进行处理,以及向组件提供绑定机制。最重要的需要重写的回调方法如下。

1) onStartCommand 方法。当其他组件通过调用 startService 方法请求 Started 方式的服务时,系统将会调用本方法。一旦执行本方法,服务就被启动,并在后台一直运行下去。可以在 onStartCommand 方法中完成服务的初始化工作。对于 Started 方式的服务,可以在服务中调用 stopSelf 方法来终止服务,也可以在服务之外调用 stopService 方法来终止服务。如果只想提供 Bound 方式的服务,则不必实现本方法。

2) onBind 方法。当其他组件需要通过 bindService 方法来绑定服务时,系统会调用本方法。在本方法的实现代码中必须返回 IBinder 来提供一个接口,客户端通过它与服务进行通信。必须确保实现本方法,不过如果不需要提供绑定,则直接返回 null 即可。

3) onCreate 方法。当服务第一次被创建时,系统会调用本方法,用于执行一次性的配置工作。如果服务已经运行,则本方法就不会再被调用。

4) onDestroy 方法。当服务被销毁时,系统会调用本方法。服务应该实现本方法来进行资源的清理工作,诸如线程、已注册的侦听器 listener 和接收器 receiver 等。这将是服务收到的最后一个调用。

(3) Service 的存续期

如果组件通过调用 startService 启动了服务,那么服务将一直保持运行,直至自行用 stopSelf 方法终止或由其他组件调用 stopService 方法来终止。

如果组件调用 bindService 方法来创建服务(此时 onStartCommand 方法就不会被调用),则服务的生存期就与被绑定的组件一致。一旦所有客户端都调用了 unbindService 方法对服务解除了绑定,系统就会销毁该服务。

当系统资源减少时,为了保障拥有用户焦点的应用程序正常运行,Android 系统会强行终止一个服务。如果服务被拥有用户焦点的 Activity 绑定,则它一般不会被杀死。如果服务声明为"在前台运行服务",则它几乎不会被杀死。然而,如果服务已被启动并且已运行了很长时间,那么系统将会随着时间的推移而降低它在后台任务列表中的级别,此类服务将很有可能会被杀死。如果系统杀死了服务,只要资源再度够用,系统就会再次启动服务(当然这还取决于 onStartCommand 方法的返回值)。关于服务永久运行的方法将在第 9 章介绍。

(4) 在配置文件中声明服务

与广播接收器和权限使用类似,必须在应用程序的配置文件中对所有使用的服务进行声明,将 <service> 元素作为子元素加入 <application> 元素中即可。例如:

```
<manifest...>
    ...
    <application...>
        <service android:name=".ExampleService" />
        ...
    </application>
</manifest>
```

在 <service> 元素中可以包含很多其他属性,比如定义启动服务所需权限、服务运行的进程之类的属性。android:name 是唯一必需的属性,它定义了服务的类名。应用程序一经发布，就不得再修改这个类名，因为这样做可能会破坏某些显式引用该服务的 Intent 对象的功能。如果包含了 android:exported 属性并且设置为 "false"，就可以确保该服务是本应用程序的私有服务。

服务中还可以定义 Intent 过滤器，使得其他组件能用隐式 Intent 来调用服务。通过声明 Intent 过滤器，任何安装在用户设备上的应用程序组件都有能力来启动该服务，只要该服务所声明的 Intent 过滤器与其他应用程序传递给 startService 方法的 Intent 相匹配即可。

4.8 启动服务的实现

在其他组件通过调用 startService 方法请求 Started 方式的服务时，系统将自动调用 onStartCommand 方法。一旦执行 startService 方法，服务就被启动。本节通过一个浮动球来说明 Started 服务的使用方法。

(1) 布局文件

新建项目 FloatingBall，采用线性布局，水平居中设置两个按钮，btStart 按钮用来启动服务，btStop 按钮用来终止服务。

```xml
<LinearLayout xmlns:android="http://schemas.android.com/apk/res/android"
    xmlns:tools="http://schemas.android.com/tools"
    android:layout_width="match_parent"
    android:layout_height="match_parent"
    android:paddingBottom="@dimen/activity_vertical_margin"
    android:paddingLeft="@dimen/activity_horizontal_margin"
    android:paddingRight="@dimen/activity_horizontal_margin"
    android:paddingTop="@dimen/activity_vertical_margin"
    android:orientation="horizontal"
    android:gravity="center_horizontal"
    tools:context="com.walkerma.floatingball.MainActivity" >
    <Button
        android:id="@+id/btStart"
        android:layout_width="wrap_content"
        android:layout_height="wrap_content"
        android:text="@string/strStart" />
    <Button
        android:id="@+id/btStop"
        android:layout_width="wrap_content"
        android:layout_height="wrap_content"
        android:text="@string/strStop" />
</LinearLayout>
```

(2) 主窗体源代码

在 btStart 事件处理方法中，生成 Intent 对象 intent，并传递字符串，然后调用 startService 方法启动服务。在 btStop 事件处理方法中，调用 stopService 方法终止服务。

```java
package com.walkerma.floatingball;
import android.app.Activity;
import android.content.Intent;
import android.os.Bundle;
```

```
import android.view.View;
import android.view.View.OnClickListener;
import android.widget.Button;
public class MainActivity extends Activity {
    @Override
    protected void onCreate(Bundle savedInstanceState) {
        super.onCreate(savedInstanceState);
        setContentView(R.layout.activity_main);
        Button btStart = (Button)findViewById(R.id.btStart);
        btStart.setOnClickListener(new OnClickListener() {
            @Override
            public void onClick(View v) {
                Intent intent =  new Intent(MainActivity.this, FloatingService.class);
                intent.putExtra("Value", "From MainActivity");
                startService(intent);
            }
        });
        Button btStop = (Button)findViewById(R.id.btStop);
        btStop.setOnClickListener(new OnClickListener() {
            @Override
            public void onClick(View v) {
                stopService(new Intent(MainActivity.this, FloatingService.class));
            }
        });
    }
}
```

(3) 启动服务源代码

由于是启动服务，因而不需要实现 onBind 方法，直接返回 null 即可。主窗体中调用 startService 方法后，首先调用 onCreate 方法对服务进行初始化（仅执行一次），生成 ImageView 对象 ball 并进行填充；然后初始化布局参数 LayoutParams 对象 params，TYPE_SYSTEM_ALERT 常量表示窗口将显示在其他应用程序的窗口之上，FLAG_NOT_FOCUSABLE 常量表示所显示的窗口不会接收焦点输入，如不会处理返回键；最后调用系统服务获得 WindowManager 对象 wm，按照设置的参数在屏幕左上角显示 ImageView 对象 ball，并注册 onTouch 事件处理方法。

完成 onCreate 方法后，接着执行 onStartCommand 方法。在服务已经存在的情况下启动服务，此时将不再执行 onCreate 方法，而是直接执行 onStartCommand 方法。onStartCommand 方法中的 intent 对象是启动组件传递过来的，在系统重启服务的时候，该 intent 对象为 null；flags 为启动请求标志，一般为 0、START_FLAG_REDELIVERY 或 START_FLAG_RETRY 三种情况；startId 是一个唯一的请求启动标识，传递给 stopSelfResult 方法以用于停止服务。onStartCommand 方法返回 Service.START_STICKY，表示如果系统杀死了服务，只要资源再度够用，系统就会再次启动服务。

当服务被终止或被系统杀死，将自动调用 onDestroy 方法，在其中移除 ImageView 对象 ball。

```
package com.walkerma.floatingball;
import android.app.Service;
import android.content.Intent;
```

```java
import android.graphics.PixelFormat;
import android.os.IBinder;
import android.view.Gravity;
import android.view.MotionEvent;
import android.view.View;
import android.view.WindowManager;
import android.view.WindowManager.LayoutParams;
import android.widget.ImageView;
import android.widget.Toast;
public class FloatingService extends Service {
    private WindowManager wm;
    private ImageView ball;
    @Override
    public int onStartCommand(Intent intent, int flags, int startId) {
        // TODO Auto-generated method stub
        String val=null;
        if(intent!=null){
            val = intent.getExtras().getString("Value");
            Toast.makeText(this, val, Toast.LENGTH_SHORT).show();
        }
        else
            Toast.makeText(this, "NULL", Toast.LENGTH_SHORT).show();
        return Service.START_STICKY;
    }
    @Override
    public IBinder onBind(Intent intent) {
        return null;
    }
    @Override
    public void onCreate() {
        super.onCreate();
        ball = new ImageView(this);
        ball.setImageResource(R.drawable.floating);
        final LayoutParams params = new LayoutParams(
                LayoutParams.WRAP_CONTENT,
                LayoutParams.WRAP_CONTENT,
                LayoutParams.TYPE_SYSTEM_ALERT,
                LayoutParams.FLAG_NOT_FOCUSABLE,
                PixelFormat.TRANSLUCENT);
        params.gravity = Gravity.TOP | Gravity.START;
        wm = (WindowManager) getSystemService(WINDOW_SERVICE);
        wm.addView(ball, params);
        try {
            ball.setOnTouchListener(new View.OnTouchListener() {
                private LayoutParams paramsF = params;
                private int initialX;
                private int initialY;
                private float initialTouchX;
                private float initialTouchY;
                @Override
                public boolean onTouch(View v, MotionEvent event) {
                    switch (event.getAction()) {
                        case MotionEvent.ACTION_DOWN:
                            initialX = paramsF.x;
                            initialY = paramsF.y;
                            initialTouchX = event.getRawX();
```

```
                    initialTouchY = event.getRawY();
                    break;
                case MotionEvent.ACTION_UP:
                    break;
                case MotionEvent.ACTION_MOVE:
                    paramsF.x = initialX +
                            (int) (event.getRawX() - initialTouchX);
                    paramsF.y = initialY +
                            (int) (event.getRawY() - initialTouchY);
                    wm.updateViewLayout(ball, paramsF);
                    break;
                }
                return false;
            }
        });
    } catch (Exception e) {
        // TODO: handle exception
    }
}
@Override
public void onDestroy() {
    super.onDestroy();
    if (ball != null) wm.removeView(ball);
}
}
```

（4）配置文件

关于配置文件，主要在 <application> 标记中添加 <service> 子标记，登记服务类名 FloatingService，设置 exported 属性为 false，即该服务仅内部使用。另外，还需要申请上层显示应用程序的权限 permission.SYSTEM_ALERT_WINDOW，使得浮动球可以覆盖其他应用程序界面。

```xml
<?xml version="1.0" encoding="utf-8"?>
<manifest xmlns:android="http://schemas.android.com/apk/res/android"
    package="com.walkerma.floatingball"
    android:versionCode="1"
    android:versionName="1.0" >
    <uses-sdk
        android:minSdkVersion="14"
        android:targetSdkVersion="21" />
    <uses-permission android:name=
        "android.permission.SYSTEM_ALERT_WINDOW" / >
    <application
        android:allowBackup="true"
        android:icon="@drawable/ic_launcher"
        android:label="@string/app_name"
        android:theme="@style/AppTheme" >
        <service
            android:name=".FloatingService"
            android:enabled="true"
            android:exported="false" />
        <activity
            android:name=".MainActivity"
            android:label="@string/app_name" >
```

```
            <intent-filter>
                <action android:name="android.intent.action.MAIN" />
                <category android:name="android.intent.category.LAUNCHER" />
            </intent-filter>
        </activity>
    </application>
</manifest>
```

（5）运行效果

FloatingBall 项目的运行效果如图 4-7 所示，点击【Start】启动服务，在屏幕左上角（状态栏之下）显示浮动球，该浮动球顶层显示，且可以被任意移动，不受主窗体是否被关闭的影响。点击【Stop】将终止服务，同时浮动球被移除。

图 4-7 FloatingBall 项目运行效果

（6）查找服务是否运行和启动应用程序

虽然本程序在 onStartCommand 方法中返回 Service.START_STICKY，理论上表示如果系统杀死了服务，只要资源再度够用，系统就会再次启动服务。但实际情况是，服务依然会被自动终止。若需要服务一直处于 started 状态，在安装时应选择"自动启动"，在编程时还需要将服务声明为"前台服务"，并经常例行启动服务，告诉操作系统这是常用的重要服务，这样服务才不会被系统杀死。

可以使用如下 isServiceRunning 函数来查询服务是否处于运行状态，第一个参数是上下文环境，第二个参数是服务的完整名称。该函数首先获得当前服务列表，然后逐个核对指定名称的服务是否处于运行状态。

```
//"com.walkerma.floatingball. FloatingService"
public  static boolean isServiceRunning(Context context, String serviceName) {
    boolean isRunning = false;
    ActivityManager am = (ActivityManager)context
            .getSystemService(Context.ACTIVITY_SERVICE);
    List<RunningServiceInfo> servicesInfo =
            am.getRunningServices(Integer.MAX_VALUE);
    if(servicesInfo!=null && servicesInfo.size()>0){
        for(RunningServiceInfo serviceInfo : servicesInfo){
            if(serviceName.equals(serviceInfo.service.getClassName())){
                isRunning=true;
                break;
            }
        }
    }
    return isRunning;
}
```

如果发现给定的服务被终止，可以调用 startAppPackage 函数启动服务对应的应用程序，从而启动该服务。

```
//"com.walkerma.floatingball"
public static void startAppPackage(Context context, String packageName){
    PackageManager packageManager = context.getPackageManager();
```

```
        Intent intent = packageManager.getLaunchIntentForPackage(packageName);
        intent.setFlags(Intent.FLAG_ACTIVITY_NEW_TASK |
                Intent.FLAG_ACTIVITY_RESET_TASK_IF_NEEDED |
                Intent.FLAG_ACTIVITY_CLEAR_TOP);
        context.startActivity(intent);
    }
```

4.9 绑定服务的实现

实现绑定服务时，需要实现 onBind 回调方法。该方法返回 IBinder 对象，它定义了客户端用来与服务交互的程序接口，即通过该对象可实现与绑定的 Service 之间的通信。本节通过一个产生随机数的绑定服务实例来说明绑定服务的实现过程及生命周期。

（1）绑定服务源代码

新建项目 ServiceBinding，布局文件（此处省略）中仅使用一个按钮来调用绑定服务中产生随机数的方法。绑定服务所实现的 onBind 回调方法中返回的 IBinder 对象通常会采用继承 Binder 的实现类的方式来实现，这里定义了一个 public 类型的 LocalBinder 类，其中定义了一个 getService 方法来返回 LocalService 类的实例。

绑定服务中提供了一个公用方法 getRandomNumber，用来产生一个随机数。

```java
package com.walkerma.servicebinding;
import java.util.Random;
import android.app.Service;
import android.content.Intent;
import android.os.Binder;
import android.os.IBinder;
public class LocalService extends Service {
    private final IBinder binder = new LocalBinder();
    private final Random generator = new Random();
    public class LocalBinder extends Binder{
        LocalService getService(){
            return LocalService.this;
        }
    }
    @Override
    public IBinder onBind(Intent intent) {
        // TODO Auto-generated method stub
        return binder;
    }
    public int getRandomNumber(){
        return generator.nextInt(100);
    }
}
```

（2）主窗体源代码

客户端运行后，onCreate 执行完毕将自动运行 onStart 方法，并在其中调用 bindService 方法绑定服务。bindService 方法立即返回，但是，当 Android 系统创建客户端与服务之间的连接时，将调用 ServiceConnection 接口的 onServiceConnected 方法来发送客户端用来与服务通信的 IBinder 对象。此时点击命令按钮即可调用绑定服务的 getRandomNumber 方法并获得随机数。

当应用程序不可见时自动调用 onStop 方法解除绑定，当程序可见时又自动调用 onStart 方法绑定服务，并可调用绑定服务的 getRandomNumber 方法来获得随机数。

```java
package com.walkerma.servicebinding;
import com.walkerma.servicebinding.LocalService.LocalBinder;
import android.app.Activity;
import android.content.ComponentName;
import android.content.Context;
import android.content.Intent;
import android.content.ServiceConnection;
import android.os.Bundle;
import android.os.IBinder;
import android.view.View;
import android.widget.Toast;
public class MainActivity extends Activity {
    LocalService localService;
    boolean bound = false;
    private ServiceConnection connection = new ServiceConnection(){
        @Override
        public void onServiceConnected(ComponentName name,
                IBinder service) {
            // TODO Auto-generated method stub
            LocalBinder binder = (LocalBinder)service;
            localService = binder.getService();
            bound = true;
        }
        @Override
        public void onServiceDisconnected(ComponentName name) {
            // TODO Auto-generated method stub
            bound = false;
        }};
    @Override
    protected void onCreate(Bundle savedInstanceState) {
        super.onCreate(savedInstanceState);
        setContentView(R.layout.activity_main);
    }
    @Override
    protected void onStart() {
        // TODO Auto-generated method stub
        Intent intent = new Intent(this, LocalService.class);
        bindService(intent, connection, Context.BIND_AUTO_CREATE);
        super.onStart();
    }
    @Override
    protected void onStop() {
        // TODO Auto-generated method stub
        if(bound){
            unbindService(connection);
            bound = false;
        }
        super.onStop();
```

```
        }
        public void onClick(View v){
            if(bound){
                int num = localService.getRandomNumber();
                Toast.makeText(this, "Random: " + num,
                    Toast.LENGTH_SHORT).show();
            }
        }
}
```

(3)配置文件

对于配置文件,主要在 <application> 标记中添加 <service> 子标记,登记服务类名 LocalService,设置 exported 属性为 false,即该服务仅在内部使用。

```
<?xml version="1.0" encoding="utf-8"?>
<manifest xmlns:android="http://schemas.android.com/apk/res/android"
    package="com.walkerma.servicebinding"
    android:versionCode="1"
    android:versionName="1.0" >
    <uses-sdk
        android:minSdkVersion="16"
        android:targetSdkVersion="21" />
    <application
        android:allowBackup="true"
        android:icon="@drawable/ic_launcher"
        android:label="@string/app_name"
        android:theme="@style/AppTheme" >
        <service
            android:name=".LocalService"
            android:enabled="true"
            android:exported="false" >
        </service>
        <activity
            android:name=".MainActivity"
            android:label="@string/app_name" >
            <intent-filter>
                <action android:name="android.intent.action.MAIN" />
                <category android:name="android.intent.category.LAUNCHER" />
            </intent-filter>
        </activity>
    </application>
</manifest>
```

(4)运行效果

程序运行后点击【Bind】按钮,将显示不同的随机数。

4.10 本章小结

Intent 可以指定对象名称和动作名称,添加数据、种类并补充附加信息以及设置标志信息。在多 Activity 操作中,可以利用 Intent 启动 Activity、传递和返回数据。广播接收器可以在后台监控系统或用户广播，本章先后通过静态注册、动态注册、独立类和嵌入类等多种方式实现了广播接收器。服务分为启动服务和绑定服务，本章通过简洁的实例展示了两种服务的实现方法，并给出了测试服务是否运行及启动其他应用程序的源代码。利用 Intent 进行前台多 Activity 操作是编写 Android 应用程序的基础，熟练掌握 Intent 在后台 BroadcastReceiver 和 Service 中的应用，可以实现功能更强大、更实用的应用程序。

第二部分

实用案例分析

第 5 章　课堂随机点名软件

第 6 章　简易英语学习软件

第 7 章　通讯录备份与恢复软件

第 8 章　服务账号登记软件

第 9 章　地址定位及辅助服务软件

第 10 章　地址查询与地图打点软件

第 5 章　课堂随机点名软件

双色球福利彩票的六个红色球号码的范围为 1~33 且不重复，一个蓝色球号码的范围为 1~16，对规则适当变通，可以用于课堂随机点名：红色球号码改为班级学生的序号，有几个班就使用几个红色球，而蓝色球号码改为班级序号。这样，如果每班随机抽取一个学生，只考虑红色球即可；如果只随机抽取一个学生，就考虑蓝色球班级的红色球序号。一位教师可能任教多门课程，每门课程分别设置学生参数，则可以实现所有课程的随机点名。

主要知识点：SQLite 数据库、菜单、ListView 的使用、自定义对话框

5.1　主要功能和技术特点

采用 Android 应用程序实现课堂随机点名，可以随时随地进行，不必依赖于台式机的系统环境。Android 课堂随机点名软件（安装后的 App 名为 Lottery，下文简称 Lottery）的主要特点为：

- 可以处理多门课程，每门课程可以对应多个班级；
- 每次选择一门课程点名时，每班随机抽取学生的序号不重复；
- 可以每班随机抽取一名学生（根据红色球数字）；
- 也可以在所有学生中仅随机抽取一名学生（结合蓝色球班号）；
- 生成双色球福利彩票数字序列。

5.2　软件操作

Lottery 软件的运行界面如图 5-1 所示。在 Course Name 编辑框中输入课程名称，点击【Add】命令按钮，可以添加新的课程。选中某课程，在 "23 39" 所在的编辑框中输入整数（以空格分隔），几个整数即表示该课程有几个班。点击【RUN】菜单项，如果数据库中没有班级学生人数，则将输入的整数存入数据库；如果数据库中有班级学生人数，则每班随机抽取一个学生，并随机抽取一个班号。如果点击【Clear】命令按钮，则会清除所有随机数及编辑框中的课程名称，同时将数据库中的班级人数清空，这样又可以通过【RUN】菜单项来重新设置班级人数。

图 5-1 中的【RUN】菜单项一直显示在 ActionBar 上；溢出菜单项【Normal】总是自动隐藏，用于产生一

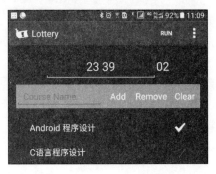

图 5-1　Lottery 软件界面

个标准的双色球福利彩票序列号。两个菜单项分开布置，主要是方便使用，以免点击时由于相隔区域较小而相互影响。随机点名与输出标准彩票号码是两个不同的功能，两者相互独立，互不影响。

在图 5-1 中选择课程列表中的课程，点击【Remove】命令按钮将会删除该门课程，此时会显示是否删除的提示对话框（自定义对话框），如图 5-2 所示。点击【Clear】按钮也会显示此对话框，但标题显示为"Clear course seed"，即清除课程的班级人数种子。点击对话框中的【Yes】按钮将删除当前课程，点击【No】按钮将放弃操作。

另外，溢出菜单 About 显示版权信息（截图略），关于版权信息的设计见下一节。

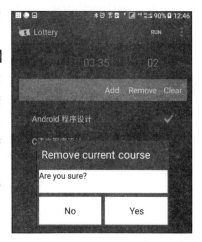

图 5-2 自定义对话框效果

5.3 界面布局与资源说明

Lottery 是一个实用的软件作品，所涉及的资源包括字符串的定义、菜单、颜色、自定义对话框和主窗体的布局文件等。大部分元素都要综合体现在主窗体布局文件中，供主窗体源代码综合调度使用。

5.3.1 字符串定义文件

字符串定义文件 strings.xml 中定义了菜单项和主窗体中组件所使用的字符串常量，其中红色球号码初始提示字符串 strRedHint 和蓝色球号码初始提示字符串 strBlueHint 都用"#"表示，指示输入数字。

```
<resources>
    <string name="app_name">Lottery</string>
    <string name="run">Run</string>
    <string name="normal">Normal</string>
    <string name="about">About</string>
    <string name="strRedHint">## ## ## ## ## ##</string>
    <string name="strBlueHint">##</string>
    <string name="strClear">Clear</string>
    <string name="strRemove">Remove</string>
    <string name="strAdd">Add</string>
    <string name="strCourseHint">Your Course Name</string>
    <string name="strCopyrights">
        Version 1.0\nAll rights reserved.\nDeveloped by </string>
</resources>
```

5.3.2 菜单项定义文件

菜单项定义文件 menu_lottery.xml 中定义了三个菜单项，其中 item_Run 总是在标题栏中显示，方便用户随机抽取或保留班级种子数字。item_Normal 用来产生标准的双色球号码，由于不是主要功能，放在溢出菜单中。item_About 用来显示版权信息，也放在溢出菜单中。

```
<?xml version="1.0" encoding="utf-8"?>
```

```xml
<menu xmlns:android="http://schemas.android.com/apk/res/android"
    xmlns:app="http://schemas.android.com/apk/res-auto">
    <item
        android:id="@+id/item_Run"
        android:orderInCategory="100"
        android:title="@string/run"
        android:showAsAction="always" />
    <item
        android:id="@+id/item_Normal"
        android:orderInCategory="101"
        android:title="@string/normal"
        android:showAsAction="never" />
    <item
        android:id="@+id/item_About"
        android:orderInCategory="102"
        android:title="@string/about"
        android:showAsAction="never" />
</menu>
```

5.3.3 颜色定义文件

颜色定义文件 colors.xml 中定义了多种颜色,这些颜色将用于后续的自定义对话框布局文件和主窗体布局文件。此文件存放于 3.3 节创建的 Library 项目的 res\values 子文件夹中,可供引用 Library 的任何项目使用,而无须再次创建。

```xml
<?xml version="1.0" encoding="utf-8"?>
<resources>
    <color name="colorPrimary">#3F51B5</color>
    <color name="colorPrimaryDark">#303F9F</color>
    <color name="colorAccent">#FF4081</color>
    <color name="black">#ff000000</color>
    <color name="dark_grey">#181818</color>
    <color name="blue">#ff1118ff</color>
    <color name="light_grey">#bababa</color>
    <color name="green">#003300</color>
    <color name="white">#ffffff</color>
</resources>
```

5.3.4 自定义对话框布局文件

自定义对话框布局文件 dialog_custom.xml 中分别定义了一个标题和消息内容的 TextView 组件,以及两个与用户交互的 Button 组件,用于删除课程或者清除班级数据。

```xml
<?xml version="1.0" encoding="utf-8"?>
<LinearLayout xmlns:android="http://schemas.android.com/apk/res/android"
    android:id="@+id/custom_root_layout"
    android:layout_width=" match_parent "
    android:layout_height="match_parent"
    android:background="@color/blue"
    android:orientation="vertical" >
    <TextView
        android:id="@+id/custom_title"
        android:layout_width="match_parent"
        android:layout_height="60dp"
```

```xml
            android:gravity="center_vertical"
            android:padding="10dp"
            android:text="New Reminder:"
            android:textColor="@color/white"
            android:textSize="24sp" />
    <TextView
            android:id="@+id/custom_message"
            android:layout_width="match_parent"
            android:layout_height="50dp"
            android:layout_margin="4dp"
            android:background="@color/light_grey"
            android:textSize="18sp"
            android:gravity="start"
            android:textColor="@color/black" />
    <LinearLayout
            android:layout_width="match_parent"
            android:layout_height="match_parent"
            android:orientation="horizontal">
        <Button
            android:id="@+id/custom_button_no"
            android:layout_margin="4dp"
            android:layout_width="0dp"
            android:layout_height="60dp"
            android:layout_weight="50"
            android:text="No"
            android:textSize="20sp"
            android:background="@color/light_grey"
            android:textColor="@color/black"/>
        <Button
            android:id="@+id/custom_button_yes"
            android:layout_margin="4dp"
            android:layout_width="0dp"
            android:layout_height="60dp"
            android:layout_weight="50"
            android:text="Yes"
            android:textSize="20sp"
            android:background="@color/light_grey"
            android:textColor="@color/black"/>
    </LinearLayout>
</LinearLayout>
```

5.3.5 ListView 列表布局文件

Android 系统自定义的 ListView 列表布局文件 android.R.layout.simple_list_item_checked 中的字体太小，显示效果欠佳，对其进行改进，将 textSize 属性设置为 18sp，得到 simple_list_item_checked.xml，如果需要继续进行个性化调整，还可以修改其他属性。

```xml
<?xml version="1.0" encoding="utf-8"?>
<CheckedTextView xmlns:android="http://schemas.android.com/apk/res/android"
    android:id="@android:id/text1"
    android:layout_width="match_parent"
    android:layout_height="?android:attr/listPreferredItemHeightSmall"
    android:checkMark="?android:attr/textCheckMark"
    android:gravity="center_vertical"
    android:paddingEnd="?android:attr/listPreferredItemPaddingEnd"
```

```
        android:paddingLeft="?android:attr/listPreferredItemPaddingLeft"
        android:paddingRight="?android:attr/listPreferredItemPaddingRight"
        android:paddingStart="?android:attr/listPreferredItemPaddingStart"
        android:textAppearance="?android:attr/textAppearanceListItemSmall"
        android:textColor="@color/white"
        android:textSize="18sp" />
```

5.3.6 版权窗体布局文件

版权窗体布局文件 dialog_about.xml 中有三个组件。txtTitle 显示软件名称，txtCopyright 显示版权信息，在程序运行时动态加载。imageView 组件显示作者图片信息。显示版权信息也是应用程序中的常用元素之一，因而可将此布局文件和相关图片信息存入 Library 项目中的相关目录中，以便共享调用。

```
<?xml version="1.0" encoding="utf-8"?>
<RelativeLayout xmlns:android="http://schemas.android.com/apk/res/android"
    android:layout_width="match_parent"
    android:layout_height="match_parent"
    android:background="@color/blue"
    android:padding="10dp">
    <TextView
        android:layout_width="wrap_content"
        android:layout_height="wrap_content"
        android:textColor="@color/white"
        android:textAppearance="?android:attr/textAppearanceLarge"
        android:text="@null"
        android:id="@+id/txtTitle"
        android:layout_alignParentTop="true"
        android:layout_centerHorizontal="true"/>
    <TextView
        android:layout_width="wrap_content"
        android:layout_height="wrap_content"
        android:textColor="@color/white"
        android:textAppearance="?android:attr/textAppearanceMedium"
        android:text="@null"
        android:id="@+id/txtCopyright"
        android:layout_below="@+id/imageView"
        android:layout_centerHorizontal="true"
        android:gravity="center"/>
    <ImageView
        android:layout_width="wrap_content"
        android:layout_height="wrap_content"
        android:id="@+id/imageView"
        android:contentDescription="@string/app_name"
        android:layout_marginTop="5dp"
        android:layout_marginBottom="5dp"
        android:layout_below="@+id/txtTitle"
        android:layout_centerHorizontal="true"
        android:src="@drawable/walker_ma"/>
</RelativeLayout>
```

5.3.7 主窗体布局文件

主窗体布局文件 activity_main.xml 的实际效果可以参考图 5-1，这里整体采用线性布局,由于课程名称具有弹性,因而将 ListView 放在最下面,layout_width 和 layout_height 都设置为"match_parent",填满剩余的屏幕空间。

```xml
<?xml version="1.0" encoding="utf-8"?>
<LinearLayout xmlns:android="http://schemas.android.com/apk/res/android"
    xmlns:tools="http://schemas.android.com/tools"
    android:id="@+id/activity_main"
    android:layout_width="match_parent"
    android:layout_height="match_parent"
    android:background="@color/black"
    android:orientation="vertical"
    android:paddingBottom="@dimen/activity_vertical_margin"
    android:paddingLeft="@dimen/activity_horizontal_margin"
    android:paddingRight="@dimen/activity_horizontal_margin"
    android:paddingTop="@dimen/activity_vertical_margin"
    tools:context="com.walkerma.lottery.MainActivity">
    <LinearLayout
        android:layout_width="wrap_content"
        android:layout_height="wrap_content"
        android:layout_gravity="center"
        android:orientation="horizontal">
        <EditText
            android:id="@+id/edRed"
            android:layout_width="wrap_content"
            android:layout_height="wrap_content"
            android:gravity="center"
            android:textSize="24sp"
            android:digits="0123456789 "
            android:hint="@string/strRedHint"
            android:inputType="textFilter"
            android:maxLength="17"
            android:padding="10dp"
            android:textAppearance="?android:attr/textAppearanceMedium"
            android:textColor="@android:color/holo_red_light">
            <!--<requestFocus />-->
        </EditText>
        <TextView
            android:id="@+id/txtBlue"
            android:layout_width="wrap_content"
            android:layout_height="wrap_content"
            android:textSize="24sp"
            android:hint="@string/strBlueHint"
            android:textAppearance="?android:attr/textAppearanceMedium"
            android:textColor="@android:color/holo_blue_light" />
    </LinearLayout>
    <LinearLayout
        android:layout_width="match_parent"
        android:layout_height="wrap_content"
        android:gravity="center"
        android:layout_marginBottom="10dp"
        android:layout_marginTop="10dp"
        android:background="@android:color/holo_green_dark"
        android:orientation="horizontal" >
        <EditText
            android:id="@+id/edCourse"
```

```xml
            android:layout_width="wrap_content"
            android:layout_height="wrap_content"
            android:ellipsize="end"
            android:hint="@string/strCourseHint"
            android:inputType="text"
            android:maxLines="1"
            android:paddingEnd="5dp"
            android:paddingStart="15dp"
            android:textAppearance="?android:attr/textAppearanceMedium"
            android:textColor="@android:color/white" />
        <Button
            android:id="@+id/btAdd"
            style="?android:attr/buttonBarButtonStyle"
            android:layout_width="wrap_content"
            android:layout_height="wrap_content"
            android:text="@string/strAdd"
            android:textColor="@android:color/white" />
        <Button
            android:id="@+id/btRemove"
            style="?android:attr/buttonBarButtonStyle"
            android:layout_width="wrap_content"
            android:layout_height="wrap_content"
            android:text="@string/strRemove"
            android:textColor="@android:color/white" />
        <Button
            android:id="@+id/btClear"
            style="?android:attr/buttonBarButtonStyle"
            android:layout_width="wrap_content"
            android:layout_height="wrap_content"
            android:text="@string/strClear"
            android:textColor="@android:color/white" />
    </LinearLayout>
    <ListView
        android:id="@+id/list"
        android:layout_width="match_parent"
        android:layout_height="match_parent"
        android:textSize="24sp"
        android:choiceMode="singleChoice"
        android:layout_marginBottom="20dp"
        android:layout_marginLeft="20dp"
        android:layout_marginRight="20dp"
        android:layout_marginTop="5dp"
        android:drawSelectorOnTop="false" />
</LinearLayout>
```

5.4 配置文件

读写 SQLite 数据库不需要任何权限。需要将 activity 标记的 configChanges 属性设置为"orientation|screenSize",以便在源代码中进行转屏事件处理;将 windowSoftInputMode 属性设置为"adjustUnspecified|stateHidden",这样在显示 EditText 组件时就不会首先显示输入软键盘,只有在点击 EditText 组件时才会显示输入软键盘。

```xml
<?xml version="1.0" encoding="utf-8"?>
```

```xml
<!DOCTYPE xml>
<manifest xmlns:android="http://schemas.android.com/apk/res/android"
    package="com.walkerma.lottery"
    android:versionCode="1"
    android:versionName="1.0" >
    <uses-sdk
        android:minSdkVersion="17"
        android:targetSdkVersion="22" />
    <application
        android:allowBackup="true"
        android:icon="@drawable/ic_launcher"
        android:label="@string/app_name"
        android:theme="@style/AppTheme" >
        <activity
            android:name=".MainActivity"
            android:label="@string/app_name"
            android:configChanges="orientation|screenSize"
            android:windowSoftInputMode="adjustUnspecified|stateHidden">
            <intent-filter>
                <action android:name="android.intent.action.MAIN" />
                <category android:name="android.intent.category.LAUNCHER" />
            </intent-filter>
        </activity>
    </application>
</manifest>
```

5.5 主窗体源代码

颜色定义文件及版权信息相关布局文件和图片文件都来自 Library，另外 DatabaseHelper 类和显示版权信息的 fireAboutDialog 函数都在 Library 中定义，因而，需要按照 3.3 节中的方法引用 Library 类库。

本程序使用的 SQLite 数据库包含两个字段：_courseName 字段存放课程名称，为主键；_classItems 字段存放班级参数，多个班则用空格分隔。getClassArray 函数以 _classItems 字符串为参数，得到一个整型数组，其中存放每个班的学生数。getRandomStringNumber 函数以整数为参数，返回一个不大于该数的随机数（转换为字符串）。getNextData 以历史随机数字符串和最大数为参数，返回下一个不重复的字符串形式的随机数，但是，在产生随机数的过程中只尝试十次，即十次以后数据还重复，就直接返回重复的数据。

getCourseLottery 以 _classItems 字符串为参数，得到所有红色球和蓝色球随机数，其中字符串数组 hitItems 中存放每个班已经抽取的历史随机数，使用 bResetHitItems 布尔变量对 hitItems 的初始化进行控制，只有当新选择一门课程时，bResetHitItems 才为 true，这时清空 hitItems 中的历史数据（设置为初值"/"）。

getNormalLottery 函数用于产生标准的福彩号码，采用 ArrayList<String> 类型的变量 classItems 来保存红色球号码，采用 Collections 类的静态方法 sort 对 classItems 中的元素排序，最后中间加空格合成一个字符串。而蓝色球号码只需调用 getRandomStringNumber 函数即可一步解决。

程序运行时首先执行 onCreate 方法，创建数据库，进行组件的初始化，接着初始化

ListView 对象 list 的适配器，注册选项点击侦听器，然后遍历数据库，将课程名称加入 list 对象，并利用 3.5 节的方法处理后退键。

在与用户交互的过程中，输入课程名称，点击【Add】，将调用 addOneCourse 函数，将课程名加入数据库和 list。同理，点击【Remove】调用 removeOneCourse 函数将课程名从数据库中删除，同时移出 list。点击【Clear】将清除编辑框中的数据，如果选中了某课程，还要调用 fireCustomDialog 函数显示对话框，提示用户是否删除班级参数，如果用户在对话框中点击【Yes】，将调用 clearSeedOfCourse 函数删除班级参数。

点击菜单项【Run】，如果选中了某课程且班级数据非空，则调用上文介绍的 getCourseLottery 以显示班级随机数；如果点击溢出菜单【Normal】，则调用 getNormalLottery 函数显示标准彩票数据；【About】菜单项调用 Library 类库中的 GeneralProcess 类的静态方法 fireAboutDialog，显示版权信息。

```
package com.walkerma.lottery;
import android.app.Activity;
import android.app.Dialog;
import android.content.ContentValues;
import android.content.res.Configuration;
import android.database.Cursor;
import android.os.Handler;
import android.os.Message;
import android.os.Bundle;
import android.text.TextUtils;
import android.view.KeyEvent;
import android.view.Menu;
import android.view.MenuItem;
import android.view.View;
import android.view.Window;
import android.widget.AdapterView;
import android.widget.ArrayAdapter;
import android.widget.Button;
import android.widget.EditText;
import android.widget.ListView;
import android.widget.TextView;
import android.widget.Toast;
import java.util.ArrayList;
import java.util.Collections;
import java.util.Random;
import com.walkerma.library.DatabaseHelper;
import com.walkerma.library.GeneralProcess;
import static java.lang.Math.abs;
public class MainActivity extends Activity implements View.OnClickListener {
    private DatabaseHelper db;
    private String strQuery = "SELECT * FROM Courses";
    private String strQueryCourse;
    private EditText edRed, edCourse;
    private TextView txtBlue;
    private ArrayAdapter<String> adapter;
    private ListView list;
    private ArrayList<String> listItems = new ArrayList<String>();
    private String[] hitItems;
    private final int TEST_TIME = 10;
```

```java
    private boolean bResetHitItems = true;
    private int nCurrentPos = -1;
    private String itemSelected;
    private enum DIALOG_SORT{Remove, Clear}
    private DIALOG_SORT nDialogSort;
    private boolean bCanExit = false;
    private Handler mHandler;
    class IncomingHandlerCallback implements Handler.Callback {
        @Override
        public boolean handleMessage(Message msg) {
            bCanExit = false;
            return true;
        }
    }
    @Override
    protected void onCreate(Bundle savedInstanceState) {
        super.onCreate(savedInstanceState);
        setContentView(R.layout.activity_main);
        db = new DatabaseHelper(this,"dbCourse", 1, "Courses",
                "_courseName TEXT PRIMARY KEY, " +
                "_classItems TEXT NULL");
        db.refreshReadDB(strQuery, null);
        initWidgets();
        initAdapter();
        addCoursesToListItems();
        mHandler = new Handler(new IncomingHandlerCallback());
    }
    private void addCoursesToListItems() {
        int nLen = db.getCount();
        if (nLen == 0) return;
        Cursor row = db.getRecords();
        row.moveToFirst();
        listItems.clear();
        for (int i = 0; i < nLen; i++) {
            listItems.add(row.getString(0));
            row.moveToNext();
        }
        adapter.notifyDataSetChanged();
    }
    private void initWidgets() {
        Button btAdd = (Button) findViewById(R.id.btAdd);
        btAdd.setOnClickListener(this);
        Button btRemove = (Button) findViewById(R.id.btRemove);
        btRemove.setOnClickListener(this);
        Button btClear = (Button) findViewById(R.id.btClear);
        btClear.setOnClickListener(this);
        edRed = (EditText) findViewById(R.id.edRed);
        edCourse = (EditText) findViewById(R.id.edCourse);
        txtBlue = (TextView) findViewById(R.id.txtBlue);
        list = (ListView) findViewById(R.id.list);
    }
    private void initAdapter() {
        // android.R.layout.simple_list_item_checked
        adapter = new ArrayAdapter<String>(this,
                R.layout.simple_list_item_checked, listItems);
        list.setAdapter(adapter);
```

```java
        list.setOnItemClickListener(new AdapterView.OnItemClickListener() {
            @Override
            public void onItemClick(AdapterView<?> parent, View view,
                                    int position, long id) {
                // TODO Auto-generated method stub
                if (nCurrentPos != position) {
                    nCurrentPos = position;
                    list.setItemChecked(position, true);
                    itemSelected = parent.getItemAtPosition(position).toString();
                    strQueryCourse = strQuery +
                            " WHERE _courseName = '" + itemSelected + "'";
                    db.refreshReadDB(strQueryCourse, null);
                    Cursor row = db.getRecords();
                    row.moveToFirst();
                    edRed.setText(row.getString(1));
                    bResetHitItems = true;
                } else {
                    nCurrentPos = -1;
                    list.setItemChecked(position, false);
                    itemSelected = "";
                    edRed.setText("");
                }
                txtBlue.setText("");
                adapter.notifyDataSetChanged();
            }
        });
    }
    @Override
    public boolean onKeyDown(int keyCode, KeyEvent event) {
        // TODO Auto-generated method stub
        switch (keyCode) {
            case KeyEvent.KEYCODE_BACK:
                if (bCanExit) {
                    mHandler.removeCallbacksAndMessages(null);
                    db.close();
                    break;
                } else {
                    bCanExit = true;
                    Toast.makeText(getApplicationContext(),
                            "Press again to quit.",
                            Toast.LENGTH_SHORT).show();
                }
                int MESSAGE_Delay = 10000;
                mHandler.sendEmptyMessageDelayed(MESSAGE_Delay, 1000);
                return true;
        }
        return super.onKeyDown(keyCode, event);
    }
    @Override
    public boolean onCreateOptionsMenu(Menu menu) {
        getMenuInflater().inflate(R.menu.menu_lottery, menu);
        return true;
    }
    @Override
    public boolean onOptionsItemSelected(MenuItem item) {
        switch (item.getItemId()) {
```

```java
            case R.id.item_Run:
                if (nCurrentPos == -1) {
                    Toast.makeText(this, "Please select a course...",
                            Toast.LENGTH_SHORT).show();
                    return true;
                }
                db.refreshReadDB(strQueryCourse, null);
                Cursor row = db.getRecords();
                row.moveToFirst();
                String strClassData = row.getString(1);
                if (TextUtils.isEmpty(strClassData))
                    addSeedToDatabase();
                else
                    getCourseLottery(strClassData);
                return true;
            case R.id.item_Normal:
                getNormalLottery();
                return true;
            case R.id.item_About:
                **GeneralProcess.fireAboutDialog(this,**
                        getString(R.string.app_name),
                        getString(R.string.strCopyrights));
                return true;
            default:
                return false;
        }
    }
    @Override
    public void onConfigurationChanged(Configuration newConfig) {
        super.onConfigurationChanged(newConfig);
        adapter.notifyDataSetChanged();
    }
    private void addSeedToDatabase() {
        String strSeed = edRed.getText().toString().trim();
        while (strSeed.contains("  ")) strSeed =
                strSeed.replaceAll("  ", " ");
        if (strSeed.isEmpty()) {
            Toast.makeText(this, "Please input seed number...",
                    Toast.LENGTH_SHORT).show();
        } else {
            edRed.setText(strSeed);
            ContentValues cv = new ContentValues();
            cv.put("_classItems", strSeed);
            db.updateRecords(cv, "_courseName = '" +
            itemSelected + "'", null);
            Toast.makeText(this, "Seed saved ...",
                    Toast.LENGTH_SHORT).show();
        }
    }
    private void getCourseLottery(String strClassData) {
        int[] nClassArray = getClassArray(strClassData);
        if (nClassArray == null) {
            Toast.makeText(this, "Class data error ...",
                    Toast.LENGTH_SHORT).show();
            return;
        }
```

```java
        String strRed = "";
        int nLen = nClassArray.length;
        if (bResetHitItems) {
            hitItems = new String[nLen];
         // contains not permit blank value
            for (int i = 0; i < nLen; i++) hitItems[i] = "/";
            bResetHitItems = false;
        }
        for (int i = 0; i < nLen; i++) {
            if (nClassArray[i] == 0) {
                Toast.makeText(this, "Class data error ...",
                        Toast.LENGTH_SHORT).show();
                return;
            }
            String strCurrent = getNextData(hitItems[i],
                    nClassArray[i]);
            if (!hitItems[i].contains(strCurrent)) hitItems[i]
                    += strCurrent + "/";
            strRed = strRed + strCurrent;
            if (i < nLen - 1) strRed = strRed + " ";
        }
        edRed.setText(strRed);
        txtBlue.setText(getRandomStringNumber(nLen));
    }
    private String getNextData(String strHistory, int nBig){
        String strNew = getRandomStringNumber(nBig);
        int nCount = 0;
        while (strHistory.contains(strNew)) {
            strNew = getRandomStringNumber(nBig);
            nCount++;
            if (nCount > TEST_TIME) {
                Toast.makeText(this, "Data redundant...",
                        Toast.LENGTH_SHORT).show();
                break;
            }
        }
        return strNew;
    }
    private int[] getClassArray(String strClassData) {
        String[] strArray = strClassData.split(" ");
        int nLen = strArray.length;
        if (nLen == 0) return null;
        int[] nArray = new int[nLen];
        for (int i = 0; i < nLen; i++)
            nArray[i] = Integer.parseInt(strArray[i]);
        return nArray;
    }
    private void getNormalLottery() {
        int nBig = 33;
        String strNumber = getRandomStringNumber(nBig);
        ArrayList<String> classItems = new ArrayList<String>();
        classItems.add(strNumber);
        // red*6, 1-33; blue*1, 1-16
        for (int i = 0; i < 5; i++) {
            while (classItems.contains(strNumber))
                strNumber = getRandomStringNumber(nBig);
```

```java
            classItems.add(strNumber);
        }
        Collections.sort(classItems);
        String strRed = "";
        for (int i = 0; i < 6; i++) {
            strRed = strRed + classItems.get(i);
            if (i < 5) strRed = strRed + " ";
        }
        edRed.setText(strRed);
        txtBlue.setText(getRandomStringNumber(16));
    }
    private String getRandomStringNumber(int nBig) {
        Random rand = new Random();
        int n = rand.nextInt() % nBig;
        n = abs(n);
        if (n == 0) n = nBig;
        String result = Integer.toString(n);
        if (result.length() == 1) result = "0" + result;
        return result;
    }
    @Override
    public void onClick(View v) {
        switch (v.getId()) {
            case R.id.btAdd:
                addOneCourse();
                break;
            case R.id.btRemove:
                if (nCurrentPos == -1) {
                    if (listItems.size() > 0)
                        Toast.makeText(this,
                                "Please select an item...",
                                Toast.LENGTH_SHORT).show();
                    else
                        Toast.makeText(this, "No any item...",
                                Toast.LENGTH_SHORT).show();
                    return;
                }
                nDialogSort = DIALOG_SORT.Remove;
                fireCustomDialog();
                break;
            case R.id.btClear:
                if ((TextUtils.isEmpty(edRed.getText().toString()) &&
                        TextUtils.isEmpty(txtBlue.getText().toString())) &&
                        TextUtils.isEmpty(edCourse.getText().toString()) &&
                        (nCurrentPos == -1)) {
                    Toast.makeText(this, "All data is empty already.",
                            Toast.LENGTH_SHORT).show();
                    return;
                }
                txtBlue.setText("");
                edCourse.setText("");
                if (nCurrentPos == -1) {
                    edRed.setText("");
                    return;
                }
                nDialogSort = DIALOG_SORT.Clear;
```

```java
                    fireCustomDialog();
                break;
            }
        }
        private void clearSeedOfCourse() {
            ContentValues cv = new ContentValues();
            cv.put("_classItems", "");
            db.updateRecords(cv, "_courseName = '" +
            itemSelected + "'", null);
            Toast.makeText(this, "Seed cleared ...",
                    Toast.LENGTH_SHORT).show();
        }
        private void fireCustomDialog() {
            // custom dialog
            final Dialog dialog = new Dialog(this);
            dialog.requestWindowFeature(Window.FEATURE_NO_TITLE);
            dialog.setContentView(R.layout.dialog_custom);
            //dialog.setCancelable(false);
            TextView titleView = (TextView) dialog.
                    findViewById(R.id.custom_title);
            if (nDialogSort == DIALOG_SORT.Clear)
                titleView.setText("Clear course seed");
            else
                titleView.setText("Remove current course");
            TextView msgCustom = (TextView) dialog.
                    findViewById(R.id.custom_message);
            msgCustom.setText("Are you sure?");
            Button btYes = (Button) dialog.
                    findViewById(R.id.custom_button_yes);
            btYes.setOnClickListener(new View.OnClickListener() {
                @Override
                public void onClick(View v) {
                    if (nDialogSort == DIALOG_SORT.Clear)
                        clearSeedOfCourse();
                    else
                        removeOneCourse();
                    edRed.setText("");
                    dialog.dismiss();
                }
            });
            Button btNo = (Button) dialog.findViewById(R.id.custom_button_no);
            btNo.setOnClickListener(new View.OnClickListener() {
                @Override
                public void onClick(View v) {
                    dialog.dismiss();
                }
            });
            dialog.show();
        }
        private void removeOneCourse() {
            listItems.remove(nCurrentPos);  // checked in onClick
            db.deleteRecords("_courseName = '" +
            itemSelected + "'", null);
            if (nCurrentPos >= 0 && nCurrentPos <=
                    listItems.size() - 1)
                list.setItemChecked(nCurrentPos, false);
```

```
            nCurrentPos = -1;
            itemSelected = "";
            adapter.notifyDataSetChanged();
    }
    private void addOneCourse() {
        String strCourse = edCourse.getText().toString();
        if (strCourse.isEmpty()) {
            Toast.makeText(this, "Data empty.",
                    Toast.LENGTH_SHORT).show();
            return;
        }
        if (listItems.contains(strCourse)) {
            Toast.makeText(this, strCourse +
                    " redundant.", Toast.LENGTH_SHORT).show();
            return; //old data, return
        }
        listItems.add(strCourse);    // checked automatically
        int nPos = listItems.indexOf(strCourse);
        if (list.isItemChecked(nPos)) list.setItemChecked(nPos, false);
        ContentValues cv = new ContentValues();
        cv.put("_courseName", strCourse);
        db.insertRecords(cv);
        adapter.notifyDataSetChanged();
        Toast.makeText(this, strCourse + " added.",
                Toast.LENGTH_SHORT).show();
        edCourse.setText("");
    }
}
```

5.6 本章小结

本章使用 3.3 节构建的 Library 类库和静态函数，以及 3.8 节的 SQLite 数据库技术实现了一个课堂随机点名程序，可以随时随地随机抽查考勤和检查教学效果，不必依赖台式机及其运行环境。另外，还向 Library 类库中添加了颜色设置、版权对话框相关的布局文件与图片等元素。构建自己的 Library 类库，将常用的代码抽象成共性的类，可以提高程序的可靠性和效率，因为修改 Library 中的类，即修改了所有调用该类的应用程序，只需重新编译即可。

第6章 简易英语学习软件

英语是一门重要的语言工具,学好了用处较多,因而设计此简易英语学习软件,既可以打发无聊时间,又能学以致用。初步设想主要用于学习英语口语,将文本复制到手机,以回车换行符进行分隔,每次显示一句话,可以前后翻页。由于孩子也需要用平板学习英语,因而对软件进行改进,使其可以适应多种屏幕的 Android 系统,而且对文本文件可以排序处理,这样方便选取以日期为文件名的文本。

主要知识点:Spinner 组件的使用、外部文本文件读取、多线程的使用、适应多屏幕

6.1 主要功能和技术特点

简易英语学习软件(安装后的 App 名为 EnglishReader,下文简称 EnglishReader)可以自行设置学习内容,既可以用来学习英语对话,也可以用来背单词。EnglishReader 的主要特点为:

- 适应多种屏幕的 Android 系统;
- 可以设置多个文本文件;
- 可以对文本文件进行排序;
- 可以显示当前文件的总条数和当前进度;
- 实现了前后翻页和开始结尾定位;
- 每个文件均可独立设置书签。

6.2 软件操作

EnglishReader 调用 getExternalStorageDirectory 方法获取外部存储目录,即读取 SD 卡,但是,不同的系统会有不同的结果,有的手机仍然读取的是手机存储。因而,需要根据手机类型将文本文件,如"English Idioms.txt",复制到手机或 SD 目录"/data/EnglishReader/"。

运行 EnglishReader,主界面如图 6-1 所示。当前文本文件名为"English Idioms",可以通过下拉列表项选取新的文件,该文件共有 145 句,当前显示第 11 句,如果现在关闭程序,重新打开,现场数据不变。屏幕底部的【Start】返回第一句,【End】返回最后一句,【Previous】返回前一句,【Next】返回下一句(为了节省图片空间,

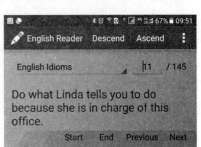

图 6-1 EnglishReader 软件界面

将这四个 Button 截取移动到靠近句子下面，实际居于屏幕底部）。为了方便选取文件，标题栏上的 Descend 菜单项可对文件名降序排序，Ascend 菜单项可对文件升序排序，另外，溢出菜单中的 About 菜单项显示版权信息。

6.3 界面布局与资源说明

第 5 章是第一个完整的应用程序作品，对界面布局与资源进行了详细的说明。本章开始只对有特色的重点部分加以介绍。EnglishReader 的文件名通过 Spinner 进行显示，相关技术已经在 2.7 节中做了介绍，这里不再赘述。另外，字符串定义文件 strings.xml 结构简单，也不再列入。

6.3.1 适应多屏幕的 dimens 文件

屏幕小，分辨率低，对应的字体就小，组件之间的距离也较近；反之，屏幕大，分辨率高，对应的字体相对较大，组件之间的距离也需要相应调大。这可以通过资源目录 res 中 values 目录的 dimens.xml 文件来控制，将距离大小和字体大小用变量来表示，变量值则在 dimens.xml 文件中定义。从图 6-1 中可以看出，句子所使用的字号最大，布局文件中就用 big_textsize 来表示，该字号到底多大，系统根据屏幕情况自行选择。例如，对于小屏幕，系统自动选择 values-small 目录下 dimens.xml 文件中定义的数据。

```xml
<resources>
    <!-- values-small -->
    <dimen name="activity_horizontal_margin">12dp</dimen>
    <dimen name="activity_vertical_margin">12dp</dimen>
    <dimen name="interval">6dp</dimen>
    <dimen name="normal_textsize">12sp</dimen>
    <dimen name="big_textsize">18sp</dimen>
</resources>
```

对于正常屏幕，系统自动选择 values-normal 目录下 dimens.xml 文件中定义的数据。

```xml
<resources>
    <!-- values-normal -->
    <dimen name="activity_horizontal_margin">16dp</dimen>
    <dimen name="activity_vertical_margin">16dp</dimen>
    <dimen name="interval">6dp</dimen>
    <dimen name="normal_textsize">18sp</dimen>
    <dimen name="big_textsize">24sp</dimen>
</resources>
```

对于大屏幕，系统自动选择 values-large 目录下 dimens.xml 文件中定义的数据。

```xml
<resources>
    <!-- values-large -->
    <dimen name="activity_horizontal_margin">16dp</dimen>
    <dimen name="activity_vertical_margin">16dp</dimen>
    <dimen name="interval">6dp</dimen>
    <dimen name="normal_textsize">24sp</dimen>
    <dimen name="big_textsize">36sp</dimen>
</resources>
```

对于特大屏幕，系统自动选择 values-xlarge 目录下 dimens.xml 文件中定义的数据。

```xml
<resources>
    <!-- values-xlarge -->
    <dimen name="activity_horizontal_margin">32dp</dimen>
    <dimen name="activity_vertical_margin">32dp</dimen>
    <dimen name="interval">12dp</dimen>
    <dimen name="normal_textsize">36sp</dimen>
    <dimen name="big_textsize">48sp</dimen>
</resources>
```

6.3.2 菜单项定义文件

菜单文件 main.xml 定义了三个菜单项，item_Descend 用于对 Spinner 中的文件名降序排序，item_Ascend 为升序排序，由于为常用选项，showAsAction 属性都设置为"ifRoom"；item_About 菜单项用于显示版权信息。菜单项对应的文本在 strings.xml 文件中定义，这里不再列出。

```xml
<menu xmlns:android="http://schemas.android.com/apk/res/android"
    xmlns:tools="http://schemas.android.com/tools"
    tools:context="com.walkerma.englishreader.MainActivity" >
    <item
        android:id="@+id/item_Descend"
        android:orderInCategory="101"
        android:showAsAction="ifRoom"
        android:title="@string/strDescend"/>
    <item
        android:id="@+id/item_Ascend"
        android:orderInCategory="102"
        android:showAsAction="ifRoom"
        android:title="@string/strAscend"/>
    <item
        android:id="@+id/item_About"
        android:orderInCategory="103"
        android:showAsAction="never"
        android:title="@string/strAbout"/>
</menu>
```

6.3.3 主窗体布局文件

主窗体布局文件 activity_main.xml 中的主体大框架采用相对布局，结合图 6-1 进行说明。Spinner 所在行采用线性布局，居中；句子所在位置由 TextView 组件承担，layout_width 属性为"match_parent"，即宽度填满父空间；layout_height 属性设置为"wrap_content"，即高度与句子的内容相匹配；singleLine 属性设置为"false"，即允许多行显示。由于 TextView 中的句子可长可短，翻页按钮如果直接放在下面，将会忽上忽下。因而，将四个翻页按钮放在子相对布局中，居于大框架的底部。

```xml
<RelativeLayout xmlns:android="http://schemas.android.com/apk/res/android"
    xmlns:tools="http://schemas.android.com/tools"
    android:layout_width="match_parent"
    android:layout_height="match_parent"
    android:orientation="vertical"
    android:paddingBottom="@dimen/activity_vertical_margin"
```

```xml
        android:paddingLeft="@dimen/activity_horizontal_margin"
        android:paddingRight="@dimen/activity_horizontal_margin"
        android:paddingTop="@dimen/activity_vertical_margin"
        tools:context="com.walkerma.englishreader.MainActivity" >
        <LinearLayout
            android:id="@+id/layoutTitle"
            android:layout_width="wrap_content"
            android:layout_height="wrap_content"
            android:layout_alignParentTop="true"
            android:layout_centerInParent="true"
            android:orientation="horizontal" >
            <Spinner
                android:id="@+id/spFiles"
                android:layout_width="wrap_content"
                android:layout_height="wrap_content"
                android:layout_marginRight="@dimen/interval"
                android:textSize="@dimen/normal_textsize" />
            <EditText
                android:id="@+id/edIndex"
                android:layout_width="wrap_content"
                android:layout_height="wrap_content"
                android:layout_marginRight="@dimen/interval"
                android:gravity="center"
                android:hint="@string/strIndex"
                android:inputType="number"
                android:textSize="@dimen/normal_textsize" />
            <TextView
                android:id="@+id/txtTotal"
                android:layout_width="wrap_content"
                android:layout_height="wrap_content"
                android:text="@null"
                android:textSize="@dimen/normal_textsize" />
        </LinearLayout>
        <TextView
            android:id="@+id/txtContent"
            android:layout_width="match_parent"
            android:layout_height="wrap_content"
            android:layout_below="@id/layoutTitle"
            android:paddingBottom="20dp"
            android:paddingTop="20dp"
            android:singleLine="false"
            android:text="@null"
            android:textSize="@dimen/big_textsize" />
        <RelativeLayout
            android:layout_width="match_parent"
            android:layout_height="wrap_content"
            android:layout_alignParentBottom="true"
            android:orientation="horizontal" >
            <Button
                android:id="@+id/btStart"
                style="?android:attr/buttonBarButtonStyle"
                android:layout_width="wrap_content"
                android:layout_height="wrap_content"
                android:layout_toLeftOf="@+id/btEnd"
                android:text="@string/strStart"
                android:textSize="@dimen/normal_textsize" />
```

```xml
<Button
    android:id="@+id/btEnd"
    style="?android:attr/buttonBarButtonStyle"
    android:layout_width="wrap_content"
    android:layout_height="wrap_content"
    android:layout_marginLeft="@dimen/interval"
    android:layout_toLeftOf="@+id/btPrevious"
    android:text="@string/strEnd"
    android:textSize="@dimen/normal_textsize" />
<Button
    android:id="@+id/btPrevious"
    style="?android:attr/buttonBarButtonStyle"
    android:layout_width="wrap_content"
    android:layout_height="wrap_content"
    android:layout_marginLeft="@dimen/interval"
    android:layout_toLeftOf="@+id/btNext"
    android:text="@string/strPrevious"
    android:textSize="@dimen/normal_textsize" />
<Button
    android:id="@+id/btNext"
    style="?android:attr/buttonBarButtonStyle"
    android:layout_width="wrap_content"
    android:layout_height="wrap_content"
    android:layout_alignBaseline="@+id/btNext"
    android:layout_alignParentRight="true"
    android:layout_marginLeft="@dimen/interval"
    android:text="@string/strNext"
    android:textSize="@dimen/normal_textsize" />
    </RelativeLayout>
</RelativeLayout>
```

6.3.4 主题设置文件

为了个性化显示应用程序的背景和菜单项，对主题设置文件 styles.xml 中的选项进行设置。windowBackground 属性可以设置窗体背景，对于多窗体的应用程序，只要设置该属性，所有窗体将采用同一个图片作为背景，这里的 background 图片放在 Library 类库的 drawable 文件夹中。

显示在标题栏中的菜单项文本自动大写，这里可以设置 textAllCaps 属性为 false，即可使得菜单项以原来的大小写方式进行显示。

```xml
<resources xmlns:android="http://schemas.android.com/apk/res/android">
    <style name="AppBaseTheme" parent="android:Theme.Light">
    </style>
    <!-- Application theme. -->
    <style name="AppTheme" parent="AppBaseTheme">
        <item name="android:windowBackground">@drawable/background</item>
        < item name="android:actionMenuTextAppearance">
            @style/menutextappearance</item>
        <item name="android:itemTextAppearance">
            @style/itemtextappearance</item >
    </style>
    <style name="menutextappearance">
        <item name="android:textAllCaps">false</item>
        <item name="android:textColor">#ffffff</item>
```

```xml
        <item name="android:textSize">18sp</item>
    </style>
    <style name="itemtextappearance">
        <item name="android:textAllCaps">false</item>
        <item name="android:textColor">#000000</item>
        <item name="android:textSize">18sp</item>
    </style>
</resources>
```

6.4 配置文件

如果编译器的版本高于 targetSdkVersion，则会显示警告信息，设置 tools:ignore 属性为 OldTargetApi，就可以忽略旧版本相关的警告信息。另外，读写文件和创建目录需要添加相关的权限，否则程序执行相关操作时要么没有效果，要么直接崩溃。

```xml
<?xml version="1.0" encoding="utf-8"?>
<!DOCTYPE xml>
<manifest xmlns:android="http://schemas.android.com/apk/res/android"
    package="com.walkerma.englishreader"
    android:versionCode="1"
    android:versionName="1.0" xmlns:tools="http://schemas.android.com/tools">
    <uses-sdk
        android:minSdkVersion="14"
        android:targetSdkVersion="22"
        tools:ignore="OldTargetApi"/>
    <uses-permission android:name=
        "android.permission.MOUNT_UNMOUNT_FILESYSTEMS" />
    <uses-permission android:name=
        "android.permission.WRITE_EXTERNAL_STORAGE" />
    <uses-permission android:name=
        "android.permission.READ_EXTERNAL_STORAGE"/>
    <application
        android:allowBackup="false"
        android:icon="@drawable/ic_launcher"
        android:label="@string/app_name"
        android:theme="@style/AppTheme" >
        <activity
            android:name=".MainActivity"
            android:configChanges="orientation|screenSize"
            android:label="@string/title_activity_main"
            android:windowSoftInputMode="adjustUnspecified|stateHidden">
            <intent-filter>
                <action android:name="android.intent.action.MAIN" />
                <category android:name="android.intent.category.LAUNCHER" />
            </intent-filter>
        </activity>
    </application>
</manifest>
```

6.5 目录与文件处理类源代码

Library 类库中创建了内部文件读写静态函数，并在 3.11 节做了介绍，关于外部文件写入在下一节介绍，这里仅介绍路径管理静态函数。isExternalStorageValid 函数用来判断 SD 卡是否有效，isFolderExists 函数判断给定的目录是否存在。createDir 函数调用 isFolderExists 函数，检查给定目录 dirSrc 之下是否存在 dirDest 子目录，如果不存在，就调用 File 对象的 mkdirs 方法创建目录。

```java
package com.walkerma.library;
import java.io.File;
import java.io.FileInputStream;
import java.io.FileNotFoundException;
import java.io.FileOutputStream;
import java.io.IOException;
import android.content.Context;
import android.os.Environment;
public class FileProcess {
    public static boolean isFolderExists(String strPath){
        File dirDest = new File(strPath);
        return dirDest.exists();
    }
    public static String createDir(String dirSrc, String dirDest){
        String strPath = dirSrc + File.separator + dirDest;
        if(isFolderExists(strPath)) return strPath;
        try{
            File file = new File(strPath);
            file.mkdirs();
        }catch(Exception e){
            e.printStackTrace();
            return null;
        }
        return strPath;
    }
    public static boolean isExternalStorageValid(){
        String strStatus = Environment.getExternalStorageState();
        return strStatus.equals(Environment.MEDIA_MOUNTED);
    }
}
```

6.6　文本读取类源代码

文本读取是一种经常性的工作，而且也比较耗时，因而，在 Library 类库中创建多线程类 ThreadReadText，以用于读取文本文件。该类的构造函数有两个变量：一个是调用函数传入的 Handler 对象 mHandler，字符串数据通过该对象以消息的形式返回调用函数的线程空间；另一个参数为完整的文件名 fileName，该类通过此文件名读取数据。

ThreadReadText 类的核心工作是对 run 方法的重载，在其中以 fileName 为参数初始化 File 对象 file，然后经过多次转换得到 BufferedReader 对象 br，接着通过 while 循环每次读取一字节以存入 StringBuffer 对象 sb 中，读取完毕则通过 mHandler 对象以消息的形式将字符串数据返回主线程空间。

```java
package com.walkerma.library;
import java.io.BufferedInputStream;
import java.io.BufferedReader;
```

```java
import java.io.File;
import java.io.FileInputStream;
import java.io.FileNotFoundException;
import java.io.IOException;
import java.io.InputStreamReader;
import android.os.Handler;
public class ThreadReadText extends Thread {
    public final static int MSG_SD_NotReady = 30000;
    public final static int MSG_FileNotFound = 30001;
    public final static int MSG_ReadError = 30002;
    public final static int MSG_ReadOK = 30003;
    private Handler mHandler;
    private String fileName;
    public ThreadReadText(Handler mHandler, String fileName) {
        this.mHandler = mHandler;
        this.fileName = fileName;
    }
    public void run(){
        if(FileProcess.isExternalStorageValid() == false) {
            mHandler.obtainMessage(MSG_SD_NotReady,
                    0, 0, null).sendToTarget();
            return;
        }
        File file = new File(fileName);
        FileInputStream fin;
        try {
            fin = new FileInputStream(file);
        } catch (FileNotFoundException e) {
            e.printStackTrace();
            mHandler.obtainMessage(MSG_FileNotFound,
                    0, 0, null).sendToTarget();
            return;
        }
        BufferedReader br = new BufferedReader(
                new InputStreamReader(new BufferedInputStream(fin)));
        StringBuffer sb = new StringBuffer();
        try{
            while(br.ready()) sb.append((char)(br.read()));
        }catch (IOException e) {
            e.printStackTrace();
            mHandler.obtainMessage(MSG_ReadError,
                    0, 0, null).sendToTarget();
            return;
        }
        try {
            br.close();
        } catch (IOException e) {
            e.printStackTrace();
            return;
        }
        mHandler.obtainMessage(MSG_ReadOK,
                0, 0, sb.toString()).sendToTarget();
    }
}
```

6.7 主窗体源代码

主窗体源代码需要保存用户学习的历史记录，即当前学习的文本文件及其书签、文本文件的排序方法。String 变量 currentName 中保存当前学习的文件名，当程序退出时，在 onDestroy 方法中调用 saveCurrentName 函数以保存文本文件名。int 变量 nCurrent 中存放当前文件的页码，当程序退出时或者用户选择另一个文件时（Spinner 对象的 onItemSelected 方法），都要调用 savePageInfo 函数，保存 currentName 和 nCurrent 键值对。boolean 变量 bReverse 为 true 表示文件名逆序排序，为 false 表示升序排序，程序退出时调用 saveReverse 函数保存 bReverse 的值，在 onCreate 方法中调用 getReverse 函数以取得保存的布尔值并存入 bReverse 变量。

程序运行时，执行 onCreate 方法进行组件和参数的初始化，然后启动多线程对象 thGetFiles，将"/data/EnglishReader/"目录下的文本文件名加入 ArrayList<String> 对象 listFiles 中，为 Spinner 对象 spFiles 提供数据。thGetFiles 添加文件名后发送消息 MSG_Files，转入消息处理函数 processMessage，在其中调用 setSpinnerOrder 函数，对文件名进行排序，调用 spFiles 对象的 setSelection 方法设置当前文件名，从而触发 spFiles 对象的事件处理方法 onItemSelected，在其中调用 prepareData 函数，启动读取文本文件内容的 ThreadReadText 对象 thRead。

文本文件读取完毕，执行 processMessage 函数中的 ThreadReadText.MSG_ReadOK 消息部分，通过 updateContent 函数将得到的字符串以回车换行符为分隔符转换成字符串数组，存入全局 String 数组 arrayContent 中，然后调用 getPageInfo 函数读取该文件的书签，最后调用 displayData 函数显示数据。displayData 函数主要完成图 6-1 中当前项和总条数的显示，以及对应文本的显示。

点击屏幕底部的四个按钮，将修改 nCurrent 的值，然后调用 displayData 函数显示新的文本内容。菜单项【Descend】和【Ascend】将改变 bReverse 的值，然后调用 setSpinnerOrder 对 spFiles 中的字符串排序。

```
package com.walkerma.englishreader;
import com.walkerma.library.FileProcess;
import com.walkerma.library.GeneralProcess;
import com.walkerma.library.ThreadReadText;
import java.io.File;
import java.util.ArrayList;
import java.util.Collections;
import java.util.Locale;
import android.app.Activity;
import android.content.SharedPreferences;
import android.content.SharedPreferences.Editor;
import android.content.res.Configuration;
import android.os.Bundle;
import android.os.Environment;
import android.os.Handler;
import android.os.Message;
import android.text.TextUtils;
import android.view.KeyEvent;
import android.view.Menu;
```

```java
import android.view.MenuItem;
import android.view.View;
import android.view.View.OnClickListener;
import android.widget.AdapterView;
import android.widget.ArrayAdapter;
import android.widget.Button;
import android.widget.EditText;
import android.widget.Spinner;
import android.widget.TextView;
import android.widget.Toast;
import android.widget.AdapterView.OnItemSelectedListener;
public class MainActivity extends Activity implements OnClickListener{
    private Spinner spFiles;
    private ArrayList<String> listFiles = new ArrayList<String>();
    private ArrayAdapter<String> adapter;
    private boolean bReverse = false;
    private EditText edIndex;
    private TextView txtTotal, txtContent;
    private Button btStart, btEnd, btPrevious, btNext;
    private String currentName = "";      // current file name
    private String currentFullName = ""; // with ".txt"
    private int nCurrent = 0;
    private int nTotal = 0;
    private String[] arrayContent;
    private boolean bDataReady = false;
    private boolean bCanExit = false;
    private Handler mHandler;
    private final int MSG_Delay = 10000;
    private final int MSG_Files = 10001;
    class IncomingHandlerCallback implements Handler.Callback{
        @Override
        public boolean handleMessage(Message msg) {
            processMessage(msg);
            return true;
        }
    }
    private void processMessage(Message msg){
        switch(msg.what){
        case MSG_Delay:
            bCanExit = false;
            break;
        case MSG_Files:
            if(msg.arg1 == 0) {
                Toast.makeText(this,  "File is empty.",
                        Toast.LENGTH_SHORT ).show();
                break;
            }
            setSpinnerOrder();
            break;
        case ThreadReadText.MSG_SD_NotReady:
            Toast.makeText(this,  "SD card is not ready.",
                    Toast.LENGTH_SHORT ).show();
            break;
        case ThreadReadText.MSG_ReadOK:
            updateContent((String)msg.obj);
            nCurrent = getPageInfo(currentName);
```

```java
            displayData();
            break;
        case ThreadReadText.MSG_ReadError:
            Toast.makeText(this,   "Read Error.",
                    Toast.LENGTH_SHORT ).show();
            break;
        }
    }
    private void updateContent(String content) {
        if (TextUtils.isEmpty(content)) {
            Toast.makeText(this, "Content empty.",
                    Toast.LENGTH_SHORT).show();
            arrayContent = null;
            bDataReady = false;
        } else {
            arrayContent = content.split("\r\n");
            bDataReady = true;
        }
    }
    private void displayData(){
        nTotal = arrayContent.length;
        txtTotal.setText("/ " + Integer.toString(nTotal));
        if(nCurrent>=0 && nCurrent<nTotal){
            txtContent.setText(arrayContent[nCurrent]);
            edIndex.setText(Integer.toString(nCurrent + 1));
        }
    }
    private int getPageInfo(String strItem){
        String strAppName = getString(R.string.app_name);
        SharedPreferences sp = getSharedPreferences(strAppName, 0);
        return sp.getInt(strItem, 0);
    }
    private void savePageInfo(String strItem){
        String strAppName = getString(R.string.app_name);
        SharedPreferences sp = getSharedPreferences(strAppName, 0);
        Editor editor = sp.edit();
        editor.putInt(strItem, nCurrent);
        editor.commit();
    }
    @Override
    public boolean onKeyDown(int keyCode, KeyEvent event) {
        // TODO Auto-generated method stub
        switch(keyCode){
        case KeyEvent.KEYCODE_BACK:
            if(bCanExit){
                break;
            }
            else{
                bCanExit = true;
                Toast.makeText(this, "Press again to quit.",
                        Toast.LENGTH_SHORT).show();
            }
            mHandler.sendEmptyMessageDelayed(MSG_Delay, 1000);
            return true;
        }
        return super.onKeyDown(keyCode, event);
```

```java
    }
    private void saveCurrentName(){
        String strAppName = getString(R.string.app_name);
        SharedPreferences sp = getSharedPreferences(strAppName, 0);
        Editor editor = sp.edit();
        editor.putString("CurrentName", currentName);
        editor.commit();
    }
    private String getCurrentName(){
        String strAppName = getString(R.string.app_name);
        SharedPreferences sp = getSharedPreferences(strAppName, 0);
        return sp.getString("CurrentName", "");
    }
    private void saveReverse(){
        String strAppName = getString(R.string.app_name);
        SharedPreferences sp = getSharedPreferences(strAppName, 0);
        Editor editor = sp.edit();
        editor.putBoolean("Reverse", bReverse);
        editor.commit();
    }
    private boolean getReverse(){
        String strAppName = getString(R.string.app_name);
        SharedPreferences sp = getSharedPreferences(strAppName, 0);
        return sp.getBoolean("Reverse", false);
    }
    @Override
    protected void onCreate(Bundle savedInstanceState) {
        super.onCreate(savedInstanceState);
        setContentView(R.layout.activity_main);
        bReverse = getReverse();
        initWidgets();
        mHandler = new Handler(new IncomingHandlerCallback());
        ThreadGetFiles thGetFiles = new ThreadGetFiles();
        thGetFiles.start();
    }
    private void initSpArticle(){
        spFiles = (Spinner)findViewById(R.id.spFiles);
        adapter= new ArrayAdapter<String>(this, R.layout.spinner_item, listFiles);
        adapter.setDropDownViewResource(R.layout.spinner_dropdown_item);
        spFiles.setAdapter(adapter);
        spFiles.setOnItemSelectedListener(new OnItemSelectedListener(){
            @Override
            public void onItemSelected(AdapterView<?> parent, View view,
                    int position, long id) {
                // TODO Auto-generated method stub
                if(!TextUtils.isEmpty(currentName))
                    savePageInfo(currentName);
                bDataReady = false;
                prepareData();
            }
            @Override
            public void onNothingSelected(AdapterView<?> parent) {
                // TODO Auto-generated method stub
            }});
    }
    private void setSpinnerOrder(){
```

```java
        if(bReverse){
            Collections.reverse(listFiles);
            Toast.makeText(this,  "Descending order.",
                    Toast.LENGTH_SHORT ).show();
        }
        else{
            Collections.sort(listFiles);
            Toast.makeText(this,  "Ascending order.",
                    Toast.LENGTH_SHORT ).show();
        }
        adapter.notifyDataSetChanged();
        spFiles.setSelection(getSpinnerItemIndex(spFiles, getCurrentName()));
    }
    private int getSpinnerItemIndex(Spinner spinner, String strItem){
        int index = 0;
        for (int i=0;i<spinner.getCount();i++){
            if (spinner.getItemAtPosition(i).equals(strItem)){
                index = i;
            }
        }
        return index;
    }
    class ThreadGetFiles extends Thread{
        public void run(){
            String strPath = getAppPath();
            File dir = new File(strPath);
            String[] strArray = dir.list();
            if(strArray == null) return;
            listFiles.clear();
            int nLen = strArray.length;
            for(int i=0; i<nLen; i++){
                if(strArray[i].endsWith(".txt")){
                    int nLoc = strArray[i].
                            toLowerCase(Locale.ENGLISH).indexOf(".txt");
                    listFiles.add(strArray[i].substring(0, nLoc));
                }
            }
            mHandler.obtainMessage(MSG_Files,
                    nLen, 0, null).sendToTarget();
        }
    }
    private String getAppPath(){
        String strPath = Environment.
                getExternalStorageDirectory().getPath();
        String strAppName = getString(R.string.app_name);
        return FileProcess.createDir(strPath, "data" +
                File.separator + strAppName) + File.separator;
    }
    private String getFullFileName(){
        currentName = spFiles.getSelectedItem().toString();
        if(TextUtils.isEmpty(currentName)) return null;
        return getAppPath() + currentName + ".txt";
    }
    private void initWidgets(){
        edIndex = (EditText)findViewById(R.id.edIndex);
```

```java
        txtTotal = (TextView)findViewById(R.id.txtTotal);
        txtContent = (TextView)findViewById(R.id.txtContent);
        btStart = (Button)findViewById(R.id.btStart);
        btStart.setOnClickListener(this);
        btEnd = (Button)findViewById(R.id.btEnd);
        btEnd.setOnClickListener(this);
        btPrevious = (Button)findViewById(R.id.btPrevious);
        btPrevious.setOnClickListener(this);
        btNext = (Button)findViewById(R.id.btNext);
        btNext.setOnClickListener(this);
        initSpArticle();
    }
    @Override
    public boolean onCreateOptionsMenu(Menu menu) {
        getMenuInflater().inflate(R.menu.main, menu);
        return true;
    }
    @Override
    public boolean onOptionsItemSelected(MenuItem item) {
        int id = item.getItemId();
        switch(id){
        case R.id.item_Descend:
            bReverse = true;
            setSpinnerOrder();
            return true;
        case R.id.item_Ascend:
            bReverse = false;
            setSpinnerOrder();
            return true;
        case R.id.item_About:
            GeneralProcess.fireAboutDialog(this,
                    getString(R.string.app_name),
                    getString(R.string.strCopyrights));
            return true;
        }
        return super.onOptionsItemSelected(item);
    }
    @Override
    public void onConfigurationChanged(Configuration newConfig) {
        // TODO Auto-generated method stub
        displayData();
        super.onConfigurationChanged(newConfig);
    }
    @Override
    protected void onDestroy() {
        // TODO Auto-generated method stub
        mHandler.removeCallbacksAndMessages(null);
        if(!TextUtils.isEmpty(currentName)) {
            saveCurrentName();
            savePageInfo(currentName);
            saveReverse();
        }
        super.onDestroy();
    }
    @Override
```

```java
public void onClick(View v) {
    // TODO Auto-generated method stub
    if(TextUtils.isEmpty(currentName) || (!bDataReady)){
        Toast.makeText(this, "Data empty...",
                Toast.LENGTH_SHORT).show();
        return;
    }
    String strCurrent = edIndex.getText().toString();
    if(TextUtils.isEmpty(strCurrent)){
        Toast.makeText(this, "Index empty.",
                Toast.LENGTH_SHORT).show();
        return;
    }
    nCurrent = Integer.parseInt(strCurrent) - 1;
    switch(v.getId()){
    case R.id.btStart:
        nCurrent = 0;
        displayData();
        break;
    case R.id.btEnd:
        nCurrent = nTotal - 1;
        displayData();
        break;
    case R.id.btPrevious:
        if(nCurrent >= 1){
            nCurrent--;
            displayData();
        }
        else
            Toast.makeText(this, "Reached head.",
                    Toast.LENGTH_SHORT).show();
        break;
    case R.id.btNext:
        if(nCurrent < nTotal - 1){
            nCurrent++;
            displayData();
        }
        else
            Toast.makeText(this, "Reached end.",
                    Toast.LENGTH_SHORT).show();
        break;
    }
}
private void prepareData(){
    currentFullName = getFullFileName();
    if(TextUtils.isEmpty(currentFullName)) {
        Toast.makeText(this, "No text file.", Toast.LENGTH_SHORT).show();
        return;
    }
    ThreadReadText thRead = new ThreadReadText(mHandler, currentFullName);
    thRead.start();
}
}
```

6.8 本章小结

本章在 Library 类库中增加了 ThreadReadText 类,可用于后台读取文本文件内容,以消息形式返回主线程空间;补充了 FileProcess 类中的目录查询与创建静态函数。所完成的简易英语学习软件可以自动读取指定目录下的文本文件,实现了文件的升序和降序排序及翻页功能,并可以较好地保存用户数据。

第 7 章　通讯录备份与恢复软件

手机通讯录是最基本的个人隐私，更换手机时需要转移通讯录，但是如果手机型号不一样，往往不能通过蓝牙一键转移，需要借助 SIM 卡作为中间介质辅助转移。手机也可能被遗失，这时就无法转移了通讯录。因此，平时做好通讯录备份是一个良好的习惯。利用第三方软件进行备份，由于植入广告和文件格式问题，往往不太方便；利用大公司的软件工具将通讯录备份到云端也不太安全，而且备份通讯录时，对于相同姓名、不同电话号码的数据能否覆盖基本定制好了，用户不一定能够修改。下面我们设计一款通讯录备份与恢复软件，将通讯录备份成外部文本文件，便于修改和保存，必要时将文本文件复制到手机的指定目录，又可以一键恢复通讯录。

主要知识点：通讯录操作、外部文件读写、多线程技术、消息机制

7.1　主要功能和技术特点

通讯录备份与恢复软件（安装后的 App 名为 BookBackup，下文简称 BookBackup）可一键完成所有通讯录的备份，一键将备份全部恢复（合并）到通讯录。BookBackup 的主要特点为：

- 一键备份到指定的外部存储，方便复制保存；
- 备份的形式为文本文件，方便阅读修改；
- 备份速度较快，只显示最终的通讯录条数；
- 将备份通讯录复制到指定外部存储，一键完成恢复（合并）；
- 备份电话号码与手机电话号码不同才加入通讯录；
- 恢复速度较慢，同步显示进度。

7.2　软件操作

BookBackup 运行后的软件界面如图 7-1 所示，当前处于通讯录恢复状态，上面的文本框显示恢复进度，此时左侧的【Backup】按钮禁用变灰。软件初始运行时，可以点击【Backup】一键完成通讯录的备份，将所有通讯录备份到"/data/BookBackup"目录下，文件名为"ContactsPhone.txt"，由于该文件采用外部写入的方式生成，因而对所有用户可见，可以复制到合适的地方保存并删除手机中的备份。

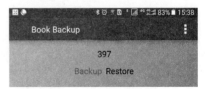

图 7-1　BookBackup 软件界面

如欲恢复通讯录，则需要将"ContactsPhone.txt"复制到"/data/BookBackup"目录下，点击【Restore】即可。由于恢复通讯录时，需要查看在手机通讯录中是否存在备份中的每个手机号码，因此比较耗时，可采用多线程技术在文本框中同步显示恢复进度。

7.3 界面布局

界面布局中，txtTotal 文本框居中显示备份或恢复的总数目，也同步显示恢复进度。两个 Button 中，btBackup 用于备份通讯录，btRestore 用于恢复通讯录。

```xml
<LinearLayout xmlns:android="http://schemas.android.com/apk/res/android"
    xmlns:tools="http://schemas.android.com/tools"
    android:id="@+id/LinearLayout1"
    android:layout_width="match_parent"
    android:layout_height="match_parent"
    android:layout_marginBottom="@dimen/activity_horizontal_margin"
    android:layout_marginLeft="@dimen/activity_horizontal_margin"
    android:layout_marginRight="@dimen/activity_horizontal_margin"
    android:layout_marginTop="@dimen/activity_horizontal_margin"
    android:orientation="vertical" >
    <TextView
        android:id="@+id/txtTotal"
        android:layout_width="wrap_content"
        android:layout_height="wrap_content"
        android:layout_gravity="center"
        android:text="@string/strTotal"
        android:textAppearance="?android:attr/textAppearanceMedium" />
    <LinearLayout
        android:layout_width="match_parent"
        android:layout_height="wrap_content"
        android:gravity="center"
        android:orientation="horizontal" >
        <Button
            android:id="@+id/btBackup"
            style="?android:attr/buttonBarButtonStyle"
            android:layout_width="wrap_content"
            android:layout_height="wrap_content"
            android:text="@string/strBackup" />
        <Button
            android:id="@+id/btRestore"
            style="?android:attr/buttonBarButtonStyle"
            android:layout_width="wrap_content"
            android:layout_height="wrap_content"
            android:text="@string/strRestore" />
    </LinearLayout>
</LinearLayout>
```

7.4 配置文件

与第 6 章一样，BookBackup 软件也需要添加读写文件和创建目录与文件的权限，另外，还需要添加读写通讯录的权限。为了避免软件运行时请求用户授权，将 targetSdkVersion

属性设置为"22"。Activity 标记中的 screenOrientation 属性设置为"portrait",即 Book-Backup 只工作在竖屏状况下。

```xml
<?xml version="1.0" encoding="utf-8"?>
<!DOCTYPE xml>
<manifest xmlns:android="http://schemas.android.com/apk/res/android"
    package="com.walkerma.bookbackup"
    android:versionCode="1"
    android:versionName="1.0" >
    <uses-sdk
        android:minSdkVersion="14"
        android:targetSdkVersion="22" />
    <uses-permission android:name=
        "android.permission.READ_EXTERNAL_STORAGE"/>
    <uses-permission android:name=
        "android.permission.WRITE_EXTERNAL_STORAGE"/>
    <uses-permission android:name=
        "android.permission.MOUNT_UNMOUNT_FILESYSTEMS"/>
    <uses-permission android:name=
        "android.permission.READ_CONTACTS"/>
    <uses-permission android:name=
        "android.permission.WRITE_CONTACTS"/>
    <application
        android:allowBackup="true"
        android:icon="@drawable/ic_launcher"
        android:label="@string/app_name"
        android:theme="@style/AppTheme" >
        <activity
            android:name=".MainActivity"
            android:screenOrientation="portrait"
            android:label="@string/title_activity_main" >
            <intent-filter>
                <action android:name="android.intent.action.MAIN" />
                <category android:name="android.intent.category.LAUNCHER" />
            </intent-filter>
        </activity>
    </application>
</manifest>
```

7.5 通讯录操作源代码

在 Library 类库中创建 PhoneBook 子类,专门处理通讯录相关事宜,并设置如下静态函数。与 3.9 节类似,getAllNamePhone 函数获取并返回所有联系人的信息,但是只取姓名和电话号码两个字段。

```
public static Cursor getAllNamePhone(Context context){
    String[] projection = new String[] {Phone.DISPLAY_NAME, Phone.NUMBER};
    Cursor cr = context.getContentResolver().query(Phone.CONTENT_URI,
            projection, null, null, null);
    return cr;
}
```

getNumberCursor 函数根据输入的电话号码(number),返回含有用户 ID、电话号码和姓名的记录。

```java
public static Cursor getNumberCursor(Context context, String number){
    Uri lookupUri = Uri.withAppendedPath(
            PhoneLookup.CONTENT_FILTER_URI, Uri.encode(number));
    String[] mPhoneNumberProjection = {
            PhoneLookup._ID, PhoneLookup.NUMBER,
            PhoneLookup.DISPLAY_NAME};
    return context.getContentResolver().query(
            lookupUri,mPhoneNumberProjection, null, null, null);
}
```

contactExists 函数根据输入的电话号码，调用 getNumberCursor 函数，查看返回的记录是否为空：如果为空，则返回 false，表示手机中无此联系人；如果非空，则返回 true，表示手机中有此联系人。

```java
public static boolean contactExists(Context context, String number){
    Cursor cr = getNumberCursor(context, number);
    if(cr == null) return false;
    if(cr.getCount()>0)
        return true;
    else
        return false;
}
```

addName 函数以 ArrayList<ContentProviderOperation> 对象 ops 和姓名 name 为参数，将姓名插入 ops 对象并返回该 ops 对象。

```java
public static ArrayList<ContentProviderOperation> addName(
        ArrayList<ContentProviderOperation> ops, String name){
    if(TextUtils.isEmpty(name)) return null;
    ops.add(ContentProviderOperation
            .newInsert(Data.CONTENT_URI)
            .withValueBackReference(Data.RAW_CONTACT_ID, 0)
            .withValue(Data.MIMETYPE,
                StructuredName.CONTENT_ITEM_TYPE)
                    .withValue(StructuredName.DISPLAY_NAME,
                        name).build());
    return ops;
}
```

addMobileNumber 函数以 ArrayList<ContentProviderOperation> 对象 ops 和电话号码 phone 为参数，将电话号码插入 ops 并返回该 ops 对象。

```java
public static ArrayList<ContentProviderOperation> addMobileNumber(
        ArrayList<ContentProviderOperation> ops, String phone){
    if(TextUtils.isEmpty(phone)) return null;
    ops.add(ContentProviderOperation
            .newInsert(Data.CONTENT_URI)
            .withValueBackReference(
                Data.RAW_CONTACT_ID, 0)
                    .withValue(Data.MIMETYPE,
                        Phone.CONTENT_ITEM_TYPE)
                            .withValue(Phone.NUMBER, phone)
                            .withValue(Phone.TYPE, Phone.TYPE_MOBILE)
```

```
                .build());
    return ops;
}
```

7.6 外部文本写入源代码

在 3.11 节直接调用 openFileOutput 方法得到 FileOutputStream 对象，即可用来写入内部文件。而外部文件写入需要以绝对路径为参数初始化 File 对象，然后以此为参数初始化 FileOutputStream 对象 fos。两者的写入方式一样，需要将字符串转换为字节数组，然后调用 fos 将其写入文件即可。程序卸载后，内部文件将随之消失，但是外部文件依然可以存取。

```
public static boolean writeText(String strPath, String strTxt){
    File file = new File(strPath);
    try {
        FileOutputStream fos = new FileOutputStream (file);
        fos.write(strTxt.getBytes());
        fos.close();
        return true;
    } catch (Exception e) {
        // TODO Auto-generated catch block
        e.printStackTrace();
        return false;
    }
}
```

7.7 主窗体源代码

在 onClick 方法中，btBackup 启动 ThreadWriteText 类的对象 threadRead 读取通讯录。threadRead 主要通过调用 7.5 节中的 getAllNamePhone 函数获得通讯录记录集，即 Cursor 对象 row，然后利用循环将 row 中的姓名和电话以 "," 分隔，每条记录之间用 "\r\n" 分隔，形成一个字符串，然后调用 7.6 节中的 writeText 函数，将通讯录字符串写入指定目录的 ContactsPhone.txt 文件中。

btRestore 启动 ThreadReadText 类的对象 threadRead 以读取 ContactsPhone.txt 文件内容，并以消息 MSG_ReadOK 的形式将读取的字符串返回主线程空间。然后接着启动 ThreadProcess 类的对象 thProcess，在其中将通讯录字符串分解成 String 数组 strArray，每一条包括姓名和电话，再进一步将 strArray 元素分解为 strRecord 数组，位置 0 为姓名，位置 1 即为电话号码。这样就可以调用 7.5 节的 contactExists 函数查询电话号码是否存在于手机通讯录中，如果不存在，则调用 addName 和 addMobileNumber 函数添加姓名及其对应的电话号码，然后调用 applyBatch 方法，将数据写入通讯录。

```
package com.walkerma.bookbackup;
import java.io.File;
import java.util.ArrayList;
import com.walkerma.library.FileProcess;
import com.walkerma.library.GeneralProcess;
```

```java
import com.walkerma.library.PhoneBook;
import com.walkerma.library.ThreadReadText;
import android.app.Activity;
import android.content.ContentProviderOperation;
import android.content.Context;
import android.database.Cursor;
import android.os.Bundle;
import android.os.Environment;
import android.os.Handler;
import android.os.Message;
import android.provider.ContactsContract;
import android.text.TextUtils;
import android.view.KeyEvent;
import android.view.Menu;
import android.view.MenuItem;
import android.view.View;
import android.view.View.OnClickListener;
import android.widget.Button;
import android.widget.TextView;
import android.widget.Toast;
public class MainActivity extends Activity implements OnClickListener{
    private TextView txtTotal;
    private Button btBackup, btRestore;
    private boolean bCanExit = false;
    private String strContacts;
    private Context context;
    private Handler mHandler;
    private final int MSG_Delay = 10000;
    private final int MSG_SD_NotReady = 10001;
    private final int MSG_WriteError = 10002;
    private final int MSG_WriteOK = 10003;
    private final int MSG_ContactsBlank = 10004;
    private final int MSG_Process = 10005;
    class IncomingHandlerCallback implements Handler.Callback{
        @Override
        public boolean handleMessage(Message msg) {
            processMessage(msg);
            return true;
        }
    }
    private void processMessage(Message msg){
        switch(msg.what){
        case MSG_Delay:
            bCanExit = false;
            break;
        case ThreadReadText.MSG_FileNotFound:
            Toast.makeText(this, "File not found.",
                    Toast.LENGTH_SHORT ).show();
            break;
        case MSG_SD_NotReady:
        case ThreadReadText.MSG_SD_NotReady:
            Toast.makeText(this, "SD card is not ready.",
                    Toast.LENGTH_SHORT ).show();
            break;
        case MSG_ContactsBlank:
            Toast.makeText(this, "Contacts empty.",
```

```
                    Toast.LENGTH_SHORT ).show();
            break;
        case MSG_WriteOK:
            txtTotal.setText(Integer.toString(msg.arg1));
            Toast.makeText(this,   "Write OK.",
                    Toast.LENGTH_SHORT ).show();
            break;
        case MSG_WriteError:
            Toast.makeText(this,   "Write Error.",
                    Toast.LENGTH_SHORT ).show();
            break;
        case ThreadReadText.MSG_ReadOK:
            strContacts = (String)msg.obj;
            ThreadProcess thProcess = new ThreadProcess();
            thProcess.start();
            Toast.makeText(this,   "Read OK.",
                    Toast.LENGTH_SHORT ).show();
            break;
        case ThreadReadText.MSG_ReadError:
            Toast.makeText(this,   "Read Error.",
                    Toast.LENGTH_SHORT ).show();
            break;
        case MSG_Process:
            txtTotal.setText(Integer.toString(msg.arg1));
            break;
        }
    }
    class ThreadProcess extends Thread {
        public void run() {
            if (TextUtils.isEmpty(strContacts)) return;
            String[] strArray = strContacts.split("\r\n");
            for (int i=0; i<strArray.length; i++) {
                mHandler.obtainMessage(MSG_Process,
                        i+1, 0, null).sendToTarget();
                String[] strRecord = strArray[i].split(",");
                String phone = strRecord[1];
                if ((TextUtils.isEmpty(phone))
                        || PhoneBook.contactExists(context, phone))
                    continue;
                try {
                    ArrayList<ContentProviderOperation> ops = PhoneBook
                            .buildOps();
                    ops = PhoneBook.addName(ops, strRecord[0]);
                    ops = PhoneBook.addMobileNumber(ops, phone);
                    getContentResolver().applyBatch(
                            ContactsContract.AUTHORITY, ops);
                } catch (Exception e) {
                    e.printStackTrace();
                }
            }
        }
    }
    @Override
    public boolean onKeyDown(int keyCode, KeyEvent event) {
        // TODO Auto-generated method stub
        switch(keyCode){
```

```java
        case KeyEvent.KEYCODE_BACK:
            if(bCanExit){
                mHandler.removeCallbacksAndMessages(null);
                break;
            }
            else{
                bCanExit = true;
                Toast.makeText(getApplicationContext(),
                        "Press again to quit.",
                        Toast.LENGTH_SHORT).show();
            }
            mHandler.sendEmptyMessageDelayed(MSG_Delay, 1000);
            return true;
        }
        return super.onKeyDown(keyCode, event);
    }
    @Override
    protected void onCreate(Bundle savedInstanceState) {
        super.onCreate(savedInstanceState);
        setContentView(R.layout.activity_main);
        context = this;
        initWidgets();
        mHandler = new Handler(new IncomingHandlerCallback());
    }
    private void initWidgets(){
        txtTotal = (TextView)findViewById(R.id.txtTotal);
        btBackup = (Button)findViewById(R.id.btBackup);
        btBackup.setOnClickListener(this);
        btRestore = (Button)findViewById(R.id.btRestore);
        btRestore.setOnClickListener(this);
    }
    @Override
    public boolean onCreateOptionsMenu(Menu menu) {
        getMenuInflater().inflate(R.menu.main, menu);
        return true;
    }
    @Override
    public boolean onOptionsItemSelected(MenuItem item) {
        int id = item.getItemId();
        switch(id){
        case R.id.item_About:
            GeneralProcess.fireAboutDialog(this,
                    getString(R.string.app_name),
                    getString(R.string.strCopyrights));
            return true;
        }
        return super.onOptionsItemSelected(item);
    }
    @Override
    public void onClick(View v) {
        // TODO Auto-generated method stub
        txtTotal.setText("#");
        switch(v.getId()){
        case R.id.btBackup:
            ThreadWriteText threadWrite = new ThreadWriteText();
            threadWrite.start();
```

```java
                break;
            case R.id.btRestore:
                ThreadReadText threadRead = new ThreadReadText(
                        mHandler, getFileName());
                threadRead.start();
                btBackup.setEnabled(false);
                break;
        }
    }
    private String getAppPath(){
        String strPath = Environment.
                getExternalStorageDirectory().getPath();
        String strAppName = getString(R.string.app_name);
        strPath = FileProcess.createDir(strPath, "data" +
                File.separator + strAppName) + File.separator;
        return strPath;
    }
    private String getFileName(){
        return getAppPath() + "ContactsPhone.txt";
    }
    class ThreadWriteText extends Thread{
        public void run(){
            if(FileProcess.isExternalStorageValid() == false) {
                mHandler.obtainMessage(MSG_SD_NotReady,
                        0, 0, null).sendToTarget();
                return;
            }
            Cursor row = PhoneBook.getAllNamePhone(context);
            if(row == null){
                mHandler.obtainMessage(MSG_ContactsBlank,
                        0, 0, null).sendToTarget();
                return;
            }
            int nLen = row.getCount();
            row.moveToFirst();
            StringBuffer sb = new StringBuffer();
            while(true){
                String name = row.getString(0);
                String phoneNumber = row.getString(1);
                sb.append(name + "," + phoneNumber + "\r\n");
                if (row.isLast())
                    break;
                else
                    row.moveToNext();
            }
            row.close();    //important
            if(FileProcess.writeText(getFileName(), sb.toString()))
                mHandler.obtainMessage(MSG_WriteOK,
                        nLen, 0, null).sendToTarget();
            else mHandler.obtainMessage(MSG_WriteError,
                    0, 0, null).sendToTarget();
        }
    }
}
```

7.8 本章小结

本章在 Library 类库中增加了 PhoneBook 类,可用于通讯录的查询和写入,另外,还在 FileProcess 类中补充了外部写入文件的静态函数 writeText。所完成的通讯录备份与恢复软件主要通过读取通讯录并写入外部文本多线程、读取外部文本多线程和处理恢复通讯录多线程为主线展开,后台多线程在完成任务的过程中,通过消息将数据传递到主线程空间。

第 8 章　服务账号登记软件

现在很多人都有各类金融账号、通信账号和生活账号等，由于要求和重要性有差异，其密码也不尽相同，而且，一个网银账号常有登录密码、查询密码和支付密码等。如何随时准确地输入账号和密码，简直是一个挑战。本章设计一个服务账号登记软件，将这些账号全部存于手机，并做好加密备份。即使手机遗失，由于手机和软件都有一定的保护功能，所以不怕信息泄露，而且可以通过加密备份随时恢复。

主要知识点：SQLite 数据库、外部文件读取、个性化 ListView、拖曳技术的实现

8.1　主要功能和技术特点

服务账号登记软件（安装后的 App 名为 ServiceRegister，下文简称 ServiceRegister）采用一个静态密码登录或采用动态密码登录，完成所有账号的录入与查询工作。采用 SQLite 数据库实现，主要字段及作用如表 8-1 所示，以 _unit 作为主键，除了 _unit 和 _service 以外，其他字段都可以为空值。

表 8-1　ServiceRegister 采用的主要字段

字段名	中文含义	用途
_unit	单位	单位名称，可用于查询
_service	服务	服务分类，相同服务的单位放在一起
_no	账号	可存放各种账号，也可以为空值
_date	日期	存放银行开户日期，或服务的最后一天
_address	地址	存放地址和电话相关信息
_remark	备注	补充信息，可以存放密码

ServiceRegister 的主要特点为：
- 可以服务门类（以银行为例）为浏览入口；
- 打开银行服务，可浏览各家银行（单位）；
- 点击某银行，即可查询和修改银行信息；
- 可以查询为入口，输入单位关键字，即可显示相关单位；
- 服务和单位可以上下移动并保存；
- 服务门类可以备份与恢复；
- 单位列表中的详细信息可以备份与恢复。

8.2 软件操作

ServiceRegister 运行后的登录界面如图 8-1 所示，Static 为固定不变的静态密码，Dynamic 是与日期相关的动态密码。输入密码后，点击【Login】，如果密码正确，则【Go】按钮有效。选择 Category 进入服务浏览界面，选择 Search 进入单位查找界面。

在图 8-1 的状态下点击【Go】，进入图 8-2 所示界面。在编辑框中输入服务种类，点击【Add】即可完成添加，选中某列表项，点击【Delete】即可删除某服务。点击【Backup】即可将所有服务种类写入外部目录，点击【Restore】即可从外部目录将备份的服务种类恢复到列表项中。

在图 8-2 所示的状态下点击【Enter】，进入如图 8-3 所示的界面，所有银行相关的单位品种全部显示出来。

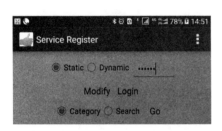

图 8-1　ServiceRegister 登录界面

图 8-2　服务浏览界面

在图 8-3 中选择"工行（北京）"，点击【Inquire】，进入图 8-4 所示界面，显示其详细信息，包括账号、开户日期和有效日期、开户地址、账号密码和银行网点号等。点击【Clear】将清除信息，仅保留 Service；点击【Save】将保存修改的数据。

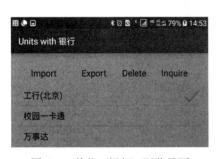

图 8-3　单位（银行）浏览界面

图 8-4　服务（银行）的详细信息

如果在图 8-3 中不选择任何选项，则【Inquire】按钮显示为"Add"，点击【Add】，将增加新的 Unit 及其相关信息。可以在 Service 中添加其他服务选项，即在多个服务种类中，可以查到同一个单位。例如，"图书馆账号"可以放在"工作"和"科研"两个服务门类下。

在图 8-1 的状态下点击【Modify】将可以修改静态密码，如图 8-5 所示，正确输入旧密码，两次输入新密码即可。密码修改完毕，点击【Enter】即返回图 8-1。静态密码也是加密密码，动态密码仅用于浏览信息，不参与加密。

在图 8-1 中选择 Search 选项，点击【Go】进入图 8-6 所示界面，如果在编辑框中输入"行"，点击【Search】则单位文本中含有"行"的选项全部列出，这样可以避免记不清服务分类无法查找具体单位的情况。在本图点击【Inquire】即可进入图 8-4 所示界面，但是，从图 8-4 将直接返回图 8-6。

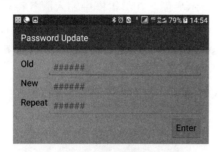

图 8-5　静态密码修改界面

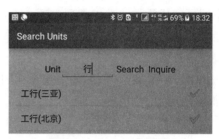

图 8-6　单位搜索界面

8.3　配置文件

配置文件中需要申请读写外部文件的权限，因为数据库中的数据需要备份到外部文件，另外还需要振动权限。application 标记中的 allowBackup 属性设置为 false，即不允许系统备份用户数据，这样即使手机丢失也能保证数据的安全；supportsRtl 属性设置为 true，表示支持从右到左的布局。同时，在 activity 布局文件中，layout_marginLeft 将用 layout_marginStart 来表示，layout_marginRight 将用 layout_marginEnd 来表示，其他情况类似。supportsRtl 属性要求 minSdkVersion 为 17 或更高。各个窗体的背景在 styles.xml 文件中统一设置，可参考 6.3.4 节。

```xml
<?xml version="1.0" encoding="utf-8"?>
<manifest xmlns:android="http://schemas.android.com/apk/res/android"
    package="com.walkerma.serviceregister"
    android:versionCode="1"
    android:versionName="1.0" >
    <uses-sdk
        android:minSdkVersion="17"
        android:targetSdkVersion="21" />
    <uses-permission android:name=
        "android.permission.MOUNT_UNMOUNT_FILESYSTEMS" />
    <uses-permission android:name=
        "android.permission.WRITE_EXTERNAL_STORAGE" />
    <uses-permission android:name=
        "android.permission.READ_EXTERNAL_STORAGE" />
    <uses-permission android:name=
        "android.permission.VIBRATE" />
    <application
        android:allowBackup="false"
        android:icon="@drawable/ic_launcher"
        android:label="@string/app_name"
        android:supportsRtl="true"
        android:theme="@style/AppTheme" >
        <activity
```

```xml
            android:name=".LoginActivity"
            android:label="@string/title_activity_login"
            android:windowSoftInputMode="adjustUnspecified|stateHidden"
            android:launchMode="standard"
            android:screenOrientation="portrait" >
            <intent-filter>
                <action android:name="android.intent.action.MAIN" />
                <category android:name="android.intent.category.LAUNCHER" />
            </intent-filter>
        </activity>
        <activity
            android:name=".MainActivity"
            android:label="@string/title_activity_main"
            android:windowSoftInputMode="adjustUnspecified|stateHidden"
            android:screenOrientation="portrait" >
        </activity>
        <activity
            android:name=".UnitsActivity"
            android:configChanges="orientation|screenSize"
            android:label="@string/title_activity_units"
            android:launchMode="singleTask" >
        </activity>
        <activity
            android:name=".InfoActivity"
            android:configChanges="orientation|screenSize"
            android:label="@string/title_activity_info" >
        </activity>
        <activity
            android:name=".PwdActivity"
            android:label="@string/title_activity_pwd" >
        </activity>
        <activity
            android:name=".SearchActivity"
            android:configChanges="orientation|screenSize"
            android:windowSoftInputMode="adjustUnspecified|stateHidden"
            android:label="@string/title_activity_search" >
        </activity>
    </application>
</manifest>
```

8.4 登录窗体

登录窗体即主窗体、软件的入口，在这里验证身份并选择服务浏览或单位搜索。其布局文件名为 activity_login.xml，对应的源代码为 LoginActivity.java。

8.4.1 布局文件

布局文件外层采用线性布局：第一行采用相对布局，用来处理密码；第二行采用线性布局，两个按钮中一个用来修改静态密码，一个用来登录；第三行也采用线性布局，用来浏览或搜索信息。初始状况下命令按钮无法与两个单选按钮对齐，可设置 Button 标记的 paddingBottom 属性调整其高度，达到命令按钮与单选按钮对齐的目的。

```xml
<LinearLayout xmlns:android="http://schemas.android.com/apk/res/android"
```

```xml
    xmlns:tools="http://schemas.android.com/tools"
    android:layout_width="match_parent"
    android:layout_height="match_parent"
    android:layout_marginLeft="@dimen/activity_horizontal_margin"
    android:layout_marginRight="@dimen/activity_horizontal_margin"
    android:layout_marginTop="@dimen/activity_horizontal_margin"
    android:orientation="vertical" >
    <RelativeLayout
        android:layout_width="wrap_content"
        android:layout_height="wrap_content"
        android:layout_gravity="center"
        android:orientation="horizontal" >
        <RadioGroup
            android:id="@+id/rg_Status"
            android:layout_width="wrap_content"
            android:layout_height="wrap_content"
            android:orientation="horizontal" >
            <RadioButton
                android:id="@+id/rbStatic"
                android:layout_width="wrap_content"
                android:layout_height="wrap_content"
                android:checked="true"
                android:onClick="onClick_RB"
                android:text="@string/strStatic" />
            <RadioButton
                android:id="@+id/rbDynamic"
                android:layout_width="wrap_content"
                android:layout_height="wrap_content"
                android:onClick="onClick_RB"
                android:text="@string/strDynamic" />
        </RadioGroup>
        <EditText
            android:id="@+id/etPwd"
            android:layout_width="wrap_content"
            android:layout_height="wrap_content"
            android:layout_marginStart="6dp"
            android:layout_toEndOf="@+id/rg_Status"
            android:hint="@string/strLogin_Hint"
            android:inputType="numberPassword"
            android:maxLength="6"
            android:textAppearance="?android:attr/textAppearanceMedium" >
            <requestFocus />
        </EditText>
    </RelativeLayout>
    <LinearLayout
        android:layout_width="match_parent"
        android:layout_height="wrap_content"
        android:gravity="center"
        android:orientation="horizontal" >
        <Button
            android:id="@+id/btModify"
            style="?android:attr/buttonBarButtonStyle"
            android:layout_width="wrap_content"
            android:layout_height="wrap_content"
            android:onClick="onClick_BT"
            android:text="@string/strModify" />
```

```xml
        <Button
            android:id="@+id/btLogin"
            style="?android:attr/buttonBarButtonStyle"
            android:layout_width="wrap_content"
            android:layout_height="wrap_content"
            android:onClick="onClick_BT"
            android:text="@string/strLogin" />
    </LinearLayout>
    <LinearLayout
        android:layout_width="wrap_content"
        android:layout_height="wrap_content"
        android:layout_gravity="center"
        android:orientation="horizontal" >
        <RadioGroup
            android:id="@+id/rg_Function"
            android:layout_width="wrap_content"
            android:layout_height="wrap_content"
            android:orientation="horizontal" >
            <RadioButton
                android:id="@+id/rbCategory"
                android:layout_width="wrap_content"
                android:layout_height="wrap_content"
                android:checked="true"
                android:onClick="onClick_RB"
                android:text="@string/strCategory" />
            <RadioButton
                android:id="@+id/rbSearch"
                android:layout_width="wrap_content"
                android:layout_height="wrap_content"
                android:onClick="onClick_RB"
                android:text="@string/strSearch" />
        </RadioGroup>
        <Button
            android:id="@+id/btGo"
            style="?android:attr/buttonBarButtonStyle"
            android:layout_width="wrap_content"
            android:layout_height="wrap_content"
            **android:paddingBottom="16dp"**
            android:onClick="onClick_BT"
            android:text="@string/strGo" />
    </LinearLayout>
</LinearLayout>
```

8.4.2 源代码

登录窗体主要完成身份认证工作，readStoragePwd 是一个静态函数，可以方便地供其他窗体调用，用来读取保存的静态密码，该密码参与数据库的加密备份。writeStoragePwd 函数用于保存静态密码。getCurrentDatePwd 函数用来获取当前的动态密码，即月份的天数加上 11 转换为文本，前面加上固定的头部，动态密码只用于浏览信息。

checkPassword 函数用来检查密码是否正确，正确返回 true，错误返回 false。首先将编辑框里面的密码保存到 strEnter 变量中，如果采用动态密码方式，则只要检查 strEnter 与动态密码是否相等即可返回布尔值。如果是静态密码方式，则要求密码长度为 6 位，如果原来的密码为空，则调用 Library 类库中 StringProcess 类的静态函数 reverseString，将密码倒序，然后再调用 writeStoragePwd 函数保存倒序后的密码；如果原来密码不为空，则检查 strEnter 与倒序后的密码是否相等。

与用户交互的 RadioButton 和 Button 原理比较简单，这里不再赘述。

```java
package com.walkerma.serviceregister;
import java.util.Calendar;
import java.util.Locale;
import com.walkerma.library.GeneralProcess;
import com.walkerma.library.StringProcess;
import android.app.Activity;
import android.content.Context;
import android.content.Intent;
import android.content.SharedPreferences;
import android.content.SharedPreferences.Editor;
import android.os.Bundle;
import android.os.Handler;
import android.os.Message;
import android.text.TextUtils;
import android.view.KeyEvent;
import android.view.Menu;
import android.view.MenuItem;
import android.view.View;
import android.widget.Button;
import android.widget.EditText;
import android.widget.RadioButton;
import android.widget.Toast;
public class LoginActivity extends Activity{
    private Button btModify, btGo;
    private RadioButton rbStatic, rbDynamic;
    private EditText edPwd;
    private String strAppName;
    private final int STATIC_MODE = 0;
    private final int DYNAMIC_MODE = 1;
    private int pwdMode;
    private enum WorkMode{Category, Search};
    private WorkMode workMode=WorkMode.Category;
    private boolean bCanExit = false;
    private Handler mHandler;
    private final int MSG_Delay = 10000;
    class IncomingHandlerCallback implements Handler.Callback{
        @Override
        public boolean handleMessage(Message msg) {
            ProcessMessage(msg);
            return true;
        }
    }
    private void ProcessMessage(Message msg){
        switch(msg.what){
        case MSG_Delay:
            bCanExit = false;
            break;
        }
    }
    @Override
    protected void onCreate(Bundle savedInstanceState) {
```

```java
        super.onCreate(savedInstanceState);
        setContentView(R.layout.activity_login);
        strAppName = getString(R.string.app_name);
        edPwd = (EditText)findViewById(R.id.etPwd);
        btModify = (Button)findViewById(R.id.btModify);
        rbStatic = (RadioButton)findViewById(R.id.rbStatic);
        rbDynamic = (RadioButton)findViewById(R.id.rbDynamic);
        btGo = (Button)findViewById(R.id.btGo);
        SharedPreferences sp = getSharedPreferences(strAppName, 0);
        pwdMode = sp.getInt("PwdMode", STATIC_MODE);
        if(pwdMode==STATIC_MODE){
            rbStatic.setChecked(true);
            btModify.setEnabled(true);
        }
        else{
            rbDynamic.setChecked(true);
            btModify.setEnabled(false);
        }
        Intent it = getIntent();
        if(it!=null)
            btGo.setEnabled(it.getBooleanExtra("Returned", false));
        else
            btGo.setEnabled(false);
        mHandler = new Handler(new IncomingHandlerCallback());
}
// This for dynamic password.
private String getCurrentDatePwd(){
    Calendar cl = Calendar.getInstance();
    int nDay = cl.get(Calendar.DAY_OF_MONTH) + 11;
    return "99" + String.format(Locale.ENGLISH, "%2d", nDay);
}
public static String readStoragePwd(Context context){
    SharedPreferences sp = context.getSharedPreferences(
            context.getString(R.string.app_name), 0);
    return sp.getString("Pwd", "Blank");
}
private void writeStoragePwd(String strPwd){
    SharedPreferences sp = getSharedPreferences(strAppName, 0);
    Editor editor = sp.edit();
    editor.putString("Pwd", strPwd);
    editor.commit();
}
private boolean checkPassword() {
    String strEnter = edPwd.getText().toString();
    if(TextUtils.isEmpty(strEnter)) return false;
    switch(pwdMode){
    case DYNAMIC_MODE:
        if(!strEnter.equals(getCurrentDatePwd()))
            return false;
        else
            return true;
    case STATIC_MODE:
        if(strEnter.length()!=6) return false;
        String strPwdDisk = readStoragePwd(this);
        if(strPwdDisk.equals("Blank")){
            // Password is reversed at first, then saved.
```

```java
                    writeStoragePwd(StringProcess.reverseString(strEnter));
                    Toast.makeText(this, "Password set.",
                            Toast.LENGTH_SHORT).show();
                    return true;
                }
                else{
                    if(!strEnter.equals(StringProcess.reverseString(strPwdDisk)))
                        return false;
                    else
                        return true;
                }
            default:
                return true;
        }
    }
    @Override
    public boolean onCreateOptionsMenu(Menu menu) {
        getMenuInflater().inflate(R.menu.login, menu);
        return true;
    }
    @Override
    public boolean onOptionsItemSelected(MenuItem item) {
        int id = item.getItemId();
        switch (id){
        case R.id.menu_about:
            GeneralProcess.fireAboutDialog(this,
                    getString(R.string.app_name),
                    getString(R.string.strCopyrights));
            break;
        }
        return super.onOptionsItemSelected(item);
    }
    public void onClick_RB(View v){
        int id = v.getId();
        switch(id){
        case R.id.rbStatic:
            pwdMode = STATIC_MODE;
            btModify.setEnabled(true);
            break;
        case R.id.rbDynamic:
            pwdMode = DYNAMIC_MODE;
            btModify.setEnabled(false);
            break;
        case R.id.rbCategory:
            workMode = WorkMode.Category;
            break;
        case R.id.rbSearch:
            workMode = WorkMode.Search;
            break;
        }
    }
    public void onClick_BT(View v){
        Intent intent;
        int id = v.getId();
        switch(id){
        case R.id.btModify:
```

```java
            String strOld_Disk = readStoragePwd(this);
            if(strOld_Disk.equals("Blank")){
                Toast.makeText(this, "Password blank.",
                        Toast.LENGTH_SHORT).show();
                break;
            }
            intent = new Intent(this, PwdActivity.class);
            startActivity(intent);
            finish();
            break;
        case R.id.btLogin:
            if(btGo.isEnabled()) {
                Toast.makeText(this, "OK", Toast.LENGTH_SHORT).show();
                break;
            }
            if(checkPassword())
                btGo.setEnabled(true);
            else{
                btGo.setEnabled(false);
                Toast.makeText(this, "Password error.",
                        Toast.LENGTH_SHORT).show();
            }
            break;
        case R.id.btGo:
            if(workMode==WorkMode.Category)
                intent = new Intent(this, MainActivity.class);
            else
                intent = new Intent(this, SearchActivity.class);
            startActivity(intent);
            finish();
            break;
        }
    }
    @Override
    public boolean onKeyDown(int keyCode, KeyEvent event) {
        // TODO Auto-generated method stub
        switch(keyCode){
        case KeyEvent.KEYCODE_BACK:
            if(bCanExit){
                break;
            }
            else{
                bCanExit = true;
                Toast.makeText(getApplicationContext(),
                        "Press again to quit.",
                        Toast.LENGTH_SHORT).show();
            }
            mHandler.sendEmptyMessageDelayed(MSG_Delay, 1000);
            return true;
        }
        return super.onKeyDown(keyCode, event);
    }
    @Override
    protected void onDestroy() {
        mHandler.removeCallbacksAndMessages(null);
        SharedPreferences sp = getSharedPreferences(strAppName, 0);
```

```
        Editor editor = sp.edit();
        editor.putInt("PwdMode", pwdMode);
        editor.commit();
        super.onDestroy();
    }
}
```

8.5 服务浏览窗体

服务浏览窗体源代码列出了所有服务种类,可以通过各选项中的箭头上下移动服务种类,离开该窗体时,移动后的服务种类的位置将得到保存。其布局文件名为 activity_main.xml,对应的窗体源代码为 MainActivity.java,所使用的适配器源代码为 MainListAdapter.java。

8.5.1 适配器布局文件

适配器布局文件包含一个 TextView 标记,用来显示服务文本;三个 ImageView 标记,第一个用来显示该选项是否被选中,最后两个用来移动选项。

```xml
<?xml version="1.0" encoding="utf-8"?>
<RelativeLayout xmlns:android="http://schemas.android.com/apk/res/android"
    android:layout_width="match_parent"
    android:layout_height="match_parent"
    android:orientation="horizontal" >
    <TextView
        android:id="@+id/txtItem"
        android:layout_width="wrap_content"
        android:layout_height="wrap_content"
        android:layout_alignParentStart="true"
        android:layout_centerVertical="true"
        android:layout_marginLeft="10dp"
        android:layout_marginRight="10dp"
        android:lines="1"
        android:text="@string/strService"
        android:textAppearance="?android:attr/textAppearanceMedium" />
    <ImageView
        android:id="@+id/check_mark"
        android:layout_width="40dp"
        android:layout_height="40dp"
        android:layout_centerVertical="true"
        android:layout_toEndOf="@id/txtItem"
        android:contentDescription="@null"
        android:src="@drawable/check_mark_pure"
        android:visibility="invisible" />
    <ImageView
        android:id="@+id/circled_up"
        android:layout_width="40dp"
        android:layout_height="40dp"
        android:layout_alignParentEnd="true"
        android:layout_centerVertical="true"
        android:layout_marginEnd="10dp"
        android:layout_marginStart="20dp"
        android:contentDescription="@null"
        android:src="@drawable/circled_up"
```

```
            android:visibility="invisible" />
        <ImageView
            android:id="@+id/circled_down"
            android:layout_width="40dp"
            android:layout_height="40dp"
            android:layout_centerVertical="true"
            android:layout_toStartOf="@id/circled_up"
            android:contentDescription="@null"
            android:src="@drawable/circled_down"
            android:visibility="invisible" />
</RelativeLayout>
```

8.5.2 窗体布局文件

窗体布局文件外层采用线性布局，内部第一行和第二行的组件也采用线性布局，底部的 ListView 的 layout_height 属性设置为 match_parent，即填满剩余的高度空间。

```
<LinearLayout xmlns:android="http://schemas.android.com/apk/res/android"
    xmlns:tools="http://schemas.android.com/tools"
    android:id="@+id/LinearLayout1"
    android:layout_width="wrap_content"
    android:layout_height="match_parent"
    android:layout_marginBottom="@dimen/activity_horizontal_margin"
    android:layout_marginLeft="@dimen/activity_horizontal_margin"
    android:layout_marginRight="@dimen/activity_horizontal_margin"
    android:layout_marginTop="@dimen/activity_horizontal_margin"
    android:orientation="vertical" >
    <LinearLayout
        android:layout_width="match_parent"
        android:layout_height="wrap_content"
        android:focusable="true"
        android:focusableInTouchMode="true"
        android:gravity="center"
        android:orientation="horizontal" >
        <TextView
            android:id="@+id/txtService"
            android:layout_width="wrap_content"
            android:layout_height="wrap_content"
            android:text="@string/strService"
            android:textAppearance="?android:attr/textAppearanceMedium" />
        <EditText
            android:id="@+id/edService"
            android:layout_width="wrap_content"
            android:layout_height="wrap_content"
            android:layout_marginStart="6dp"
            android:hint="@string/strCategory"
            android:inputType="text" />
        <Button
            android:id="@+id/btRestore"
            style="?android:attr/buttonBarButtonStyle"
            android:layout_width="wrap_content"
            android:layout_height="wrap_content"
            android:text="@string/strRestore" />
        <Button
            android:id="@+id/btBackup"
```

```xml
            style="?android:attr/buttonBarButtonStyle"
            android:layout_width="wrap_content"
            android:layout_height="wrap_content"
            android:text="@string/strBackup" />
    </LinearLayout>
    <LinearLayout
        android:layout_width="match_parent"
        android:layout_height="wrap_content"
        android:gravity="center"
        android:orientation="horizontal" >
        <Button
            android:id="@+id/btAdd"
            style="?android:attr/buttonBarButtonStyle"
            android:layout_width="wrap_content"
            android:layout_height="wrap_content"
            android:text="@string/strAdd" />
        <Button
            android:id="@+id/btDelete"
            style="?android:attr/buttonBarButtonStyle"
            android:layout_width="wrap_content"
            android:layout_height="wrap_content"
            android:text="@string/strDelete" />
        <Button
            android:id="@+id/btEnter"
            style="?android:attr/buttonBarButtonStyle"
            android:layout_width="wrap_content"
            android:layout_height="wrap_content"
            android:text="@string/strEnter" />
    </LinearLayout>
    <ListView
        android:id="@+id/list"
        android:layout_width="wrap_content"
        android:layout_height="match_parent"
        android:drawSelectorOnTop="false" >
    </ListView>
</LinearLayout>
```

8.5.3 适配器源代码

在适配器源代码中，布尔变量 bItemMoved 是一个选项有无移动的标志（包括上下移动或插入与删除）。在 getView 方法中定义了点击上下移动箭头所执行的函数，而且，如果选项当前位置为最前面，则隐藏向上箭头；如果选项的当前位置为最后面，则隐藏向下箭头。

```java
package com.walkerma.serviceregister;
import java.util.ArrayList;
import android.app.Activity;
import android.content.Context;
import android.os.Vibrator;
import android.view.LayoutInflater;
import android.view.View;
import android.view.ViewGroup;
import android.widget.ArrayAdapter;
import android.widget.ImageView;
import android.widget.TextView;
public class MainListAdapter extends ArrayAdapter<String>{
```

```java
private int nCurrentPosition = -1;
private boolean bItemMoved = false;
private boolean bCanVibrate = false;
private final int nVibrateTime = 30;
private final Activity context;
private ArrayList<String> listItems;
private int mLayout;
public MainListAdapter(Activity context,
        int mLayout, ArrayList<String> listItems) {
    super(context, mLayout, listItems);
    this.context = context;
    this.mLayout = mLayout;    // row xml
    this.listItems = listItems;
}
public View getView(int position,View convertView,ViewGroup parent) {
    View rowItem = convertView;
    final ViewHolder holder;
    if(rowItem == null){
        LayoutInflater inflater = context.getLayoutInflater();
        rowItem = inflater.inflate(mLayout, parent, false);
        holder = new ViewHolder();
        holder.txtItem = (TextView)
                rowItem.findViewById(R.id.txtItem);
        holder.imgCheckMark = (ImageView)
                rowItem.findViewById(R.id.check_mark);
        holder.imgCircledDown = (ImageView)
                rowItem.findViewById(R.id.circled_down);
        holder.imgCircledUp = (ImageView)
                rowItem.findViewById(R.id.circled_up);
        rowItem.setTag(holder);
    }
    else holder = (ViewHolder) rowItem.getTag();
    holder.imgCircledDown.setOnClickListener(
            new View.OnClickListener() {
        @Override
        public void onClick(View view) {
            downMoveItem(nCurrentPoistion); }});
    holder.imgCircledUp.setOnClickListener(
            new View.OnClickListener() {
        @Override
        public void onClick(View view) {
            upMoveItem(nCurrentPosition); }});
    String strData = listItems.get(position);
    holder.txtItem.setText(strData);
    if(nCurrentPosition>=0 && nCurrentPosition <listItems.size()){
        if(strData.equals(listItems.get(nCurrentPosition))){
            holder.imgCheckMark.setVisibility(View.VISIBLE);
            if(nCurrentPosition < listItems.size() - 1)
                holder.imgCircledDown.setVisibility(View.VISIBLE);
            if(nCurrentPosition > 0)
                holder.imgCircledUp.setVisibility(View.VISIBLE);
        }
        else{
            holder.imgCheckMark.setVisibility(View.INVISIBLE);
            holder.imgCircledDown.setVisibility(View.INVISIBLE);
            holder.imgCircledUp.setVisibility(View.INVISIBLE);
```

```java
            }
        }
        if(nCurrentPosition == -1){
            holder.imgCheckMark.setVisibility(View.INVISIBLE);
            holder.imgCircledDown.setVisibility(View.INVISIBLE);
            holder.imgCircledUp.setVisibility(View.INVISIBLE);
        }
        return rowItem;
    }
    public void setCurrentPosition(int nPos){
        nCurrentPosition = nPos;
    }
    public void setVibrateStatus(boolean bCanVibrate){
        this.bCanVibrate = bCanVibrate;
    }
    public void vibrateCell(){
        //long pattern[] = {50,100,100,250,150,350};
        Vibrator v = (Vibrator)context.
                getSystemService(Context.VIBRATOR_SERVICE);
        //v.vibrate(pattern,3);
        v.vibrate(nVibrateTime);
    }
    public int getCurrentPosition(){
        return nCurrentPosition;
    }
    public String getItem() {
        if(nCurrentPosition != -1)
            return listItems.get(nCurrentPosition);
        else
            return null;
    }
    public void removeItem(int position){
        if(position < 0 || position >= listItems.size()) return;
        listItems.remove(position);
        nCurrentPosition = -1;
        notifyDataSetChanged();
        bItemMoved = true;
    }
    public void addItem(int pos, String strItem){
        listItems.add(pos, strItem);
        bItemMoved = true;
    }
    public int getItemIndex(String strItem){
        int i, nLen;
        nLen = listItems.size();
        for(i=0; i<nLen; i++){
            if(strItem.equals(listItems.get(i))) break;
        }
        if (i != nLen)
            return i;
        else
            return -1;
    }
    public boolean getItemMovedStatus(){
        return bItemMoved;
    }
```

```
    private void upMoveItem(int which){
        if(which <= 0 || which >= listItems.size()) return;
        String strTmp = listItems.get(which);
        listItems.remove(which);
        listItems.add(which - 1, strTmp);
        nCurrentPosition = which - 1;    // adjust current position
        notifyDataSetChanged();
        bItemMoved = true;
        if(bCanVibrate) vibrateCell();
    }
    private void downMoveItem(int which){
        if(which < 0 || which >= listItems.size() -1 ) return;
        String strTmp = listItems.get(which);
        listItems.remove(which);
        listItems.add(which+1, strTmp);
        nCurrentPosition = which + 1;    // adjust current position
        notifyDataSetChanged();
        bItemMoved = true;
        if(bCanVibrate) vibrateCell();
    }
    static class ViewHolder {
        TextView txtItem;
        ImageView imgCheckMark;
        ImageView imgCircledDown;
        ImageView imgCircledUp;
    }
}
```

8.5.4　窗体源代码

窗体源代码主要实现服务选项的显示、增加和删除，以及对选项进行排序并保存。在 onCreate 方法中，readStorage 函数读取保存在 SD 卡上的服务数据，服务选项之间用"/"进行分隔；readServiceData 函数则将读取的数据转换为字符串数组，然后添加到 ArrayList<String> 对象 listItems 中。restoreServiceStatus 函数用来恢复服务的状态，即如果从当前选项（图 8-2）进入单位查询界面（图 8-3），则从图 8-3 返回图 8-2 时，将恢复图 8-2 中选项的选中状态。

在 onResume 方法中生成数据库对象 db。调用 addItem 函数将需要新增的服务种类加入 listItems 中；调用 deleteItem 将选中的服务删除，这时需要与用户交互，实际删除选项时执行 deleteSelectedService 函数，将凡是只包含该服务种类的所有记录删除，或者在多服务种类的记录中删除该服务。

需要从 SD 卡读取备份的数据时，在 onClick 方法中启动多线程对象 threadRead，然后在 processMessage 消息 ThreadReadText.MSG_ReadOK 中调用 updateItems 函数，将 listItems 中不存在的新的服务加入其中，同时，布尔变量 bRestored 设置为 true，以便保存数据。

需要将 ListView 对象所列服务种类存入 SD 卡时，在 onClick 方法中启动多线程对象 threadExport，在其中调用 getServiceString 函数，将 listItems 中的所有选项转换为一个以"/"分隔的字符串，然后调用 Library 类库中的 ByteProcess 类（第 11 章将详细介绍）中的静态函数 stringToUnicodeHexChars，将该字符串转换为 Unicode 字符串，最后调用 FileProcess 类中的静态函数 writeText 以保存。

命令按钮 btEnter 调用 enterItem 函数，向 Intent 对象传入服务种类，从而启动单位浏览窗体。

窗体销毁时，在 onDestroy 方法中调用适配器的 getItemMovedStatus 函数，检查 ListView 对象中的选项有无变化，如果有变化或者 bRestored 为 true，则调用 writeServiceData 函数，在其中通过 SharedPreferences 对象将 listItems 中的所有选项保存在 Service-Register 的内部目录。

```java
package com.walkerma.serviceregister;
import java.io.File;
import java.util.ArrayList;
import com.walkerma.library.ThreadReadText;
import com.walkerma.library.ByteProcess;
import com.walkerma.library.FileProcess;
import com.walkerma.library.DatabaseHelper;
import android.app.ActionBar;
import android.app.Activity;
import android.app.AlertDialog;
import android.content.ContentValues;
import android.content.DialogInterface;
import android.content.Intent;
import android.content.SharedPreferences;
import android.content.SharedPreferences.Editor;
import android.database.Cursor;
import android.os.Bundle;
import android.os.Environment;
import android.os.Handler;
import android.os.Message;
import android.text.TextUtils;
import android.view.KeyEvent;
import android.view.View;
import android.view.View.OnClickListener;
import android.widget.AdapterView;
import android.widget.AdapterView.OnItemClickListener;
import android.widget.Button;
import android.widget.EditText;
import android.widget.ListView;
import android.widget.Toast;
public class MainActivity extends Activity implements OnClickListener{
    private ListView list;
    private EditText edService;
    private Button btEnter, btRestore, btBackup, btAdd, btDelete;
    private boolean bRestored = false;
    private DatabaseHelper db;
    public static String DB_NAME = "dbService";
    public static String TABLE_NAME = "services";
    public static String FIELDS_NAME =
            "_num INTEGER DEFAULT 0, " +
                "_unit TEXT PRIMARY KEY, " +
                "_service TEXT NOT NULL, " +
                "_no TEXT NULL, " +
                "_date TEXT NULL, " +
                "_address TEXT NULL, " +
                "_remark TEXT NULL";
```

```java
    private ArrayList<String> listItems = new ArrayList<String>();
    private MainListAdapter adapter;
    private Handler mHandler;
    private final int MSG_SD_NotReady = 10001;
    private final int MSG_ExportOK = 10002;
    private final int MSG_ExportError = 10003;
    private final int MSG_ServiceEmpty = 10004;
    class IncomingHandlerCallback implements Handler.Callback{
        @Override
        public boolean handleMessage(Message msg) {
            processMessage(msg);
            return true;
        }
    }
    private void processMessage(Message msg){
        switch(msg.what){
        case ThreadReadText.MSG_FileNotFound:
            Toast.makeText(this,  "File not found.",
                    Toast.LENGTH_SHORT ).show();
            break;
        case MSG_SD_NotReady:
        case ThreadReadText.MSG_SD_NotReady:
            Toast.makeText(this,  "SD card is not ready.",
                    Toast.LENGTH_SHORT ).show();
            break;
        case ThreadReadText.MSG_ReadOK:
            updateItems((String)msg.obj);
            Toast.makeText(this,  "Read OK.",
                    Toast.LENGTH_SHORT ).show();
            bRestored = true;
            break;
        case ThreadReadText.MSG_ReadError:
            Toast.makeText(this,  "Read Error.",
                    Toast.LENGTH_SHORT ).show();
            break;
        case MSG_ExportOK:
            Toast.makeText(this,  "Write OK.",
                    Toast.LENGTH_SHORT ).show();
            break;
        case MSG_ExportError:
            Toast.makeText(this,  "Write Error.",
                    Toast.LENGTH_SHORT ).show();
        }
        return;
    }
    @Override
    protected void onCreate(Bundle savedInstanceState) {
        super.onCreate(savedInstanceState);
        setContentView(R.layout.activity_main);
        initWidgets();
        list = (ListView)findViewById(R.id.list);
        adapter = new MainListAdapter(this, R.layout.list_main, listItems);
        adapter.setVibrateStatus(true);
        list.setAdapter(adapter);
        list.setOnItemClickListener(new OnItemClickListener(){
            @Override
```

```java
            public void onItemClick(AdapterView<?> parent, View view,
                    int position, long id) {
                if(adapter.getCurrentPosition() != position){
                    //setItemChecked
                    adapter.setCurrentPosition(position);
                }
                else{
                    //setItemUnChecked
                    adapter.setCurrentPosition(-1);
                }
                adapter.notifyDataSetChanged();     // call getView automatically
            }});
        readServiceData();      // from SD
        restoreServiceStatus();         //from intent
        mHandler = new Handler(new IncomingHandlerCallback());
    }
    @Override
    protected void onDestroy() {
        // TODO Auto-generated method stub
        mHandler.removeCallbacksAndMessages(null);
        db.close();
        if(adapter.getItemMovedStatus() || bRestored) writeServiceData();
        super.onDestroy();
    }
    private void updateItems(String strData){
        if(strData.isEmpty()) return;
        String[] strArray = strData.split("/");
        for(int i=0; i<strArray.length; i++)
            if (!listItems.contains(strArray[i]))
                listItems.add(strArray[i]);
        adapter.notifyDataSetChanged();
    }
    class ThreadExport extends Thread{
        public void run(){
            if(FileProcess.isExternalStorageValid() == false) {
                mHandler.obtainMessage(MSG_SD_NotReady,
                        0, 0, null).sendToTarget();
                return;
            }
            String strData = getServiceString();
            if(TextUtils.isEmpty(strData)){
                mHandler.obtainMessage(MSG_ServiceEmpty,
                        0, 0, null).sendToTarget();
                return;
            }
            strData = ByteProcess.stringToUnicodeHexChars(strData, false);
            if(FileProcess.writeText(getFileName(), strData))
                mHandler.obtainMessage(MSG_ExportOK,
                        0, 0, null).sendToTarget();
            else mHandler.obtainMessage(MSG_ExportError,
                    0, 0, null).sendToTarget();
        }
    }
    private String getAppPath(){
        String strPath = Environment.
                getExternalStorageDirectory().getPath();
```

```java
        String strAppName = getString(R.string.app_name);
        strPath = FileProcess.createDir(strPath,
                "data" + File.separator +
                strAppName) + File.separator;
        return strPath;
    }
    private String getFileName(){
        return getAppPath() + "056E5553.dat"; //Unicode only
    }
    private void initWidgets(){
        ActionBar actionBar = getActionBar();
        actionBar.setDisplayShowHomeEnabled(false);
        actionBar.setDisplayShowTitleEnabled(true);
        edService = (EditText)findViewById(R.id.edService);
        btEnter = (Button)findViewById(R.id.btEnter);
        btEnter.setOnClickListener(this);
        btRestore = (Button)findViewById(R.id.btRestore);
        btRestore.setOnClickListener(this);
        btBackup = (Button)findViewById(R.id.btBackup);
        btBackup.setOnClickListener(this);
        btAdd = (Button)findViewById(R.id.btAdd);
        btAdd.setOnClickListener(this);
        btDelete = (Button)findViewById(R.id.btDelete);
        btDelete.setOnClickListener(this);
    }
    private void restoreServiceStatus(){
        Intent it = getIntent();
        String strRet = it.getStringExtra("Service");
        if(TextUtils.isEmpty(strRet)) return;
        int i = adapter.getItemIndex(strRet);
        adapter.setCurrentPosition(i);
        adapter.notifyDataSetChanged();
    }
    private String readStorage(){
        String strAppName = getString(R.string.app_name);
        SharedPreferences sp = getSharedPreferences
                (strAppName, MODE_PRIVATE);
        return sp.getString("Services", "");
    }
    private void readServiceData(){
        String strDisk = readStorage();
        if(TextUtils.isEmpty(strDisk)) return;
        String[] strArray = strDisk.split("/");
        listItems.clear();
        for(int i=0; i<strArray.length; i++){
            listItems.add(strArray[i]);
        }
        adapter.notifyDataSetChanged();
    }
    @Override
    public boolean onKeyDown(int keyCode, KeyEvent event){
        if(keyCode == KeyEvent.KEYCODE_BACK){
            Intent it = new Intent(this, LoginActivity.class);
            it.putExtra("Returned", true);
            startActivity(it);
            finish();
```

```java
            return true;
        }
        return super.onKeyDown(keyCode, event);
    }
    @Override
    protected void onResume() {
        // TODO Auto-generated method stub
        db = new DatabaseHelper(this, DB_NAME, 1,
                TABLE_NAME, FIELDS_NAME);
        super.onResume();
    }
    private String getServiceString(){
        String strTmp = "";
        for(int i=0; i<listItems.size(); i++){
            strTmp += listItems.get(i);
            if(i!=listItems.size()-1) strTmp += "/";
        }
        return strTmp;
    }
    private void writeServiceData(){
        String strAppName = getString(R.string.app_name);
        SharedPreferences sp = getSharedPreferences(
                strAppName, MODE_PRIVATE);
        Editor editor = sp.edit();
        if(listItems.isEmpty()) return;
        String strInput = getServiceString();
        editor.putString("Services", strInput);
        editor.commit();
    }
    private void deleteSelectedService(String strSvc){
        String strTmpSvc = "";
        String strInquery = "SELECT _unit, _service FROM " +
                TABLE_NAME + " WHERE _service LIKE '%" + strSvc + "%'";
        db.refreshReadDB(strInquery, null);
        if(db.getCount()==0) return;
        Cursor row = db.getRecords();
        row.moveToFirst();
        for(int i=0; i<db.getCount(); i++){
            strTmpSvc = row.getString(
                    row.getColumnIndex("_service"));
            if(strTmpSvc.equals(strSvc)){
                db.deleteRecords("_unit = '" + row.getString(
                        row.getColumnIndex("_unit")) + "'", null);
            }
            else{
                strTmpSvc = removeSubItem(strTmpSvc, "|", strSvc);
                ContentValues cv = new ContentValues();
                cv.put("_service", strTmpSvc);
                db.updateRecords(cv, "_unit = '" + row.getString(
                        row.getColumnIndex("_unit")) + "'", null);
            }
            row.moveToNext();
        }
        Toast.makeText(MainActivity.this, Integer.toString(
                db.getCount()) + " item(s) updated.",
                Toast.LENGTH_SHORT).show();
```

```java
    }
// "abcdWXabWXcd", "WX", "ab" ==> "abcdWXcd"
// "ab|cd|xy|uv", "|", "xy" ==> "ab|cd|uv"
    public static String removeSubItem(String strSrc,
            String strSeg, String strSub){
        String strSegOld = strSeg;
        if((strSeg.length()==0) || (strSub.length()==0))
            return strSrc;
        if(strSeg.equals("|")) {
            strSeg = "/";
            strSrc = strSrc.replace('|','/');
        }
        String[]  strArray = strSrc.split(strSeg);
        String strDest=""; //'|' cannot be split
        for(int i=0; i<strArray.length; i++){
            if(strArray[i].equals(strSub) != true){
                strDest += strArray[i] + strSeg;
            }
        }
        int nLen = strDest.length();
        if (strSeg.equals(strSegOld)==false)
            strDest=strDest.replace('/', '|');
        if(nLen < strSeg.length())
            return "";
        else
            return strDest.substring(0, nLen-strSeg.length());
    }
    private void GeneralAlert(int nResource,
            String strTitle, String strMessage){
        AlertDialog.Builder builder = new AlertDialog.Builder(this);
        builder.setIcon(nResource)
        .setTitle(strTitle)
        .setMessage(strMessage)
        .setPositiveButton("OK", null);
        AlertDialog dialog = builder.create();
        dialog.show();
    }
    private void tipForDelete(String strItem){
        AlertDialog alert = new AlertDialog.Builder(this).create();
        alert.setIcon(R.drawable.questionmark);
        alert.setTitle(strItem);
        alert.setMessage("Delete this item?");
        // Add "No"
        alert.setButton(DialogInterface.BUTTON_NEGATIVE, "No",
                new DialogInterface.OnClickListener() {
            @Override
            public void onClick(DialogInterface dialog, int which) {
                return;
            }
        });
        // Add "Yes"
        alert.setButton(DialogInterface.BUTTON_POSITIVE,"Yes",
                new DialogInterface.OnClickListener() {
            @Override
            public void onClick(DialogInterface arg0, int arg1) {
                deleteSelectedService(adapter.getItem());
```

```java
                adapter.removeItem(adapter.getCurrentPosition());
                adapter.setCurrentPosition(-1);
            }
        });
        alert.show();
    }
    @Override
    public void onClick(View v) {
        // TODO Auto-generated method stub
        switch(v.getId()){
        case R.id.btRestore:
            ThreadReadText threadRead = new ThreadReadText(
                    mHandler, getFileName());
            threadRead.start();
            break;
        case R.id.btBackup:
            if(listItems.size()==0){
                Toast.makeText(this, "No any item...",
                        Toast.LENGTH_SHORT).show();
                return;
            }
            ThreadExport threadExport = new ThreadExport();
            threadExport.start();
            break;
        case R.id.btAdd:
            addItem();
            break;
        case R.id.btDelete:
            deleteItem();
            break;
        case R.id.btEnter:
            enterItem();
            break;
        }
    }
    private void addItem(){
        String strInput = edService.getText().toString();
        if(TextUtils.isEmpty(strInput)) {
            GeneralAlert(R.drawable.info, "Add",
                    "Please input Service...");
            return;
        }
        if(adapter.getItemIndex(strInput)!=-1) {
            GeneralAlert(R.drawable.info, "Add",
                    "【" + strInput + "】 has been added...");
            return; //old data, return
        }
        adapter.addItem(0, strInput);     // last added in the front
        adapter.setCurrentPosition(0);
        adapter.notifyDataSetChanged();
        edService.setText("");
    }
    private void enterItem(){
        if(adapter.getCurrentPosition()==-1) {
            Toast.makeText(this, "Please select an item...",
                    Toast.LENGTH_SHORT).show();
```

```
            return;
        }
        Intent it = new Intent(this, UnitsActivity.class);
        it.putExtra("Service", adapter.getItem());
        startActivity(it);
        finish(); //!!
    }
    private void deleteItem(){
        if(adapter.getCurrentPosition()!=-1)
            tipForDelete(adapter.getItem());
        else if(listItems.size() > 0)
            GeneralAlert(R.drawable.info,
                    "Delete", "Please select an item...");
        else
            Toast.makeText(this, "No any item...",
                    Toast.LENGTH_SHORT).show();
    }
}
```

8.6 单位浏览窗体

服务浏览窗体通过 Intent 对象传入服务种类，单位浏览窗体则将包含该服务的所有单位列举显示。为了更全面地学习技术，这里使用拖曳来调整单位顺序。其布局文件名为 activity_units.xml，对应的窗体源代码为 UnitsActivity.java，所使用的适配器源代码为 UnitsDragListAdapter.java，源代码 UnitsDragShadowBuilder.java 用于绘制拖曳阴影。

8.6.1 适配器布局文件

适配器布局文件包含一个 TextView 标记，用来显示单位文本；一个 ImageView 标记，用来显示该选项是否被选中，也用于拖曳。

```xml
<?xml version="1.0" encoding="utf-8"?>
<RelativeLayout xmlns:android="http://schemas.android.com/apk/res/android"
    android:layout_width="match_parent"
    android:layout_height="match_parent"
    android:orientation="horizontal" >
    <TextView
        android:id="@+id/txtItem"
        android:layout_width="wrap_content"
        android:layout_height="wrap_content"
        android:layout_marginLeft="10dp"
        android:layout_marginRight="10dp"
        android:layout_alignParentStart="true"
        android:layout_centerVertical="true"
        android:lines="1"
        android:textAppearance="?android:attr/textAppearanceMedium"
        android:text="@string/strService" />
    <ImageView
        android:id="@+id/check_mark"
        android:layout_width="40dp"
        android:layout_height="40dp"
        android:layout_alignParentEnd="true"
```

```
            android:layout_centerVertical="true"
            android:src="@drawable/check_mark_pure"
            android:visibility="invisible"
            android:contentDescription="@null"/>
</RelativeLayout>
```

8.6.2 窗体布局文件

窗体布局文件外层采用线性布局,内部第一行也采用线性布局,最后一个 Button 标记中的 id 为 btAdd_Inquire,该按钮既用来添加新的数据,也用来查询详细信息,具体操作根据上下文环境进行。

```
<LinearLayout xmlns:android="http://schemas.android.com/apk/res/android"
    xmlns:tools="http://schemas.android.com/tools"
    android:id="@+id/LinearLayout1"
    android:layout_width="wrap_content"
    android:layout_height="match_parent"
    android:layout_marginBottom="@dimen/activity_horizontal_margin"
    android:layout_marginLeft="@dimen/activity_horizontal_margin"
    android:layout_marginRight="@dimen/activity_horizontal_margin"
    android:layout_marginTop="@dimen/activity_horizontal_margin"
    android:orientation="vertical" >
    <LinearLayout
        android:layout_width="match_parent"
        android:layout_height="wrap_content"
        android:gravity="center"
        android:orientation="horizontal" >
        <Button
            android:id="@+id/btImport"
            style="?android:attr/buttonBarButtonStyle"
            android:layout_width="0dp"
            android:layout_height="wrap_content"
            android:layout_weight="1"
            android:text="@string/strImport" />
        <Button
            android:id="@+id/btExport"
            style="?android:attr/buttonBarButtonStyle"
            android:layout_width="0dp"
            android:layout_height="wrap_content"
            android:layout_weight="1"
            android:text="@string/strExport" />
        <Button
            android:id="@+id/btDelete"
            style="?android:attr/buttonBarButtonStyle"
            android:layout_width="wrap_content"
            android:layout_height="wrap_content"
            android:text="@string/strDelete" />
        <Button
            android:id="@+id/btAdd_Inquire"
            style="?android:attr/buttonBarButtonStyle"
            android:layout_width="0dp"
            android:layout_height="wrap_content"
            android:layout_marginEnd="@dimen/interval"
            android:layout_weight="1"
```

```xml
            android:text="@string/strAdd_Inquire" />
    </LinearLayout>
    <ListView
        android:id="@+id/list"
        android:layout_width="wrap_content"
        android:layout_height="match_parent"
        android:drawSelectorOnTop="false" >
    </ListView>
</LinearLayout>
```

8.6.3 拖放阴影源代码

在适配器布局文件中没有使用背景,因而在 getBitmapFromView 方法中直接绘制白色背景,用来表示被拖曳对象正在被移动。

```java
package com.walkerma.serviceregister;
import android.graphics.Bitmap;
import android.graphics.Canvas;
import android.graphics.Color;
import android.graphics.Point;
import android.graphics.drawable.Drawable;
import android.view.View;
import android.view.View.DragShadowBuilder;
public class UnitsDragShadowBuilder extends DragShadowBuilder {
    private View v;
    public UnitsDragShadowBuilder(View v) {
        super(v);
        this.v=v;
    }
    @Override
    public void onDrawShadow(Canvas canvas) {
        super.onDrawShadow(canvas);
        canvas.drawBitmap(getBitmapFromView(v), 0, 0, null);
    }
    private Bitmap getBitmapFromView(View view) {
        int nWidth = view.getWidth();
        int nHeight = view.getHeight();
        Bitmap returnedBitmap = Bitmap.createBitmap(
                nWidth, nHeight, Bitmap.Config.ARGB_8888);
        Canvas canvas = new Canvas(returnedBitmap);
        Drawable bgDrawable =view.getBackground();
        if (bgDrawable!=null)
            bgDrawable.draw(canvas);
        else
            canvas.drawColor(Color.WHITE);
        view.draw(canvas);
        return returnedBitmap;
    }
    @Override
    public void onProvideShadowMetrics(Point shadowSize, Point touchPoint) {
        shadowSize.set(v.getWidth(),v.getHeight());
        touchPoint.set(v.getWidth(), v.getHeight());
    }
}
```

8.6.4 适配器源代码

在适配器源代码中设置了一个内部类 ImageOnTouchListener，用来侦听 imgCheckMark 的 OnTouch 事件，当被按下时，得到行对象 rowItem，开始绘制背景；被放下时，执行 performClick 的默认动作。在 getView 方法中通过调用 imgCheckMark 对象的 setOnTouch-Listener 方法来注册 OnTouch 侦听器。

```java
package com.walkerma.serviceregister;
import java.util.ArrayList;
import android.app.Activity;
import android.content.ClipData;
import android.content.Context;
import android.os.Vibrator;
import android.view.LayoutInflater;
import android.view.MotionEvent;
import android.view.View;
import android.view.View.OnTouchListener;
import android.view.ViewGroup;
import android.widget.ArrayAdapter;
import android.widget.ImageView;
import android.widget.TextView;
public class UnitsDragListAdapter extends ArrayAdapter<String>{
    private int nCurrentPosition = -1;
    private boolean bItemMoved = false;
    private boolean bCanVibrate = false;
    private final int nVibrateTime = 30;
    private final Activity context;
    private ArrayList<String> listItems;
    private int mLayout;
    public UnitsDragListAdapter(Activity context, int mLayout,
            ArrayList<String> listItems) {
        super(context, mLayout, listItems);
        // TODO Auto-generated constructor stub
        this.context = context;
        this.mLayout = mLayout;    // row xml
        this.listItems = listItems;
    }
    public View getView(int position, View convertView,
            ViewGroup parent) {
        View rowItem = convertView;
        final ViewHolder holder;
        if(rowItem == null){
            LayoutInflater inflater = context.getLayoutInflater();
            rowItem = inflater.inflate(mLayout, parent, false);
            holder = new ViewHolder();
            holder.txtItem = (TextView)
                    rowItem.findViewById(R.id.txtItem);
            holder.imgCheckMark = (ImageView)
                    rowItem.findViewById(R.id.check_mark);
            rowItem.setTag(holder);
        }
        else holder = (ViewHolder) rowItem.getTag();
        String strData = listItems.get(position);
        holder.txtItem.setText(strData);
```

```java
        if(nCurrentPosition>=0 && nCurrentPosition <listItems.size()){
            if(strData.equals(listItems.get(nCurrentPosition))){
                holder.imgCheckMark.setVisibility(View.VISIBLE);
            }
            else{
                holder.imgCheckMark.setVisibility(View.INVISIBLE);
            }
        }
        if(nCurrentPosition == -1)
            holder.imgCheckMark.setVisibility(View.INVISIBLE);
        holder.imgCheckMark.setTag(strData);
        holder.imgCheckMark.setOnTouchListener(new ImageOnTouchListener());
        return rowItem;
    }
    private class ImageOnTouchListener implements OnTouchListener{
        @Override
        public boolean onTouch(View v, MotionEvent event) {
            // TODO Auto-generated method stub
            switch (event.getAction()) {
            case MotionEvent.ACTION_DOWN:
                ClipData data = ClipData.newPlainText("",
                        v.getTag().toString());
                View rowItem = (View)v.getParent();
                UnitsDragShadowBuilder shadowBuilder =
                        new UnitsDragShadowBuilder(rowItem);
                v.startDrag(data, shadowBuilder, v, 0);
                if(bCanVibrate) vibrateCell();
                break;
            case MotionEvent.ACTION_UP:
                v.performClick();
                break;
            default:
                break;
            }
            return true;
        }
    }
    public void setCurrentPosition(int nPos){
        nCurrentPosition = nPos;
    }
    public void setVibrateStatus(boolean bCanVibrate){
        this.bCanVibrate = bCanVibrate;
    }
    public void vibrateCell(){
        //long pattern[] = {50,100,100,250,150,350};
        Vibrator v = (Vibrator)context.getSystemService(
                Context.VIBRATOR_SERVICE);
        //v.vibrate(pattern,3);
        v.vibrate(nVibrateTime);
    }
    public int getCurrentPosition(){
        return nCurrentPosition;
    }
    public String getItem() {
        if(nCurrentPosition != -1)
            return listItems.get(nCurrentPosition);
```

```
            else
                return "";
        }
        public void removeItem(int position){
            if(position < 0 || position >= listItems.size()) return;
            listItems.remove(position);
            nCurrentPosition = -1;
            notifyDataSetChanged();
        }
        public int getItemIndex(String strItem){
            int i, nLen;
            nLen = listItems.size();
            for(i=0; i<nLen; i++){
                if(strItem.equals(listItems.get(i))) break;
            }
            if (i != nLen)
                return i;
            else
                return -1;
        }
        public boolean getItemMovedStatus(){
            return bItemMoved;
        }
        static class ViewHolder {
            TextView txtItem;
            ImageView imgCheckMark;
        }
    }
```

8.6.5 窗体源代码

本窗体主要用来备份所有单位的详细数据，或者从备份中读取数据并加入数据库。单位顺序的调整通过拖曳进行，在窗体源代码中定义了一个 ListOnDragListener 类，处理单位选项的拖曳，关键在其中处理 ACTION_DROP 事件，通过坐标得到选项的新位置 posNew，然后通过拖曳的内容调用适配器的 getItemIndex 函数以得到原位置 posOld，最后通过 swap 函数调整位置，即删除原位置的选项，再在新位置插入选项。在窗体的 onCreate 方法中初始化各组件，注册 ListView 对象 list 的拖曳事件侦听器。在 onResume 方法中初始化数据库，并调用 refreshUnits 函数向 list 中添加单位选项。

函数 getDataAsString 将相同服务类型的所有单位记录的内部字段用 segFieldInterval（"\n"）分隔，记录之间用 segRecords（"\r\n"）分隔，最后得到一个完整的字符串。对象 btExport 启动多线程对象 threadExport，调用 getDataAsString 函数以得到数据库记录的字符串文本，再通过 encodeText 函数对字符串进行加密保存，文件名为服务名称的 Unicode 编码。

函数 getEncodeKey 将中文常量 CN_KEY 和静态密码都转换为字节数组，然后进行异或加密，最后再调用 ByteProcess 类中的静态函数 unicodeBytesToString 将 Unicode 字节数组转换为普通字符串，参与 encodeText 函数的加密工作，同时也参与 decodeText 函数的解密工作。

与 btExport 功能相反，btImport 启动多线程对象 threadRead，读取服务的文本字符串，然后调用 updateDatabase 函数将记录添加到数据库中（新单位），或者更新数据库中的内容（已有此单位）；接着，调用 refreshUnits 函数将数据库中的单位更新到 ListView 中，最后再调用 restoreUnitStatus 函数对当前单位进行标记，并调整 btAdd_Inquire 的文本内容，即如果某选项被选中，则该按钮显示"Inquire"，否则显示"Add"。

Button 对象 btDelete 删除选中的单位选项，具体工作通过函数 deleteItem 进行，即从数据库和 list 中都删除该选项。btAdd_Inquire 用于打开单位详细信息窗体，用于增加新的单位或者编辑旧的单位。

当窗体退出时，执行 onDestroy 方法，调用 adjustNumber 函数，根据 list 中所列单位的顺序，更新数据库中序号字段的内容，以便下次按照新的序号来显示单位。

```java
package com.walkerma.serviceregister;
import java.io.File;
import java.util.ArrayList;
import com.walkerma.library.ByteProcess;
import com.walkerma.library.FileProcess;
import com.walkerma.library.StringProcess;
import com.walkerma.library.DatabaseHelper;
import com.walkerma.library.ThreadReadText;
import android.app.ActionBar;
import android.app.Activity;
import android.app.AlertDialog;
import android.content.ClipData;
import android.content.ClipDescription;
import android.content.ContentValues;
import android.content.DialogInterface;
import android.content.Intent;
import android.content.res.Configuration;
import android.database.Cursor;
import android.os.Bundle;
import android.os.Environment;
import android.os.Handler;
import android.os.Message;
import android.text.TextUtils;
import android.view.DragEvent;
import android.view.KeyEvent;
import android.view.View;
import android.view.View.OnClickListener;
import android.view.View.OnDragListener;
import android.widget.AdapterView;
import android.widget.Button;
import android.widget.ListView;
import android.widget.Toast;
import android.widget.AdapterView.OnItemClickListener;
public class UnitsActivity extends Activity implements OnClickListener{
    private static String segRecords = "\r\n";
    private static String segFieldInterval = "\n";
    private static String segFieldBlank = "null";
    private String strInquery, strService, strUnit;
    private String strPwd_PlainDisk;
    private ListView list;
    boolean bNewOne = false; //if add new item from "Service Information"
    boolean bSwaped = false;
```

```java
    private Button btAdd_Inquire, btImport, btExport, btDelete;
    private ArrayList<String> listItems = new ArrayList<String>();
    private UnitsDragListAdapter adapter;
    private DatabaseHelper db;
    private final String CN_KEY =
            "Android在计算机监控系统中的应用";
    private Handler mHandler;
    private final int MSG_SD_NotReady = 10001;
    private final int MSG_ExportOK = 10002;
    private final int MSG_ExportError = 10003;
    @Override
    protected void onCreate(Bundle savedInstanceState) {
        super.onCreate(savedInstanceState);
        setContentView(R.layout.activity_units);
        initWidgets(); // get strService & strUnit
        list = (ListView)findViewById(R.id.list);
        adapter = new UnitsDragListAdapter(this,
                R.layout.list_drag, listItems);
        adapter.setVibrateStatus(true);
        list.setAdapter(adapter);
        list.setOnItemClickListener(new ListOnItemClickListener());
        list.setOnDragListener(new ListOnDragListener());
        mHandler = new Handler(new IncomingHandlerCallback());
        strPwd_PlainDisk = StringProcess.reverseString(
                LoginActivity.readStoragePwd(this));
    }
    private class ListOnItemClickListener implements OnItemClickListener{
        @Override
        // TODO Auto-generated method stub
        public void onItemClick(AdapterView<?> parent, View view,
                int position, long id) {
            // TODO Auto-generated method stub
            if(adapter.getCurrentPosition() != position){
                adapter.setCurrentPosition(position);     //setItemChecked
                btAdd_Inquire.setText("Inquire");
            }
            else{
                adapter.setCurrentPosition(-1);     //setItemUnChecked
                btAdd_Inquire.setText("Add");
            }
            adapter.notifyDataSetChanged();
        }
    }
    private class ListOnDragListener implements OnDragListener{
        @Override
        public boolean onDrag(View v, DragEvent event) {
            // TODO Auto-generated method stub
            int action = event.getAction();
            switch (action) {
            case DragEvent.ACTION_DRAG_STARTED:
                bSwaped = false;
                //确定此视图是否能够接收拖放的数据
                if (event.getClipDescription().hasMimeType(
                        ClipDescription.MIMETYPE_TEXT_PLAIN)) {
                    return true;
                }
```

```
                return false;
            case DragEvent.ACTION_DRAG_ENTERED:
                return true;
            case DragEvent.ACTION_DRAG_LOCATION:
                return true;
            case DragEvent.ACTION_DRAG_EXITED:
                return true;
            case DragEvent.ACTION_DROP:
                ClipData.Item item = event.getClipData().getItemAt(0);
                String dragData = item.getText().toString();
                int x = (int)event.getX();
                int y = (int)event.getY();
                int posNew = list.pointToPosition(x, y);
                int posOld = adapter.getItemIndex(dragData);
                swap(posNew, posOld);
                return true;
            case DragEvent.ACTION_DRAG_ENDED:
                return true;
            default:
                break;
        }
        return false;
    }
}
private void swap(int posNew, int posOld){
    if((posNew == posOld) || (posNew >= listItems.size())
            || (posNew < 0)) return;
    String strData = adapter.getItem(posOld);
    adapter.removeItem(posOld);
    if(posOld > posNew)
        listItems.add(posNew, strData);
    else{
        if(posNew == listItems.size())
            listItems.add(strData);
        else
            listItems.add(posNew, strData);
    }
    adapter.setCurrentPosition(adapter.getItemIndex(strData));
    adapter.notifyDataSetChanged();
    bSwaped = true;
    return;
}
// list ==> db
private void adjustNumber(){
    db.refreshReadDB(strInquery, null);
    if(db.getCount() == 0) return;
    for(int i=0; i<listItems.size(); i++){
        String strUnit = listItems.get(i);
        Cursor row = db.getRecords();
        row.moveToFirst();
        for(int j=0; j<db.getCount(); j++){
            if(strUnit.equals(row.getString(
                    row.getColumnIndex("_unit")))){
                ContentValues cv = new ContentValues();
                cv.put("_num", i+1);
                db.updateRecords(cv,
```

```java
                            "_unit = '" + strUnit + "'", null);
                }
                row.moveToNext();
            }
        }
    }
    @Override
    protected void onDestroy() {
        // TODO Auto-generated method stub
        mHandler.removeCallbacksAndMessages(null);
        if((adapter.getItemMovedStatus() || bNewOne || bSwaped))
            adjustNumber();
        db.close();
        super.onDestroy();
    }
    class IncomingHandlerCallback implements Handler.Callback{
        @Override
        public boolean handleMessage(Message msg) {
            processMessage(msg);
            return true;
        }
    }
    private void processMessage(Message msg){
        switch(msg.what){
        case ThreadReadText.MSG_FileNotFound:
            Toast.makeText(this,  strService + ".dat not found.",
                    Toast.LENGTH_SHORT ).show();
            break;
        case MSG_SD_NotReady:
        case ThreadReadText.MSG_SD_NotReady:
            Toast.makeText(this,  "SD card is not ready.",
                    Toast.LENGTH_SHORT ).show();
            break;
        case ThreadReadText.MSG_ReadOK:
            updateDatabase(decodeText((String)msg.obj));
            refreshUnits();
            Toast.makeText(this,  "Import OK.",
                    Toast.LENGTH_SHORT ).show();
            break;
        case ThreadReadText.MSG_ReadError:
            Toast.makeText(this,  "Import Error.",
                    Toast.LENGTH_SHORT ).show();
            break;
        case MSG_ExportOK:
            Toast.makeText(this,  "Export OK.",
                    Toast.LENGTH_SHORT ).show();
            break;
        case MSG_ExportError:
            Toast.makeText(this,  "Export Error.",
                    Toast.LENGTH_SHORT ).show();
        }
    }
    private void initWidgets(){
        btImport = (Button)findViewById(R.id.btImport);
        btImport.setOnClickListener(this);
        btExport = (Button)findViewById(R.id.btExport);
```

```java
        btExport.setOnClickListener(this);
        btAdd_Inquire = (Button)findViewById(R.id.btAdd_Inquire);
        btAdd_Inquire.setText("Add");
        btAdd_Inquire.setOnClickListener(this);
        btDelete = (Button)findViewById(R.id.btDelete);
        btDelete.setOnClickListener(this);
        ActionBar actionBar = getActionBar();
        actionBar.setDisplayShowHomeEnabled(false);
        actionBar.setDisplayShowTitleEnabled(true);
        Intent it = getIntent();
        strService = it.getStringExtra("Service");
        strUnit = it.getStringExtra("Unit");
        actionBar.setTitle("Units with " + strService);
        strInquery = "SELECT * FROM services WHERE _service LIKE '%"
                + strService + "%'" + " ORDER BY _num";
    }
    @Override
    public boolean onKeyDown(int keyCode, KeyEvent event) {
        if(keyCode == KeyEvent.KEYCODE_BACK){
            Intent it = new Intent(this, MainActivity.class);
            it.putExtra("Service", strService);
            startActivity(it);
            finish();
            return true;
        }
        return super.onKeyDown(keyCode, event);
    }
    private void restoreUnitStatus(){
        if(TextUtils.isEmpty(strUnit))
            strUnit = adapter.getItem();
        if(TextUtils.isEmpty(strUnit)) return;
        int i = adapter.getItemIndex(strUnit);
        adapter.setCurrentPosition(i);
        adapter.notifyDataSetChanged();
        btAdd_Inquire.setText("Inquire");
    }
    private void refreshUnits(){
        db.refreshReadDB(strInquery, null);
        if(db.getCount()==0)        return;
        listItems.clear();
        Cursor row = db.getRecords();
        row.moveToFirst();
        for(int i=0; i<db.getCount(); i++){
            if((i == 0) && (row.getInt(0) == 0)) bNewOne = true;
            listItems.add(row.getString(row.getColumnIndex("_unit")));
            row.moveToNext();
        }
        adapter.notifyDataSetChanged();
        restoreUnitStatus();
    }
    @Override
    protected void onResume() {
        // TODO Auto-generated method stub
        super.onResume();
        db = new DatabaseHelper(this, MainActivity.DB_NAME, 1,
                MainActivity.TABLE_NAME, MainActivity.FIELDS_NAME);
```

```java
        refreshUnits();
    }
    @Override
    public void onConfigurationChanged(Configuration newConfig) {
        super.onConfigurationChanged(newConfig);
        // AndroidManifest.xml
        // android:configChanges="orientation|screenSize"
        adapter.notifyDataSetChanged();    // update ListView
    }
    private ContentValues getContentValues(String[] strItem){
        ContentValues cv = new ContentValues();
        cv.put("_unit", strItem[0]);
        cv.put("_service", strItem[1]);
        if(strItem[2].equals(segFieldBlank))
            cv.put("_no", "");
        else
            cv.put("_no", strItem[2]);
        if(strItem[3].equals(segFieldBlank))
            cv.put("_date", "");
        else
            cv.put("_date", strItem[3]);
        if(strItem[4].equals(segFieldBlank))
            cv.put("_address", "");
        else
            cv.put("_address", strItem[4]);
        if(strItem[5].equals(segFieldBlank))
            cv.put("_remark", "");
        else
            cv.put("_remark", strItem[5]);
        return cv;
    }
    private void updateDatabase(String strSrc){
        int i;
        String[] strRecords = strSrc.split(segRecords);
        if(strRecords==null) return;
        int nLen = strRecords.length;
        if(nLen == 0) return;
        for(i=0; i<nLen; i++){
            String[] strItems = strRecords[i].split(segFieldInterval);
            ContentValues cv = getContentValues(strItems);
            if(db.isNewItem("_unit", strItems[0])){
                // add new records to the end
                cv.put("_num", listItems.size()+ i + 1);
                db.insertRecords(cv);
            }
            else
                db.updateRecords(cv,
                        "_unit = '" + strItems[0] + "'", null);
        }
    }
    private String decodeText(String strSrc){
        String strKey = getEncodeKey();
        strSrc = ByteProcess.base64ToEnString(strSrc);
        strSrc = ByteProcess.decodeUnicode(strSrc, strKey, false);
        int n = strSrc.indexOf("B:");
        strSrc = strSrc.substring(n+2);
```

```
            return strSrc;
    }
    private String getDataAsString(){
        String strData="B:";     //start mark
        String strTmp="";
        int i, j, n;
        db.refreshReadDB(strInquery, null);
        n = db.getCount();
        Cursor row = db.getRecords();
        row.moveToFirst();
        for(i=0; i<n; i++){
            for(j=1; j<7; j++){
                strTmp = row.getString(j);
                if(strTmp.length()==0)
                    strData += segFieldBlank;
                else
                    strData += strTmp;
                if(j<6) strData += segFieldInterval;
            }
            if(i<n-1) strData += segRecords;
            row.moveToNext();
        }
        return strData;
    }
    private String getEncodeKey(){
        if(strPwd_PlainDisk.isEmpty()) return CN_KEY;
        byte[] bKey = ByteProcess.digitsTwoToBytes(strPwd_PlainDisk);
        byte[] bMain = ByteProcess.stringToUnicodeBytes(CN_KEY, false);
        int i, j=0;
        for(i=0; i < bMain.length; i++){
            bMain[i] ^= bKey[j];
            j++;
            if(j==3) j=0;
        }
        return ByteProcess.unicodeBytesToString(bMain, false);
    }
    private String encodeText(String strSrc){
        String strKey = getEncodeKey();
        String strData = ByteProcess.encodeUnicode(strSrc, strKey, false);
        return ByteProcess.enStringToBase64(strData);
    }
    private String getAppPath(){
        String strPath = Environment.getExternalStorageDirectory().getPath();
        String strAppName = getString(R.string.app_name);
        strPath = FileProcess.createDir(strPath,
                "data" + File.separator + strAppName) + File.separator;
        return strPath;
    }
    private String getFileName(){
        return getAppPath() + ByteProcess.stringToUnicodeHexChars(
                strService, false) + ".dat";
    }
    class ThreadExport extends Thread{
        public void run(){
            if(FileProcess.isExternalStorageValid() == false) {
                mHandler.obtainMessage(MSG_SD_NotReady,
```

```java
                        0, 0, null).sendToTarget();
                return;
            }
            String strData = encodeText(getDataAsString());
            if(FileProcess.writeText(getFileName(), strData))
                mHandler.obtainMessage(MSG_ExportOK,
                        0, 0, null).sendToTarget();
            else mHandler.obtainMessage(MSG_ExportError,
                    0, 0, null).sendToTarget();
        }
    }
    @Override
    public void onClick(View v) {
        // TODO Auto-generated method stub
        Intent it;
        switch(v.getId()){
        case R.id.btImport:
            ThreadReadText threadRead = new ThreadReadText(
                    mHandler, getFileName());
            threadRead.start();
            break;
        case R.id.btExport:
            db.refreshReadDB(strInquery, null);
            if(db.getCount()==0){
                Toast.makeText(this, "There is no records.",
                        Toast.LENGTH_SHORT).show();
                return;
            }
            ThreadExport threadExport = new ThreadExport();
            threadExport.start();
            break;
        case R.id.btDelete:
            if(adapter.getCurrentPosition()!=-1)
                tipForDelete(adapter.getItem());
            else if(listItems.size() > 0)
                Toast.makeText(this, "Please select an item.",
                        Toast.LENGTH_SHORT).show();
            else
                Toast.makeText(this, "No any item.",
                        Toast.LENGTH_SHORT).show();
            break;
        case R.id.btAdd_Inquire:
            it = new Intent(this, InfoActivity.class);
            it.putExtra("Service", strService);
            it.putExtra("Unit", adapter.getItem());
            startActivity(it);
            finish();
            break;
        }
    }
    private void tipForDelete(String strItem){
        AlertDialog alert = new AlertDialog.Builder(this).create();
        alert.setIcon(R.drawable.questionmark);
        alert.setTitle(strItem);
        alert.setMessage("Delete this item?");
        // Add "No"
```

```java
        alert.setButton(DialogInterface.BUTTON_NEGATIVE,
                "No", new DialogInterface.OnClickListener() {
            @Override
            public void onClick(DialogInterface dialog, int which) {
                // TODO Auto-generated method stub
                return;
            }
        });
        // Add "Yes"
        alert.setButton(DialogInterface.BUTTON_POSITIVE,
                "Yes", new DialogInterface.OnClickListener() {
            @Override
            public void onClick(DialogInterface arg0, int arg1) {
                deleteItem(adapter.getItem());
            }
        });
        alert.show();
    }
    private void deleteItem(String strUnit){
        db.deleteRecords("_unit = '" + strUnit + "'", null);
        adapter.removeItem(adapter.getPosition(strUnit));
        adapter.notifyDataSetChanged();
    }
}
```

8.7 单位详细信息窗体

单位浏览窗体通过 Intent 对象传入单位名称，单位详细信息窗体则将显示包含该单位的详细信息。其布局文件名为 activity_info.xml，对应的窗体源代码为 InfoActivity.java，所使用的适配器源代码为 InfoListAdapter.java。

8.7.1 适配器布局文件

适配器布局文件采用相对布局，左侧的 TextView 标记用来显示面向用户的字段名称，右侧的 EditText 标记用来显示对应字段的具体内容。

```xml
<?xml version="1.0" encoding="utf-8"?>
<RelativeLayout xmlns:android="http://schemas.android.com/apk/res/android"
    android:layout_width="match_parent"
    android:layout_height="wrap_content"
    android:orientation="horizontal" >
    <TextView
        android:id="@+id/txtItem"
        android:layout_width="wrap_content"
        android:layout_height="wrap_content"
        android:layout_marginLeft="10dp"
        android:layout_marginRight="5dp"
        android:lines="1"
        android:layout_alignParentLeft="true"
        android:layout_centerVertical="true"
        android:textAppearance="?android:attr/textAppearanceMedium"
        android:text="@string/strService" />
    <EditText
```

```
        android:id="@+id/edItem"
        android:layout_width="match_parent"
        android:layout_height="wrap_content"
        android:layout_marginRight="10dp"
        android:lines="1"
        android:layout_centerVertical="true"
        android:layout_toRightOf="@id/txtItem"
        android:hint=""
        android:textAppearance="?android:attr/textAppearanceMedium"
        android:inputType="text" />
</RelativeLayout>
```

8.7.2 窗体布局文件

窗体布局文件外层采用线性布局，第一行也采用线性布局，Spinner 标记用来存放服务选项，其数据可以通过服务浏览窗体增加和删除；Button 标记中的 btClear 用来清除编辑框中的内容，但保留服务编辑框中的内容，btSave 则用来将当前数据保存到数据库。

```
<LinearLayout xmlns:android="http://schemas.android.com/apk/res/android"
    android:id="@+id/LinearLayout1"
    android:layout_width="match_parent"
    android:layout_height="match_parent"
    android:layout_marginLeft="@dimen/activity_horizontal_margin"
    android:layout_marginRight="@dimen/activity_horizontal_margin"
    android:layout_marginTop="@dimen/activity_horizontal_margin"
    android:layout_marginBottom="@dimen/activity_horizontal_margin"
    android:gravity="center"
    android:orientation="vertical" >
    <LinearLayout
        android:layout_width="match_parent"
        android:layout_height="wrap_content"
        android:gravity="center" >
        <Spinner
            android:id="@+id/spServices"
            android:layout_width="wrap_content"
            android:layout_height="wrap_content"
            android:gravity="center_horizontal"
            android:layout_marginRight="@dimen/interval" />
        <Button
            android:id="@+id/btClear"
            android:layout_width="wrap_content"
            android:layout_height="wrap_content"
            style="?android:attr/buttonBarButtonStyle"
            android:text="@string/strClear" />
        <Button
            android:id="@+id/btSave"
            android:layout_width="wrap_content"
            android:layout_height="wrap_content"
            style="?android:attr/buttonBarButtonStyle"
            android:text="@string/strSave" />
    </LinearLayout>
    <ListView
        android:id="@+id/list"
        android:layout_width="wrap_content"
        android:layout_height="match_parent"
```

```
            android:divider="@null"
            android:dividerHeight="0dp"
            android:drawSelectorOnTop="false" >
    </ListView>
</LinearLayout>
```

8.7.3 适配器源代码

在适配器源代码的构造方法中，字符串数组 strItemName 存放面向用户的字段名称，字符串数组 strItemEdit 用来获取编辑框中的字段内容，采用 2.2 节的技术注册 EditText 组件的 onTextChanged 侦听器即可。

```
package com.walkerma.serviceregister;
import android.app.Activity;
import android.text.Editable;
import android.text.TextWatcher;
import android.view.LayoutInflater;
import android.view.View;
import android.view.ViewGroup;
import android.widget.ArrayAdapter;
import android.widget.EditText;
import android.widget.TextView;
public class InfoListAdapter extends ArrayAdapter<String> {
    private final Activity context;
    private int mLayout;
    private String[] strItemName;
    private String[] strItemEdit;
    public InfoListAdapter(Activity context, int mLayout,
                           String[] strItemName, String[] strItemEdit) {
        super(context, mLayout, strItemName); //strItemName cannot be ignored
        // TODO Auto-generated constructor stub
        this.context = context;
        this.mLayout = mLayout;
        this.strItemName = strItemName;
        this.strItemEdit = strItemEdit;
    }
    @Override
    public View getView(int position,View convertView,ViewGroup parent) {
        View rowItem = convertView;
        final ViewHolder holder;
        if(rowItem == null){
            LayoutInflater inflater = context.getLayoutInflater();
            rowItem = inflater.inflate(mLayout, parent, false);
            holder = new ViewHolder();
            holder.txtItem = (TextView) rowItem.findViewById(R.id.txtItem);
            holder.edItem = (EditText) rowItem.findViewById(R.id.edItem);
            rowItem.setTag(holder);
        }
        else holder = (ViewHolder) rowItem.getTag();
        holder.edItem.setTag(position);
        holder.edItem.addTextChangedListener(new TextWatcher() {
            @Override
            public void onTextChanged(CharSequence s,
                    int start, int before,  int count) {
                int pos = (Integer) holder.edItem.getTag();
```

```
                strItemEdit[pos] = holder.edItem.getText().toString();
            }
            @Override
            public void beforeTextChanged(CharSequence s,
                    int start, int count, int after) {
            }
            @Override
            public void afterTextChanged(Editable s) {
            }
        });
        if(position == 1){
            holder.edItem.setFocusable(false);
        }
        holder.txtItem.setText(strItemName[position]);
        holder.edItem.setText(strItemEdit[position]);
        return rowItem;
    }
    static class ViewHolder {
        TextView txtItem;
        EditText edItem;
    }
}
```

8.7.4 窗体源代码

从单位浏览窗体和单位搜索窗体都能进入单位详细信息窗体,而从单位详细信息窗体返回上一个窗体时,需要原路返回。因而,在 onCreate 方法中,除了基本的组件初始化工作外,还需要通过 Intent 对象获取服务、单位和是否来自单位搜索窗体等信息。本窗体还是通过窗体的 onResume 方法初始化数据库,并向界面填充数据。

Spinner 对象 spServices 的初始化工作在函数 initSpinner 中进行,主要读取服务数据,解析到字符串数组 strArray 中,然后在 spServices 中显示。在 spServices 的 onItemSelected 事件处理方法中,如果选择了新的服务,则叠加到服务字段中,如果选择了已有的服务,则从服务字段中删除该服务(当前服务不能删除)。

在 onClick 方法中,btClear 调用 clearData 函数清除数据;btSave 调用 saveItem 保存数据,这里需要调用数据库的 isNewItem 函数来验证是否为新的单位,如果是新单位,则插入记录,否则更新记录内容。

```
package com.walkerma.serviceregister;
import java.text.SimpleDateFormat;
import java.util.Calendar;
import java.util.Date;
import java.util.Locale;
import com.walkerma.library.DatabaseHelper;
import android.app.ActionBar;
import android.app.Activity;
import android.app.AlertDialog;
import android.content.ContentValues;
import android.content.Intent;
import android.content.SharedPreferences;
import android.content.res.Configuration;
import android.database.Cursor;
```

```java
import android.os.Bundle;
import android.text.TextUtils;
import android.view.KeyEvent;
import android.view.View;
import android.view.View.OnClickListener;
import android.widget.AdapterView;
import android.widget.AdapterView.OnItemSelectedListener;
import android.widget.ArrayAdapter;
import android.widget.Button;
import android.widget.ListView;
import android.widget.Spinner;
import android.widget.Toast;
public class InfoActivity extends Activity implements OnClickListener{
    private String strServiceOriginal;
    private String[] strItemName;    // defined in R.array
    private String[] strItemEdit = new String[6];
    private ListView list;
    private InfoListAdapter adapter;
    private Spinner spServices;
    private Button btSave, btClear;
    private DatabaseHelper db;
    private boolean fromSearch = false;
    @Override
    protected void onCreate(Bundle savedInstanceState) {
        super.onCreate(savedInstanceState);
        setContentView(R.layout.activity_info);
        ActionBar actionBar = getActionBar();
        actionBar.setDisplayShowHomeEnabled(false);
        actionBar.setDisplayShowTitleEnabled(true);
        initWidgets();
        Intent it = getIntent();
        strItemEdit[0] = it.getStringExtra("Unit");
        strServiceOriginal = it.getStringExtra("Service");
        fromSearch = it.getBooleanExtra("fromSearch", false);
        initSpinner();
        list = (ListView)findViewById(R.id.list);
        strItemName = getResources().
            getStringArray(R.array.strItemName);
        adapter = new InfoListAdapter(this,
            R.layout.list_info, strItemName, strItemEdit);
        list.setAdapter(adapter);
    }
    private void initWidgets(){
        spServices = (Spinner)findViewById(R.id.spServices);
        spServices.setOnItemSelectedListener(new OnItemSelectedListener(){
            @Override
            public void onItemSelected(AdapterView<?> parent, View view,
                    int position, long id) {
                String result = parent.getItemAtPosition(position).toString();
                // cannot delete basic service
                if(result.equals(strServiceOriginal)) return;
                if(strItemEdit[1].contains(result))
                    strItemEdit[1] = MainActivity.
                    removeSubItem(strItemEdit[1], "|", result);
                else
                    strItemEdit[1] = strItemEdit[1] + "|" + result;
```

```java
                adapter.notifyDataSetChanged();
            }
            @Override
            public void onNothingSelected(AdapterView<?> parent) {
            }});
        btSave = (Button)findViewById(R.id.btSave);
        btSave.setOnClickListener(this);
        btClear = (Button)findViewById(R.id.btClear);
        btClear.setOnClickListener(this);
    }
    private void initSpinner(){
        String strAppName = getString(R.string.app_name);
        SharedPreferences sp = getSharedPreferences(
                strAppName, MODE_PRIVATE);
        String strDisk = sp.getString("Services", "");
        if(TextUtils.isEmpty(strDisk)) return;
        String[] strArray = strDisk.split("/");
        ArrayAdapter<String> adapter= new ArrayAdapter<String>(
                this, android.R.layout.simple_spinner_item, strArray);
        adapter.setDropDownViewResource(
                android.R.layout.simple_spinner_dropdown_item);
        spServices.setAdapter(adapter);
        spServices.setSelection(
                getSpinnerItemIndex(spServices, strServiceOriginal));
    }
    private int getSpinnerItemIndex(Spinner spinner, String strService){
        int index = 0;
        for (int i=0;i<spinner.getCount();i++){
            if (spinner.getItemAtPosition(i).equals(strService)){
                index = i;
            }
        }
        return index;
    }
    @Override
    public boolean onKeyDown(int keyCode, KeyEvent event) {
        // TODO Auto-generated method stub
        if(keyCode == KeyEvent.KEYCODE_BACK){
            Intent it;
            if(fromSearch)
                it = new Intent(this, SearchActivity.class);
            else
                it = new Intent(this, UnitsActivity.class);
            it.putExtra("Service", strServiceOriginal);
            it.putExtra("Unit", strItemEdit[0]);
            startActivity(it);
            finish();
            return true;
        }
        return super.onKeyDown(keyCode, event);
    }
    private void refreshFormalDatabase(){
        if(TextUtils.isEmpty(strItemEdit[0])) return;
        String strQuery =
                "SELECT * FROM services WHERE _unit = '" +
                        strItemEdit[0] + "' AND _service LIKE '%" +
```

```java
                        strServiceOriginal +"%'";
        db.refreshReadDB(strQuery, null);
    }
    private void refreshInfo(){
        refreshFormalDatabase();
        if(db.getCount() == 0) {
            strItemEdit[1] = strServiceOriginal;
            return;
        }
        Cursor rowRecords = db.getRecords();
        rowRecords.moveToFirst();
        strItemEdit[0] = rowRecords.getString(
                rowRecords.getColumnIndex("_unit"));
        strItemEdit[1] = rowRecords.getString(
                rowRecords.getColumnIndex("_service"));
        strItemEdit[2] = rowRecords.getString(
                rowRecords.getColumnIndex("_no"));
        strItemEdit[3] = rowRecords.getString(
                rowRecords.getColumnIndex("_date"));
        strItemEdit[4] = rowRecords.getString(
                rowRecords.getColumnIndex("_address"));
        strItemEdit[5] = rowRecords.getString(
                rowRecords.getColumnIndex("_remark"));
    }
    @Override
    protected void onResume() {
        // TODO Auto-generated method stub
        super.onResume();
        db = new DatabaseHelper(this, MainActivity.DB_NAME,
                1, MainActivity.TABLE_NAME, MainActivity.FIELDS_NAME);
        refreshInfo();
        adapter.notifyDataSetChanged();
    }
    @Override
    public void onConfigurationChanged(Configuration newConfig) {
        // TODO Auto-generated method stub
        super.onConfigurationChanged(newConfig);
        adapter.notifyDataSetChanged();
    }
    @Override
    protected void onDestroy() {
        // TODO Auto-generated method stub
        db.close();
        super.onDestroy();
    }
    private void clearData(){
        strItemEdit[0] = "";
        strItemEdit[1] = strServiceOriginal;
        strItemEdit[2] = "";
        strItemEdit[3] = "";
        strItemEdit[4] = "";
        strItemEdit[5] = "";
    }
    private String getCurrentDateString(){
        SimpleDateFormat sdf =
                new SimpleDateFormat("MM/dd/yyyy", Locale.US);
```

```java
            Date today = Calendar.getInstance().getTime(); //java.util.
            return sdf.format(today);
        }
        private ContentValues getContentValues(){
            ContentValues cv = new ContentValues();
            cv.put("_unit", strItemEdit[0]);
            cv.put("_service", strItemEdit[1]);
            cv.put("_no", strItemEdit[2]);
            if(strItemEdit[3].isEmpty()){
                strItemEdit[3] = getCurrentDateString();
                adapter.notifyDataSetChanged();
            }
            cv.put("_date", strItemEdit[3]);
            cv.put("_address", strItemEdit[4]);
            cv.put("_remark", strItemEdit[5]);
            return cv;
        }
        public void GeneralAlert(int nResource,
                String strTitle, String strMessage){
            AlertDialog.Builder builder =
                    new AlertDialog.Builder(InfoActivity.this);
            builder.setIcon(nResource);
            builder.setTitle(strTitle);
            builder.setMessage(strMessage);
            builder.setPositiveButton("OK", null);
            AlertDialog dialog = builder.create();
            dialog.show();
        }
        private void saveItem(){
            if(TextUtils.isEmpty(strItemEdit[0])){
                GeneralAlert(R.drawable.info,
                        "Save", "Please input Unit...");
                return;
            }
            ContentValues cv = getContentValues();
            if(db.isNewItem("_unit", strItemEdit[0])){
                cv.put("_num", 0);     // add info number
                db.insertRecords(cv);
                Toast.makeText(this, "Data saved.",
                        Toast.LENGTH_SHORT).show();
            }
            else{
                db.updateRecords(cv,
                        "_unit = '" + strItemEdit[0] + "'", null);
                Toast.makeText(this, "Data updated.",
                        Toast.LENGTH_SHORT).show();
            }
        }
        @Override
        public void onClick(View v) {
            // TODO Auto-generated method stub
            switch(v.getId()){
            case R.id.btClear:
                clearData();
                adapter.notifyDataSetChanged();
                break;
```

```
            case R.id.btSave:
                saveItem();
                break;
        }
    }
}
```

8.8 单位搜索窗体

在单位搜索窗体中,只需在编辑框中输入单位名称,即可对单位名称进行模糊搜索并列出所有相关单位。从这里可以直接跳转到单位详细信息窗体,也可从单位详细信息窗体返回本窗体。本窗体的布局文件名为 activity_search.xml,对应的窗体源代码为 SearchActivity.java。

8.8.1 布局文件

在本布局文件中,EditText 标记用于输入单位信息;两个 Button 标记,前者用来执行搜索功能,后者用来查询详细信息。

```
<LinearLayout xmlns:android="http://schemas.android.com/apk/res/android"
    xmlns:tools="http://schemas.android.com/tools"
    android:layout_width="match_parent"
    android:layout_height="match_parent"
    android:layout_marginLeft="@dimen/activity_horizontal_margin"
    android:layout_marginRight="@dimen/activity_horizontal_margin"
    android:layout_marginTop="@dimen/activity_horizontal_margin"
    android:orientation="vertical"
    tools:context="com.walkerma.serviceregister.SearchActivity" >
    <LinearLayout
        android:layout_width="match_parent"
        android:layout_height="wrap_content"
        android:gravity="center"
        android:orientation="horizontal" >
        <TextView
            android:id="@+id/txtName_Label"
            android:layout_width="wrap_content"
            android:layout_height="wrap_content"
            android:text="@string/strUnit"
            android:textAppearance="?android:attr/textAppearanceMedium" />
        <EditText
            android:id="@+id/edUnit"
            android:layout_width="wrap_content"
            android:layout_height="wrap_content"
            android:gravity="center_horizontal"
            android:hint="@string/strUnitName"
            android:inputType="text"
            android:textAppearance="?android:attr/textAppearanceMedium" >
            <requestFocus />
        </EditText>
        <Button
            android:id="@+id/btSearch"
            style="?android:attr/buttonBarButtonStyle"
            android:layout_width="wrap_content"
            android:layout_height="wrap_content"
```

```xml
                android:onClick="onClick"
                android:text="@string/strSearch" />
        <Button
                android:id="@+id/btQuery"
                style="?android:attr/buttonBarButtonStyle"
                android:layout_width="wrap_content"
                android:layout_height="wrap_content"
                android:onClick="onClick"
                android:text="@string/strInquire" />
    </LinearLayout>
    <ListView
        android:id="@+id/list"
        android:layout_width="match_parent"
        android:layout_height="match_parent"
        android:choiceMode="singleChoice"
        android:drawSelectorOnTop="false" >
    </ListView>
</LinearLayout>
```

8.8.2 源代码

在 onClick 方法中，btSearch 按钮执行 addUnitItems 函数，在其中调用 refreshDatabase 函数以初始化数据库并执行查询功能，然后将相关信息添加到 ListView 对象中。对于从单位详细信息返回的情况，则在 onCreate 中通过 Intent 对象获取当前单位信息，然后再执行 refreshDatabase 函数以添加单位信息。按钮 btQuery 执行 startInfoActivity 函数，通过 Intent 对象传入服务、单位和 fromSearch 参数，从而通过单位详细信息窗体显示相关信息。

```java
package com.walkerma.serviceregister;
import java.util.ArrayList;
import com.walkerma.library.DatabaseHelper;
import com.walkerma.library.GeneralProcess;
import android.app.ActionBar;
import android.app.Activity;
import android.content.Intent;
import android.content.SharedPreferences;
import android.content.SharedPreferences.Editor;
import android.content.res.Configuration;
import android.database.Cursor;
import android.os.Bundle;
import android.text.TextUtils;
import android.view.KeyEvent;
import android.view.View;
import android.widget.AdapterView;
import android.widget.ArrayAdapter;
import android.widget.EditText;
import android.widget.ListView;
import android.widget.AdapterView.OnItemClickListener;
import android.widget.Toast;
public class SearchActivity extends Activity {
    private EditText edUnit;
    private ListView list;
    private ArrayList<String> listItems = new ArrayList<String>();
    private ArrayAdapter<String> adapter;
    private DatabaseHelper db;
```

```java
    private String strService, strUnit, strUnitIntent;
    private int nCurrent = -1;
    @Override
    protected void onCreate(Bundle savedInstanceState) {
        super.onCreate(savedInstanceState);
        setContentView(R.layout.activity_search);
        ActionBar actionBar = getActionBar();
        actionBar.setDisplayShowHomeEnabled(false);
        actionBar.setDisplayShowTitleEnabled(true);
        initWidgets();
        db = new DatabaseHelper(this, MainActivity.DB_NAME, 1,
                MainActivity.TABLE_NAME, MainActivity.FIELDS_NAME);
        strUnitIntent = getIntent().getStringExtra("Unit");
        if(!TextUtils.isEmpty(strUnitIntent))
            refreshListView();    // from InfoActivity
    }
    private void initWidgets(){
        edUnit = (EditText)findViewById(R.id.edUnit);
        list = (ListView)findViewById(R.id.list);
        adapter = new ArrayAdapter<String>(this,
                android.R.layout.simple_list_item_checked, listItems);
        list.setAdapter(adapter);
        list.setOnItemClickListener(new OnItemClickListener(){
            @Override
            public void onItemClick(AdapterView<?> parent, View view,
                    int position, long id) {
                // TODO Auto-generated method stub
                if(nCurrent != position){
                    list.setItemChecked(position, true);
                    nCurrent = position;
                }
                else{
                    list.setItemChecked(position, false);
                    nCurrent = -1;
                }
                adapter.notifyDataSetChanged();    // call getView automatically
            }});
    }
    private void refreshListView(){
        SharedPreferences sp = getPreferences(MODE_PRIVATE);
        edUnit.setText(sp.getString("SearchWord", null));
        addUnitItems();
    }
    public void onClick(View v){
        switch(v.getId()){
        case R.id.btSearch:
            addUnitItems();
            break;
        case R.id.btQuery:
            if(listItems.isEmpty()){
                Toast.makeText(this, "Item(s) empty.",
                        Toast.LENGTH_SHORT).show();
                return;
            }
            if(nCurrent==-1){
                Toast.makeText(this, "Please select an item.",
```

```java
                        Toast.LENGTH_SHORT).show();
                return;
            }
            startInfoActivity();
            finish();
            break;
        }
    }
    private void startInfoActivity(){
        strUnit = listItems.get(nCurrent);
        strService = fromUnitToService(strUnit);
        Intent it = new Intent(this, InfoActivity.class);
        it.putExtra("Service", strService);
        it.putExtra("Unit", strUnit);
        it.putExtra("fromSearch", true);
        startActivity(it);
    }
    private String fromUnitToService(String strUnit){
        Cursor row = db.getRecords();
        row.moveToFirst();
        while(true){
            String data = row.getString(
                    row.getColumnIndex("_unit"));
            if(strUnit.equals(data))
                return row.getString(
                        row.getColumnIndex("_service"));
            if(!row.moveToNext()) break;
        }
        return null;
    }
    private void addUnitItems(){
        refreshDatabase();
        if(db.getCount()==0){
            Toast.makeText(this, "Unit(s) empty...",
                    Toast.LENGTH_SHORT).show();
            return;
        }
        listItems.clear();
        Cursor row = db.getRecords();
        row.moveToFirst();
        for(int i=0; i<db.getCount(); i++){
            String strItem = row.getString(
                    row.getColumnIndex("_unit"));
            listItems.add(strItem);
            if(strItem.equals(strUnitIntent)){
                nCurrent = i;      //from InfoActivity
                list.setItemChecked(i, true);
            }
            row.moveToNext();
        }
        adapter.notifyDataSetChanged();
        GeneralProcess.closeKeyboard(this, edUnit);
    }
    private void refreshDatabase(){
        String strUnit = edUnit.getText().toString();
        if(TextUtils.isEmpty(strUnit)) return;
```

```
        strUnit = DatabaseHelper.getGeneralMatch(strUnit);
        String strInquery = "SELECT _unit, _service FROM " +
                MainActivity.TABLE_NAME +
                " WHERE _unit LIKE '" + strUnit +"'";
        db.refreshReadDB(strInquery, null);
    }
    @Override
    public boolean onKeyDown(int keyCode, KeyEvent event){
        if(keyCode == KeyEvent.KEYCODE_BACK){
            Intent it = new Intent(this, LoginActivity.class);
            it.putExtra("Returned", true);
            startActivity(it);
            finish();
            return true;
        }
        return super.onKeyDown(keyCode, event);
    }
    @Override
    public void onConfigurationChanged(Configuration newConfig) {
        adapter.notifyDataSetChanged();
        super.onConfigurationChanged(newConfig);
    }
    @Override
    protected void onDestroy() {
        db.close();
        saveSearchStatus();
        super.onDestroy();
    }
    private void saveSearchStatus(){
        if(listItems.size()==0) return;
        SharedPreferences sp = getPreferences(MODE_PRIVATE);
        Editor editor = sp.edit();
        editor.putString("SearchWord", edUnit.getText().toString());
        editor.commit();
    }
}
```

8.9 本章小结

本章完善了 3.8 节的 SQLite 数据库，并在 Library 类库中增加了 DatabaseHelper 类，专门用来简单高效地处理数据库相关事宜。ServiceRegister 软件利用此数据库技术和 4.2 节的 Activity 之间传递数据技术，实现了一个综合的个人账户保护软件，既可以通过服务种类浏览信息，也可以直接通过单位关键字搜索信息，服务种类可以通过箭头调整顺序，单位顺序则用拖曳来进行调整。ServiceRegister 软件可以为本科毕业设计提供一个较好的参考，熟练掌握相关技术也可以为 Android 应用软件研发打好基础。

第 9 章 地址定位及辅助服务软件

家有老人,我们需要知道老人所在位置,希望拨打电话老人能够接听,也希望能够跟老人微信视频聊天。但是,由于老人视力不好,经常不小心将情景模式设置为静音并关闭 WiFi,导致无法接听电话和微信视频。因而,本章介绍设计一款应用软件,将情景模式强制设置为正常模式,只要屏幕解锁,就自动检查 WiFi 是否打开,如果处于关闭状态就提醒打开 WiFi。另外,如果收到运营商发送的手机余额不足短信,就转发给指定手机号码,以便及时给老人的手机充值。

主要知识点:百度地图、开机启动、启动服务、来电处理、情景模式检测、屏幕状态检测、短信处理、WiFi 设置状态检测

9.1 主要功能和技术特点

地址定位及辅助服务软件(安装后的 App 名为 AddressServer,下文简称 AddressServer)可以短信形式返回地址定位,能够跟踪情景模式和 WiFi 的设置。AddressServer 的主要特点为:

- 以服务形式常驻内存运行;
- 开机自动启动;
- 通过通讯录来电刷新服务;
- 接到白名单来电自动开启正常情景模式,确保白名单来电响铃;
- 侦听情景模式变化,自动跳转到正常模式;
- 屏幕解锁自动提醒开启 WiFi;
- 将运营商的余额不足提醒自动转发到第一个白名单手机号码;
- 侦听 WiFi 设置,自动提醒开启 WiFi。

9.2 软件操作

AddressServer 运行后的软件界面如图 9-1 所示,可以在编辑框中输入电话号码(即白名单),其中电话号码 "000000" 有特殊的含义。电话号码可以通过箭头上下移动,当 "000000" 处于顶部时(称为非控制模式),可以将手机设置为静音模式而不会有任何提示和动作。同理,关闭 WiFi 时也不会有任何提示。

当 "000000" 不在顶部时(称为控制模式),如果将手机设置为静音或振动模式,将自动立即返回正常模式;如果关闭 WiFi,则立即出现图 9-1 所示的提醒打开 WiFi 的对话框。

如果手机的 WiFi 处于关闭状态，没有及时打开，则锁屏后重新解锁，将继续出现提醒打开 WiFi 的对话框。如果手机处于非正常模式（静音或振动），在控制模式下有白名单电话呼入，将立即将情景模式设置为大音量的正常模式，确保手机能够振铃。

无论是在控制模式还是非控制模式下，如果收到查询地址的短信"bd09ll"，且发送者为白名单，则将手机所在的经纬度信息加密后返回。如果收到移动运营商的客服短信，提示余额不足，将自动抽取数据，并向图 9-1 中的第一个手机白名单（"000000"除外）发送如"余额：30 元"的提醒短信。

关闭手机，重新启动手机，AddressServer 将自动启动，并在状态栏出现如图 9-1 所示的通知图标（顶部左侧），表示软件正在后台执行。

图 9-1　AddressServer 软件界面

9.3　配置文件

AddressServer 软件需要开机自动启动服务、收发短信、获取地理位置、处理来电与情景模式及无线设置等变化，因而需要设置对应的权限与广播接收器。开机启动完成的广播接收器必须在配置文件中注册方有效；屏幕的打开与关闭的广播接收器必须通过代码注册方有效，因而配置文件中没有此广播接收器。关于获取地理位置的 meta-data 标记内容，请参考 3.12 节。

```xml
<?xml version="1.0" encoding="utf-8"?>
<manifest xmlns:android="http://schemas.android.com/apk/res/android"
    package="com.walkerma.addressserver"
    android:versionCode="1"
    android:versionName="1.0" >
    <uses-sdk
        android:minSdkVersion="16"
        android:targetSdkVersion="22" />
    <uses-permission android:name=
        "android.permission.RECEIVE_BOOT_COMPLETED" />
    <uses-permission android:name="android.permission.SEND_SMS" />
    <uses-permission android:name="android.permission.READ_SMS" />
    <uses-permission android:name="android.permission.RECEIVE_SMS" />
    <uses-permission android:name=
        "android.permission.READ_PHONE_STATE" />
    <!-- WRITE_SMS is not valid after KitKat -->
    <uses-permission android:name="android.permission.WRITE_SMS" />
    <uses-permission android:name=
        "android.permission.READ_CONTACTS" />
    <uses-permission android:name=
        "android.permission.WRITE_SETTINGS" />
    <uses-permission android:name=
        "android.permission.WRITE_SECURE_SETTINGS" />
```

```xml
<uses-permission android:name=
    "android.permission.MODIFY_AUDIO_SETTINGS" />
<uses-permission android:name="android.permission.INTERNET" />
<uses-permission android:name=
    "android.permission.ACCESS_COARSE_LOCATION" />
<uses-permission android:name=
    "android.permission.ACCESS_FINE_LOCATION" />
<uses-permission android:name=
    "android.permission.ACCESS_NETWORK_STATE" />
<uses-permission android:name=
    "android.permission.ACCESS_WIFI_STATE" />
<uses-permission android:name=
    "android.permission.CHANGE_WIFI_STATE" />
<uses-permission android:name=
    "android.permission.VIBRATE" />
<uses-permission android:name=
    "com.android.launcher.permission.READ_SETTINGS" />
<uses-permission android:name="android.permission.GET_TASKS" />
<uses-permission android:name=
    "android.permission.WRITE_EXTERNAL_STORAGE" />
<uses-permission android:name=
    "android.permission.WRITE_SETTINGS" />
<uses-permission android:name=
    "android.permission.READ_EXTERNAL_STORAGE" />
<application
    android:allowBackup="false"
    android:icon="@drawable/ic_launcher"
    android:label="@string/app_name"
    android:theme="@style/AppTheme" >
    <meta-data
        android:name="com.baidu.lbsapi.API_KEY"
        android:value="填写你申请的AK " />
    <service
        android:name="com.baidu.location.f"
        android:enabled="true"
        android:process=":remote" >
    </service>
    <service
        android:name=".AutostartService"
        android:enabled="true"
        android:exported="false" >
    </service>
    <receiver
        android:name=".WiFiChangeReceiver"
        android:enabled="true"
        android:exported="true"
        android:label="NetworkConnection" >
        <intent-filter>
            <action android:name=
                "android.net.wifi.WIFI_STATE_CHANGED" />
        </intent-filter>
    </receiver>
    <receiver
        android:name=".RingModeChangeReceiver"
        android:enabled="true"
        android:exported="true" >
```

```xml
            <intent-filter>
                <action android:name=
                    "android.media.RINGER_MODE_CHANGED" />
            </intent-filter>
        </receiver>
        <receiver
            android:name=".BootCompleteReceiver"
            android:enabled="true"
            android:exported="true" >
            <!-- priority>=-1000, <=1000, default=0 -->
            <intent-filter android:priority="100" >
                <action android:name=
                    "android.intent.action.BOOT_COMPLETED" />
            </intent-filter>
        </receiver>
        <receiver
            android:name=".PhoneStateReceiver"
            android:enabled="true"
            android:exported="true" >
            <intent-filter android:priority="100" >
                <action android:name=
                    "android.intent.action.PHONE_STATE" />
            </intent-filter>
        </receiver>
        <receiver
            android:name=".SmsReceiver"
            android:enabled="true"
            android:exported="true" >
            <intent-filter android:priority="100" >
                <action android:name=
                    "android.provider.Telephony.SMS_RECEIVED" />
            </intent-filter>
        </receiver>
        <activity
            android:name=".MainActivity"
            android:label="@string/title_activity_main"
            android:screenOrientation="portrait"
            android:windowSoftInputMode="adjustUnspecified|stateHidden" >
            <intent-filter>
                <action android:name="android.intent.action.MAIN" />
                <category android:name="android.intent.category.LAUNCHER" />
            </intent-filter>
        </activity>
    </application>
</manifest>
```

9.4 广播接收器源代码

AddressServer 软件采用 6 个广播接收器来检测配置变化或进行广播事件处理。

9.4.1 启动完成

广播接收器类 BootCompleteReceiver 在启动完成后执行，主要调用主窗体中的 startBootService 函数以启动服务，确保 AddressServer 软件能够开机自动运行。

```
package com.walkerma.addressserver;
import android.content.BroadcastReceiver;
import android.content.Context;
import android.content.Intent;
public class BootCompleteReceiver extends BroadcastReceiver {
    @Override
    public void onReceive(Context context, Intent intent) {
        MainActivity.startBootService(context);
    }
}
```

9.4.2 来电处理

服务长时间处于非活动状态会被系统自动清除，因而，在来电处理广播接收器获得来电号码后，如果该号码是通讯录中的号码，则重新启动服务，这样达到刷新服务的目的。如果当前电话号码在白名单变量 strPhoneSet 中，并且手机处于控制模式，则检查情景模式：如果为非正常模式，则调用 Sound 类的静态函数 setSoundMode 将情景模式自动设置为正常模式，随后调用静态函数 setMaxInCallVolume 将通话音量设置为最大。对于情景模式的改变，有的手机在本次通话立即生效，有的手机则需要在下一次通话时才生效。

```
package com.walkerma.addressserver;
import com.walkerma.library.Sound;
import com.walkerma.library.PhoneBook;
import android.content.BroadcastReceiver;
import android.content.Context;
import android.content.Intent;
import android.media.AudioManager;
import android.telephony.TelephonyManager;
import android.text.TextUtils;
public class PhoneStateReceiver extends BroadcastReceiver {
    Context context;
    String strPhoneSet;
    private String inPhone="";
    public void onReceive(Context context, Intent intent) {
        this.context = context;
        String state = intent.getStringExtra(TelephonyManager.EXTRA_STATE);
        if(TextUtils.isEmpty(state)) return;
        if (!state.equals(TelephonyManager.EXTRA_STATE_RINGING)) return;
        strPhoneSet = MainActivity.readStorage();
        inPhone = intent.getStringExtra(
                TelephonyManager.EXTRA_INCOMING_NUMBER);
        if(TextUtils.isEmpty(inPhone)) return;
        if(inPhone.startsWith("+86")) inPhone = inPhone.substring(3);
        // keep service alive
        if(PhoneBook.contactExists(context, inPhone))
            MainActivity.startBootService(context);
        if((!strPhoneSet.contains(inPhone)) ||
                strPhoneSet.startsWith("000000")) return;
        Sound.setSoundMode(context, AudioManager.RINGER_MODE_NORMAL);
        Sound.setMaxInCallVolume(context);
    }
}
```

9.4.3 情景模式改变

如果图 9-1 中,"000000" 位于 ListView 的顶部,即处于非控制模式,则直接返回。否则,如果情景模式为振动或静音模式,则将音量设置为最大值减 1,确保手机持有者能够听到手机来电振铃。

```java
package com.walkerma.addressserver;
import android.content.BroadcastReceiver;
import android.content.Context;
import android.content.Intent;
import android.media.AudioManager;
public class RingModeChangeReceiver extends BroadcastReceiver {
    @Override
    public void onReceive(Context context, Intent intent) {
        if(MainActivity.readStorage().startsWith("000000")) return;
        AudioManager am = (AudioManager)context.
                getSystemService(Context.AUDIO_SERVICE);
        final int ringerMode = am.getRingerMode();
        switch (ringerMode) {
        case AudioManager.RINGER_MODE_VIBRATE:
        case AudioManager.RINGER_MODE_SILENT:
            am.setMode(AudioManager.MODE_IN_CALL);
            am.setStreamVolume(AudioManager.STREAM_RING,
                    am.getStreamMaxVolume(AudioManager.STREAM_RING)-1, 0);
            break;
        case AudioManager.RINGER_MODE_NORMAL:
        default:
            break;
        }
    }
}
```

9.4.4 屏幕状态变化

与情景模式的设置一样,屏幕变化的检测也只有当手机处于控制模式时才执行相关动作。函数 isScreenUnlocked 用来检测屏幕是否处于锁屏状态,只有处于解锁状态,才有必要提醒打开无线设置。函数 setWiFiOn 用于打开无线设置,因为这是系统认为的危险动作,虽然调用 setWifiEnabled 方法传入 true,仍然需要用户授权方可打开无线设置。本广播接收器正是如此检测屏幕亮且处于解锁状态,调用 setWiFiOn 函数打开无线设置,从而出现图 9-1 中的授权对话框。

```java
package com.walkerma.addressserver;
import android.app.KeyguardManager;
import android.content.BroadcastReceiver;
import android.content.Context;
import android.content.Intent;
import android.net.wifi.WifiManager;
public class ScreenBroadcastReceiver extends BroadcastReceiver {
    @Override
    public void onReceive(Context context, Intent intent) {
        if(MainActivity.readStorage().startsWith("000000")) return;
        String action = intent.getAction();
        if (Intent.ACTION_SCREEN_ON.equals(action)) {
```

```
            if(isScreenUnlocked(context))
                setWiFiOn(context);
        } else if (Intent.ACTION_SCREEN_OFF.equals(action)) {
            // Screen off
        } else if (Intent.ACTION_USER_PRESENT.equals(action)) {
            setWiFiOn(context);
        }
    }
    private boolean isScreenUnlocked(Context context){
        KeyguardManager mKeyguardManager = (KeyguardManager)
                context.getSystemService(Context.KEYGUARD_SERVICE);
        return !mKeyguardManager.inKeyguardRestrictedInputMode();
    }
    private void setWiFiOn(Context context){
        WifiManager wifiManager = (WifiManager)
                context.getSystemService(Context.WIFI_SERVICE);
        int status =wifiManager.getWifiState();
        if(status == WifiManager.WIFI_STATE_DISABLED)
            wifiManager.setWifiEnabled(true);
    }
}
```

9.4.5 短信接收

短信接收仅处理余额提醒与位置查询相关的功能，来电号码只能是运营商的客服号码（以"100"开头）、白名单号码或者两个家庭后门号码（读者自行根据需要替换 BACK_HOME1 与 BACK_HOME2），这体现在 onReceive 方法的条件语句中。

满足来电条件调用 processMsg 函数处理短信时，如果短信内容以"bd09ll"开头，表明这是查询地理位置的短信，则调用 adjustRingMode 函数关闭短信提醒声音，等待短信完全到达后，通过消息控制再打开声音提醒。接着调用 startBaiduMap 函数获取地理位置，在接口 MyLocationListener 中，将得到的经纬度信息加密后以短信的形式发送到查询手机中。短信的加密通过调用 ByteProcess 类中的静态函数 enStringToBase64 完成，即只对短信文本进行简单的 BASE64 编码。

如果短信来自运营商的客服号码，且短信内容中含有"余额"，则通过 tipRemainder 函数进行余额提醒，即调用 Library 类库中 StringProcess 类的静态函数 getDecimalString，利用正则表达式抽取"余额"与"元"之间的数据，假如该数大于或等于 MONEY_MIN（这里定义为 30 元）就直接返回，否则，通过循环选取图 9-1 中的第一个手机号码，向其发送类似短信"余额：18.24 元"。

```
package com.walkerma.addressserver;
import com.baidu.location.BDLocation;
import com.baidu.location.BDLocationListener;
import com.baidu.location.LocationClient;
import com.baidu.location.LocationClientOption;
import com.baidu.location.LocationClientOption.LocationMode;
import com.walkerma.library.ByteProcess;
import com.walkerma.library.SmsProcess;
import com.walkerma.library.NetworkProcess;
import com.walkerma.library.PhoneBook;
import com.walkerma.library.StringProcess;
```

```java
import com.walkerma.library.Sound;
import android.content.BroadcastReceiver;
import android.content.Context;
import android.content.Intent;
import android.media.AudioManager;
import android.os.Handler;
import android.os.Message;
import android.telephony.SmsMessage;
import android.text.TextUtils;
public class SmsReceiver extends BroadcastReceiver {
    private Context context;
    private String inPhone="";
    private Handler mHandler;
    private final int MSG_Delay = 10000;
    private final long INTERVAL_MSG = 9000;
    private boolean bSetSilent = false;
    private final float MONEY_MIN = 30;
    private int nMessages = 0;
    private final String BACK_HOME1 = "13901234567";
    private final String BACK_HOME2 = "15301234567";
    public MyLocationListener myListener=null;
    public LocationClient mLocClient=null;
    class IncomingHandlerCallback implements Handler.Callback{
        @Override
        public boolean handleMessage(Message msg) {
            switch(msg.what){
            case MSG_Delay:
                Sound.setSoundMode(context,
                        AudioManager.RINGER_MODE_NORMAL);
                bSetSilent = false;
                break;
            }
            return true;
        }
    }
    @Override
    public void onReceive(Context context, Intent intent) {
        this.context = context;
        SmsMessage[] inMsgs = SmsProcess.getIncomingMessage(intent);
        nMessages = inMsgs.length;
        inPhone = inMsgs[0].getOriginatingAddress().trim();
        if(TextUtils.isEmpty(inPhone)) return;
        if(inPhone.startsWith("+86")) inPhone = inPhone.substring(3);
        if(inPhone.equals(BACK_HOME1) ||
                inPhone.equals(BACK_HOME2) ||
                MainActivity.readStorage().contains(inPhone) ||
                inPhone.startsWith("100"))
            processMsg(inMsgs[0]);
    }
    private void processMsg(SmsMessage msg){
        String body = msg.getMessageBody();
        if(body.startsWith("bd0911")){
            adjustRingMode();
            startBaiduMap();
        }
        else if((inPhone.startsWith("10086") ||
```

```java
                inPhone.startsWith("10010") ||
                inPhone.startsWith("10001")) &&
                body.contains("余额")){
            adjustRingMode();
            tipRemainder(body);
        }
    }
    private void tipRemainder(String body){
        String strValue = StringProcess.getDecimalString(body, "余额","元");
        float fVal = Float.parseFloat(strValue);
        if(fVal >= MONEY_MIN) return;
        String strPhoneSet = MainActivity.readStorage();
        String strArray[] = strPhoneSet.split("/");
        for(int i=0; i<strArray.length; i++){
            if(PhoneBook.isMobileNumber(strArray[i])){
                String content = "余额: " + Float.toString(fVal) + "元";
                SmsProcess.sendMessage(content, strArray[i]);
                break;
            }
        }
    }
    private void startBaiduMap(){
        if(!NetworkProcess.isNetworkAvailable(context)){
            SmsProcess.sendMessage("Network disconnected ...", inPhone);
            return;
        }
        mLocClient = new LocationClient(context);
        myListener = new MyLocationListenner();
        mLocClient.registerLocationListener(myListener);
        LocationClientOption option = new LocationClientOption();
        option.setLocationMode(LocationMode.Hight_Accuracy);
        option.setOpenGps(true);    // open GPS
        option.setCoorType("bd09ll");
        option.setScanSpan(1000);
        option.setIsNeedAddress(true);
        option.setNeedDeviceDirect(true);
        mLocClient.setLocOption(option);
        mLocClient.start();
    }
    private void adjustRingMode(){
        int nMode = Sound.getSoundMode(context);
        if(nMode != AudioManager.RINGER_MODE_NORMAL) return;
        Sound.setSoundMode(context, AudioManager.RINGER_MODE_SILENT);
        if(bSetSilent) return;
        if(mHandler!=null)
            mHandler.removeCallbacksAndMessages(null);
        mHandler = new Handler(new IncomingHandlerCallback());
        mHandler.sendEmptyMessageDelayed(
                MSG_Delay, INTERVAL_MSG * nMessages);
        bSetSilent = true;
    }
    public class MyLocationListenner implements BDLocationListener {
        @Override
        public void onReceiveLocation(BDLocation location) {
            // map view销毁后不再处理新接收的位置
            if (location == null) return;
```

```
            //定位数据，accuracy单位为米
            String strHead = "lld/";
            String strAddress = Double.toString(location.getLongitude()) + "/";
            strAddress += Double.toString(location.getLatitude());
            SmsProcess.sendMessage(strHead +
                    ByteProcess.enStringToBase64(strAddress), inPhone);
            mLocClient.unRegisterLocationListener(myListener);
            mLocClient.stop();
            myListener = null;
        }
        public void onReceivePoi(BDLocation poiLocation) {
            return;
        }
    }
}
```

9.4.6　WiFi 设置变化

对于非控制模式，本广播接收器不执行任何动作。在控制模式下，则检测到 WiFi 处于状态"WIFI_STATE_DISABLED"时，调用 setWifiEnabled 方法传入 true，提示用户授权打开无线设置。

```
// ACCESS_NETWORK_STATE + ACCESS_WIFI_STATE
package com.walkerma.addressserver;
import android.content.BroadcastReceiver;
import android.content.Context;
import android.content.Intent;
import android.net.wifi.WifiManager;
import android.widget.Toast;
public class WiFiChangeReceiver extends BroadcastReceiver {
    @Override
    public void onReceive(Context context, Intent intent) {
        if(MainActivity.readStorage().startsWith("000000")) return;
        WifiManager wifiManager = (WifiManager)
                context.getSystemService(Context.WIFI_SERVICE);
        int status =wifiManager.getWifiState();
        switch(status){
        case WifiManager.WIFI_STATE_DISABLED:
            Toast.makeText(context, "WIFI_STATE_DISABLED",
                    Toast.LENGTH_SHORT).show();
            wifiManager.setWifiEnabled(true);
            break;
        case WifiManager.WIFI_STATE_ENABLED:
            Toast.makeText(context, "WIFI_STATE_ENABLED",
                    Toast.LENGTH_SHORT).show();
            break;
        case WifiManager.WIFI_STATE_ENABLING:
        case WifiManager.WIFI_STATE_DISABLING:
        case WifiManager.WIFI_STATE_UNKNOWN:
        default:
            break;
        }
    }
}
```

9.5 服务源代码

服务的基础知识可以参考 4.7~4.8 节，这里主要介绍特殊的关键代码。由于屏幕变化的广播接收器只能通过代码进行注册，否则无效，因而，在服务的 onCreate 方法中通过调用 registerScreenChange 函数完成其注册。

函数 buildForegroundNotification 构建一个通知栏图标 Notification 的对象 notify。PendingIntent 对象 pi 指向主窗体，通过调用 notify 的 setContentIntent 方法并传入该 pi 对象，即可通过点击通知图标打开如图 9-1 所示的主窗体。方法 setOngoing 中传入 true，表示有正在执行的后台任务；方法 setPriority 中传入参数 Notification.PRIORITY_MAX，使得该图标被置于最前列。

在服务的 onCreate 方法中调用 startForeground 方法将服务放到前台执行，使其不会被 Android 内存管理系统销毁，同时，在 9.4.2 节的来电处理代码中又不断刷新服务，就保证了服务永远不会被销毁（软件出现错误崩溃除外）。

```java
package com.walkerma.addressserver;
import java.util.Calendar;
import java.util.Date;
import android.app.Notification;
import android.app.PendingIntent;
import android.app.Service;
import android.content.Intent;
import android.content.IntentFilter;
import android.os.IBinder;
public class AutostartService extends Service {
    private ScreenBroadcastReceiver mScreenStateReceiver;
    private static int FOREGROUND_ID = 10001;
    @Override
    public void onCreate() {
        super.onCreate();
        registerScreenChange();
        Calendar cl = Calendar.getInstance();
        Date dt = cl.getTime();
        String strDate = dt.toString();
        startForeground(FOREGROUND_ID,
                buildForegroundNotification(strDate));
    }
    private void registerScreenChange(){
        // should be registered in Java code
        mScreenStateReceiver = new ScreenBroadcastReceiver();
        IntentFilter screenStateFilter = new IntentFilter();
        screenStateFilter.addAction(Intent.ACTION_SCREEN_ON);
        screenStateFilter.addAction(Intent.ACTION_SCREEN_OFF);
        screenStateFilter.addAction(Intent.ACTION_USER_PRESENT);
        registerReceiver(mScreenStateReceiver, screenStateFilter);
    }
    private Notification buildForegroundNotification(String strContent) {
        Intent intent = new Intent(this, MainActivity.class);
        intent.setFlags(Intent.FLAG_ACTIVITY_NEW_TASK);
        PendingIntent pi = PendingIntent.getActivity(this, 0, intent,
                PendingIntent.FLAG_UPDATE_CURRENT);
        Notification notify = new Notification.Builder(this)
```

```
            .setOngoing(true)
            .setPriority(Notification.PRIORITY_MAX)
            .setContentIntent(pi)
            .setContentTitle("Address Server")
            .setContentText(strContent)
            .setSmallIcon(R.drawable.ic_launcher)
            .build();
        return notify;       // API level 16
    }
    @Override
    public int onStartCommand(Intent intent, int flags, int startId) {
        return Service.START_STICKY;
    }
    @Override
    public void onDestroy() {
        unregisterReceiver(mScreenStateReceiver);
        super.onDestroy();
    }
    @Override
    public IBinder onBind(Intent intent) {
        // TODO Auto-generated method stub
        return null;
    }
}
```

9.6 适配器源代码

第 8 章已经介绍了若干相似的适配器布局文件，这里不再赘述。为了在主窗体中增加、删除或移动 ListView 选项后能够立即生效，方便广播接收器判断手机是否处于控制模式，因而在适配器源代码中，删除选项函数 removeItem、增加选项函数 addItem、下移选项函数 downMoveItem 和上移选项函数 upMoveItem 都调用了 writePhoneNumber 函数，继而在其中调用 getPhoneString 函数，将 ListView 中的各个选项电话号码合并成以"/"分隔的字符串，最后写入手机。

```
package com.walkerma.addressserver;
import java.util.ArrayList;
import android.app.Activity;
import android.content.Context;
import android.content.SharedPreferences;
import android.content.SharedPreferences.Editor;
import android.os.Vibrator;
import android.view.LayoutInflater;
import android.view.View;
import android.view.ViewGroup;
import android.widget.ArrayAdapter;
import android.widget.ImageView;
import android.widget.TextView;
public class MainListAdapter extends ArrayAdapter<String>{
    private int nCurrentPosition = -1;
    //"android.permission.VIBRATE"
    private boolean bCanVibrate = false;
    private final int nVibrateTime = 30;
```

```java
        private final Activity context;
        private ArrayList<String> listItems;
        private int mLayout;
        public MainListAdapter(Activity context,
                int mLayout, ArrayList<String> listItems) {
            super(context, mLayout, listItems);
            // TODO Auto-generated constructor stub
            this.context = context;
            this.mLayout = mLayout;     // row xml
            this.listItems = listItems;
        }
        public View getView(int position,
                View convertView, ViewGroup parent) {
            View rowItem = convertView;
            final ViewHolder holder;
            if(rowItem == null){
                LayoutInflater inflater = context.getLayoutInflater();
                rowItem = inflater.inflate(mLayout, parent, false);
                holder = new ViewHolder();
                holder.txtItem = (TextView)
                        rowItem.findViewById(R.id.txtItem);
                holder.imgCheckMark = (ImageView)
                        rowItem.findViewById(R.id.check_mark);
                holder.imgCircledDown = (ImageView)
                        rowItem.findViewById(R.id.circled_down);
                holder.imgCircledUp = (ImageView)
                        rowItem.findViewById(R.id.circled_up);
                rowItem.setTag(holder);
            }
            else holder = (ViewHolder) rowItem.getTag();
            holder.imgCircledDown.setOnClickListener(new View.OnClickListener() {
                @Override
                public void onClick(View view) {
                    downMoveItem(nCurrentPosition); }});
            holder.imgCircledUp.setOnClickListener(new View.OnClickListener() {
                @Override
                public void onClick(View view) {
                    upMoveItem(nCurrentPosition); }});
            String strData = listItems.get(position);
            holder.txtItem.setText(strData);
            if(nCurrentPosition>=0 && nCurrentPosition <listItems.size()){
                if(strData.equals(listItems.get(nCurrentPosition))){
                    holder.imgCheckMark.setVisibility(View.VISIBLE);
                    if(nCurrentPosition < listItems.size() - 1)
                        holder.imgCircledDown.setVisibility(View.VISIBLE);
                    if(nCurrentPosition > 0)
                        holder.imgCircledUp.setVisibility(View.VISIBLE);
                }
                else{
                    holder.imgCheckMark.setVisibility(View.INVISIBLE);
                    holder.imgCircledDown.setVisibility(View.INVISIBLE);
                    holder.imgCircledUp.setVisibility(View.INVISIBLE);
                }
            }
            if(nCurrentPosition == -1){
                holder.imgCheckMark.setVisibility(View.INVISIBLE);
```

```java
            holder.imgCircledDown.setVisibility(View.INVISIBLE);
            holder.imgCircledUp.setVisibility(View.INVISIBLE);
        }
        return rowItem;
    }
    public void setCurrentPosition(int nPos){
        nCurrentPosition = nPos;
    }
    public int getItemIndex(String strItem){
        int i, nLen;
        nLen = listItems.size();
        for(i=0; i<nLen; i++){
            if(strItem.equals(listItems.get(i))) break;
        }
        if (i != nLen)
            return i;
        else
            return -1;
    }
    public void setVibrateStatus(boolean bCanVibrate){
        this.bCanVibrate = bCanVibrate;
    }
    public void vibrateCell(){
        //long pattern[] = {50,100,100,250,150,350};
        Vibrator v = (Vibrator)context.
                getSystemService(Context.VIBRATOR_SERVICE);
        //v.vibrate(pattern,3);
        v.vibrate(nVibrateTime);
    }
    public int getCurrentPosition(){
        return nCurrentPosition;
    }
    public void removeItem(int position){
        if(position < 0 || position >= listItems.size()) return;
        listItems.remove(position);
        nCurrentPosition = -1;
        notifyDataSetChanged();
        writePhoneNumber();
    }
    public void addItem(String strItem){
        listItems.add(strItem);
        notifyDataSetChanged();
        writePhoneNumber();
    }
    private void upMoveItem(int which){
        if(which <= 0 || which >= listItems.size()) return;
        String strTmp = listItems.get(which);
        listItems.remove(which);
        listItems.add(which - 1, strTmp);
        nCurrentPosition = which - 1;    // adjust current position
        notifyDataSetChanged();
        writePhoneNumber();
        if(bCanVibrate) vibrateCell();
    }
    private void downMoveItem(int which){
        if(which < 0 || which >= listItems.size() -1 ) return;
```

```
            String strTmp = listItems.get(which);
            listItems.remove(which);
            listItems.add(which+1, strTmp);
            nCurrentPosition = which + 1;    // adjust current position
            notifyDataSetChanged();
            writePhoneNumber();
            if(bCanVibrate) vibrateCell();
        }
        private void writePhoneNumber(){
            String strAppName = context.getString(R.string.app_name);
            SharedPreferences sp = context.getSharedPreferences(
                    strAppName, 0);
            Editor editor = sp.edit();
            String strInput = getPhoneString();
            editor.putString(strAppName, strInput);
            editor.commit();
        }
        private String getPhoneString(){
            String strTmp = "";
            for(int i=0; i<listItems.size(); i++){
                strTmp += listItems.get(i);
                if(i!=listItems.size()-1) strTmp += "/";
            }
            return strTmp;
        }
        static class ViewHolder {
            TextView txtItem;
            ImageView imgCheckMark;
            ImageView imgCircledDown;
            ImageView imgCircledUp;
        }
    }
```

9.7 窗体源代码

本窗体主要用于添加、删除或移动电话号码，同时提供了静态函数 readStorage，将适配器源代码中保存的电话号码字符串读取出来，该静态函数可以在各个广播接收器中被调用。另外，静态函数 startBootService 用于启动服务，在窗体源代码的 onCreate 方法中调用了该函数，在 9.4.2 节的来电处理中也调用了该静态函数来刷新服务。

```
package com.walkerma.addressserver;
import java.util.ArrayList;
import com.walkerma.library.GeneralProcess;
import android.app.Activity;
import android.content.Context;
import android.content.Intent;
import android.content.SharedPreferences;
import android.os.Bundle;
import android.os.Handler;
import android.os.Message;
import android.text.TextUtils;
import android.view.KeyEvent;
import android.view.Menu;
```

```java
import android.view.MenuItem;
import android.view.View;
import android.view.View.OnClickListener;
import android.widget.AdapterView;
import android.widget.Button;
import android.widget.EditText;
import android.widget.ListView;
import android.widget.Toast;
import android.widget.AdapterView.OnItemClickListener;
public class MainActivity extends Activity implements OnClickListener{
    private Button btAdd, btRemove;
    private ListView list;
    private EditText edPhone;
    private static Context context;
    private ArrayList<String> listItems = new ArrayList<String>();
    private MainListAdapter adapter;
    private boolean bCanExit = false;
    private Handler mHandler;
    private final int MESSAGE_Delay = 10000;

    class IncomingHandlerCallback implements Handler.Callback{
        @Override
        public boolean handleMessage(Message msg) {
            bCanExit = false;
            return true;
        }
    }
    @Override
    protected void onCreate(Bundle savedInstanceState) {
        super.onCreate(savedInstanceState);
        setContentView(R.layout.activity_main);
        context = this;
        initWidgets();
        initListView();
        restoreData();
        mHandler = new Handler(new IncomingHandlerCallback());
        startBootService(context);
    }
    public static void startBootService(Context context){
        Intent service = new Intent(context, AutostartService.class);
        context.startService(service);
    }
    private void initWidgets(){
        edPhone = (EditText)findViewById(R.id.edPhone);
        btAdd = (Button)findViewById(R.id.btAdd);
        btAdd.setOnClickListener(this);
        btRemove = (Button)findViewById(R.id.btRemove);
        btRemove.setOnClickListener(this);
    }
    private void initListView(){
        list = (ListView)findViewById(R.id.list);
        adapter = new MainListAdapter(this,
                R.layout.main_list, listItems);
        adapter.setVibrateStatus(true);
        list.setAdapter(adapter);
        list.setOnItemClickListener(new OnItemClickListener(){
```

```java
            @Override
            public void onItemClick(AdapterView<?> parent,
                    View view, int position, long id) {
                if(adapter.getCurrentPosition() != position){
                    adapter.setCurrentPosition(position);
                }
                else{
                    adapter.setCurrentPosition(-1);
                }
                // call getView automatically
                adapter.notifyDataSetChanged();
            }});
}
private void restoreData(){
    String strDisk = readStorage();
    if(TextUtils.isEmpty(strDisk)) return;
    String[] strArray = strDisk.split("/");
    listItems.clear();
    for(int i=0; i<strArray.length; i++){
        listItems.add(strArray[i]);
    }
    adapter.notifyDataSetChanged();
}
public static String readStorage(){
    String strAppName = context.getString(R.string.app_name);
    SharedPreferences sp = context.
            getSharedPreferences(strAppName, MODE_PRIVATE);
    return sp.getString(strAppName, "").trim();
}
@Override
public boolean onKeyDown(int keyCode, KeyEvent event) {
    switch(keyCode){
    case KeyEvent.KEYCODE_BACK:
        if(bCanExit){
            mHandler.removeCallbacksAndMessages(null);
            break;
        }
        else{
            bCanExit = true;
            Toast.makeText(getApplicationContext(),
                    "Press again to quit.", Toast.LENGTH_SHORT).show();
        }
        mHandler.sendEmptyMessageDelayed(MESSAGE_Delay, 1000);
        return true;
    }
    return super.onKeyDown(keyCode, event);
}
@Override
public boolean onCreateOptionsMenu(Menu menu) {
    getMenuInflater().inflate(R.menu.address_server, menu);
    return true;
}
@Override
public boolean onOptionsItemSelected(MenuItem item) {
    if(item.getItemId()==R.id.menu_about)
        GeneralProcess.fireAboutDialog(this,
```

```
                getString(R.string.app_name),
                getString(R.string.strCopyrights));
        return super.onOptionsItemSelected(item);
    }
    @Override
    public void onClick(View v) {
        // TODO Auto-generated method stub
        String strPhone = edPhone.getText().toString();
        switch(v.getId()){
        case R.id.btAdd:
            if(TextUtils.isEmpty(strPhone)){
                Toast.makeText(this, "Data empty.",
                        Toast.LENGTH_SHORT).show();
                return;
            }
            if(adapter.getItemIndex(strPhone)!=-1) {
                Toast.makeText(this, strPhone +" redundant.",
                        Toast.LENGTH_SHORT).show();
                return; //old data, return
            }
            adapter.addItem(strPhone);
            Toast.makeText(this, strPhone +" added.",
                    Toast.LENGTH_SHORT).show();
            edPhone.setText("");
            break;
        case R.id.btRemove:
            int nPos = adapter.getCurrentPosition();
            if(nPos!=-1)
                adapter.removeItem(nPos);
            else if(listItems.size() > 0)
                Toast.makeText(this, "Please select an item...",
                        Toast.LENGTH_SHORT).show();
            else
                Toast.makeText(this, "No any item...",
                        Toast.LENGTH_SHORT).show();
            break;
        }
    }
}
```

9.8 本章小结

本章完成的AddressServer软件集成了多个广播接收器，并实现了一个原理上永远不会被销毁的服务，因而可以可靠地提供相应的辅助提醒服务，通过短信提供余额和位置信息。在软件安装过程中必须赋予给定的权限，否则不能正常工作，特别需要允许开机启动和自动启动。地图服务的AK码，以及获取位置的特权后门电话号码，读者可以根据需要自行调整。

第 10 章 地址查询与地图打点软件

地址查询与地图打点软件与地址定位及辅助服务软件（AddressServer）相配套，它通过主动向目标手机发送查询地址的短信，接收目标手机返回的经纬度信息，然后调用百度地图显示并可实现导航。

主要知识点：短信收发、调用百度地图的方法

10.1 主要功能和技术特点

地址查询与地图打点软件（安装后的 App 名为 QueryLocation，下文简称 QueryLocation）向目标手机发送查询地址指令"bd09ll"，目标手机返回以"lld"开头的加密经纬度信息。QueryLocation 解密地址信息，并调用百度地图在其中完成打点标注和导航。QueryLocation 的主要特点为：

- 调用百度地图，本身不需要地图开发包；
- 通过短信查询地址，简单方便；
- 通过地址信息直接进行地图打点和显示导航信息；
- 目标（打点）位置与本地位置一目了然。

10.2 软件操作

QueryLocation 运行后的软件界面如图 10-1 所示，可以在编辑框中输入通讯录名单，点击【Add】即可将名单加入列表。选中列表中某项，点击【Remove】则可将该选项从列表中删除。点击【Export】则可以将名单列表存入手机外部目录 \data\QueryLocation 下的 White_List.txt 文件中；点击【Import】则可将该文本文件中的名单导入列表。

在图 10-1 中选中某列表项，点击【Location】，则向该选项的手机发送短信"bd09ll"，如果目标手机中安装了 AddressServer，则回复以"lld"开头的加密经纬度信息。这里将 QueryLocation 和 AddressServer

图 10-1 QueryLocation 软件界面

都安装在一部手机中，短信对话如图 10-2 所示，右侧第一则短信为 QueryLocation 发出，左侧第一则短信为 AddressServer 接收；右侧第二则短信为 AddressServer 发送，左侧第二则短信为 QueryLocation 接收。

QueryLocation 收到经纬度信息后，进行解密得到具体的经纬度信息，然后调用百度地图，如图 10-3 所示，打点处为经纬度信息的标注，下面为当前位置，两者相距 13 米。

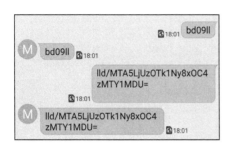

图 10-2　查询地理位置的短信对话

图 10-3　百度地图打点界面

10.3　配置文件

QueryLocation 无须百度地图开发包，只需要收发短信的权限和接收短信的广播接收器，以及读写外部文本的相关权限和读取通讯录的权限，以便通过名单得到手机号码。

```xml
<?xml version="1.0" encoding="utf-8"?>
<!DOCTYPE xml>
<manifest xmlns:android="http://schemas.android.com/apk/res/android"
    package="com.walkerma.querylocation"
    android:versionCode="1"
    android:versionName="1.0" >
    <uses-sdk
        android:minSdkVersion="17"
        android:targetSdkVersion="21" />
    <uses-permission android:name="android.permission.SEND_SMS" />
    <uses-permission android:name="android.permission.READ_SMS" />
    <uses-permission android:name="android.permission.RECEIVE_SMS" />
    <uses-permission android:name="android.permission.READ_CONTACTS" />
    <uses-permission android:name=
        "android.permission.MOUNT_UNMOUNT_FILESYSTEMS" />
    <uses-permission android:name=
        "android.permission.WRITE_EXTERNAL_STORAGE" />
    <uses-permission android:name=
        "android.permission.READ_EXTERNAL_STORAGE" />
    <application
        android:allowBackup="true"
        android:icon="@drawable/ic_launcher"
        android:label="@string/app_name"
        android:theme="@style/AppTheme" >
        <receiver
            android:name=".SmsReceiver"
            android:enabled="true"
            android:exported="true" >
            <intent-filter android:priority="1000" >
                <action android:name=
                    "android.provider.Telephony.SMS_RECEIVED" />
```

```xml
            </intent-filter>
        </receiver>
        <activity
            android:name=".MainActivity"
            android:label="@string/app_name"
            android:screenOrientation="portrait"
            android:windowSoftInputMode="adjustUnspecified|stateHidden" >
            <intent-filter>
                <action android:name="android.intent.action.MAIN" />
                <category android:name="android.intent.category.LAUNCHER" />
            </intent-filter>
        </activity>
    </application>
</manifest>
```

10.4 短信接收与处理源代码

短信广播接收器收到短信后，需要调用 ByteProcess 类中的静态函数 base64ToEnString 对文本进行解码，得到电话号码 phone 和短信内容 body。如果不是匿名电话，且短信以"lld"开头，则调用 prepareForMap 函数，按照表 10-1 填写相关内容，从而通过 Intent 对象启动百度地图。

表 10-1　百度地图打点参数说明

参数	说明	举例
location	经纬度	at,lng（先纬度，后经度）
title	打点标题	Location of M_S
content	打点内容	From QueryLocation
traffic	是否开启路况，on 表示开启，off 表示关闭	on/off

```java
package com.walkerma.querylocation;
import com.walkerma.library.ByteProcess;
import com.walkerma.library.PhoneBook;
import com.walkerma.library.SmsProcess;
import com.walkerma.library.StringProcess;
import android.content.BroadcastReceiver;
import android.content.Context;
import android.content.Intent;
import android.net.Uri;
import android.telephony.SmsMessage;
import android.text.TextUtils;
public class SmsReceiver extends BroadcastReceiver {
    private Context context;
    private String phone="";
    public void onReceive(Context context, Intent intent) {
        this.context = context;
        SmsMessage[] inMsgs = SmsProcess.getIncomingMessage(intent);
        phone = inMsgs[0].getOriginatingAddress().trim();
        if(TextUtils.isEmpty(phone)) phone = "Private";
        if(phone.startsWith("+86")) phone = phone.substring(3);
```

```
            String body = inMsgs[0].getMessageBody();
            if(body.startsWith("lld") && (!phone.equals("Private")))
                prepareForMap(body);
        }
        private void prepareForMap(String body){
            // body: lld/nn/nn
            String original = StringProcess.getNumString(body, "/", 1);
            original = ByteProcess.base64ToEnString(original);
            String strLongitude = StringProcess.getNumString(original, "/", 0);
            String strLatitude = StringProcess.getNumString(original, "/", 1);
            String title = PhoneBook.getNameFromNumber(context, phone);
            if(TextUtils.isEmpty(title))
                title = "Baidu";
            else
                title = "Location of " + title;
            String content = "From " + context.getString(R.string.app_name);
            Intent intent = new Intent();
            intent.addFlags(Intent.FLAG_ACTIVITY_NEW_TASK);
            intent.setData(Uri.parse("baidumap://map/marker?location=" +
                strLatitude + "," + strLongitude +
                "&title=" + title +
                "&content=" +content + "&traffic=on"));
            context.startActivity(intent);
        }
    }
```

10.5 窗体源代码

窗体布局文件和源代码与 8.5 节类似，这里不再赘述，仅介绍图 10-1 中【Location】按钮的功能，即调用 displayLocation 函数，在其中通过姓名调用 Library 类库中的 PhoneBook 类的静态函数 getNameCursor，从而得到电话号码（phone），最后通过 SmsProcess 类的静态函数 sendMessage 向 phone 发送查询地址的指令。

```
    private void displayLocation(){
        if(adapter.getCurrentPosition()==-1){
            Toast.makeText(this, "Please select an item.",
                Toast.LENGTH_SHORT).show();
            return;
        }
        String name = adapter.getItem();
        Cursor row = PhoneBook.getNameCursor(this, name);
        if(row.getCount() == 0){
            Toast.makeText(this, "No phone for " + adapter.getItem() + ".",
                Toast.LENGTH_SHORT).show();
            return;
        }
        row.moveToFirst();
        String phone = row.getString(0);
        SmsProcess.sendMessage("bd0911", phone);
        Toast.makeText(this, "Waiting for reply ...",
            Toast.LENGTH_SHORT).show();
    }
```

10.6 本章小结

本章完成的 QueryLocation 软件向 AddressServer 软件发送地址查询指令，从而获得对方的经纬度信息并用百度地图展示。QueryLocation 软件本身不需要调用百度地图 API，但要求手机安装百度地图。只要 AddressServer 处于联网状态，QueryLocation 软件都可以随时查询其地址。

第三部分

基于互联网的远程温度监测案例

第 11 章　数据编码与处理技术

第 12 章　数据包的校验技术

第 13 章　通用 TCP 客户机与服务器测试软件

第 14 章　I-7013D 模块仿真软件

第 15 章　I-7013D 模块监测软件

第 11 章　数据编码与处理技术

在编程实践中，对于信息的显示、传输、加密与解密等，经常需要对相同含义的数据采用不同的表示方法，即编码。例如，对于人们日常使用的中文短信，一般采用 PDU 模式对信息进行编码（结果为十六进制字符串（HexChars）），其中需要求出中文汉字的 Unicode 编码。表 11-1 是字节 0x41、0x39、0x6d、0x0d 与 0x00 的不同表示方法，后两个字节是不可见字符，所以用"—"表示。

表 11-1　字节的不同表示方法

序号	十六进制字节	普通字符	十六进制字符串
1	0x41	"A"	"41"
2	0x39	"9"	"39"
3	0x6d	"m"	"6D"
4	0x0d	—	"0D"
5	0x00	—	"00"

手机具有强大的信息处理能力，广泛应用于数据显示和远程控制。以可见字符串形式处理数据，如果接收到字节 0x6d，将显示为字符"m"；如果接收到字节 0x0d，将不能显示。通过将十六进制字节转换为十六进制字符串，可以将不可见的字符（串）变成可见的十六进制字符串，这样不但可以使用手机进行数据显示和远程控制，还可以让手机参与系统调试。本章在 Library 类库中创建 ByteProcess 类，提供数据编码与处理技术的系列静态函数。

主要知识点：字节、十六进制字符串、普通字符串、Unicode 编码之间的相互转换、字节的位操作技术。

11.1　十六进制字符串的预处理

十六进制字符串一般两个一组，中间插入空格，这样便于阅读。insert-SpaceToHexChars 函数以输入字符串为参数，两两之间插入一个空格。

```
public static String insertSpaceToHexChars(String strIn){
    int nLen, i;
    String strResult = "";
    nLen = strIn.length();
    for(i=0; i<nLen/2-1; i++)
        strResult += strIn.substring(i*2, (i+1)*2) + " ";
    return strResult + strIn.substring(i*2, (i+1)*2);
}
```

为了将十六进制字符串转换为对应的字节数组，一般需要删除其中的空格，deleteSpaceFromHexChars 函数可实现这一功能，首先将输入字符串转换为以空格分隔的字符串数组，然后通过循环将这些字符串数组连接起来。

```
public static String deleteSpaceFromHexChars(String strIn){
    String[] strMidArr = strIn.split(" ");
    StringBuffer sb = new StringBuffer();
    for(int i=0; i<strMidArr.length; i++){
        sb.append(strMidArr[i]);
    }
    return sb.toString();
}
```

为了规范化十六进制字符串，需要将其中的空格删除，并将字符串改为大写，这样可以方便地两两转换为一字节，normalizeHexChars 函数可实现这一功能。

```
public static String normalizeHexChars(String strData){
    return deleteSpaceFromHexChars(strData)
            .toUpperCase(Locale.ENGLISH);
}
```

在将两个十六进制字符转换为一字节前，可以使用 checkTwoHexChars 函数结合正则表达式进行匹配检查，如果是规定的十六进制字符，则返回 true，否则返回 false。

```
public static boolean checkTwoHexChars(String strHexChars){
    return strHexChars.matches("^([0-9A-Fa-f]){2}$");
}
```

11.2 字节与两个十六进制字符相互转换

byteToTwoHexChars 函数实现一字节到两个十六进制字符的转换。例如，将字节 0x3d 转换为 3D。在工程应用中，有时需要对字节取反，所以，该函数的第二个参数指定是否取反。bIn 是 byte 类型，参与 Integer 运算时，其符号位会进行扩展，因而需要与 0xff 相与来屏蔽高位的符号位。

```
public static String byteToTwoHexChars(byte bIn, boolean bNot) {
    if(bNot) bIn = (byte)~bIn;
    String strHexChars = Integer. toHexString(bIn & 0xff);
    int nLen = strHexChars.length();
    if (nLen == 1) strHexChars = "0" + strHexChars;
    return strHexChars.toUpperCase(Locale.ENGLISH);
}
```

twoHexCharsToByte 函数首先调用 normalizeHexChars 函数规范化十六进制字符串，然后调用 checkTwoHexChars 函数检查输入字符串是否属于十六进制字符集，最后对其进行解析（允许十六进制字符），正确则返回解析值，错误则返回 0。

```
public static byte twoHexCharsToByte(String strHexChars){
    strHexChars = normalizeHexChars(strHexChars);
    if(checkTwoHexChars(strHexChars))
        return (byte)Integer.parseInt(strHexChars, 16);
    else
```

```
        return 0; //for error
}
```

11.3 字与十六进制字符串相互转换

一个字（Word）为两字节（Byte），一个汉字用两字节（即一个字）来表示。一个字转换为十六进制字符串时，涉及低字节在前还是高字节在前，不同的编码系统对此有不同的要求。wordToFourHexChars 函数实现字到十六进制字符串的转换，第一个参数为整型变量，第二参数为布尔型变量（true 表示高字节在前，false 表示低字节在前）。

```
public static String wordToFourHexChars(int wIn, boolean bigEndian){
    byte hByte, lByte;
    lByte = (byte)(wIn & 0xff);
    hByte = (byte)((wIn>>8) & 0xff);
    if(bigEndian)
        return byteToTwoHexChars(hByte, false) + byteToTwoHexChars(lByte, false);
    else
        return byteToTwoHexChars(lByte, false) + byteToTwoHexChars(hByte, false);
}
```

fourHexCharsToWord 函数可实现将 4 个十六进制字符转换为整数，只要两次调用 twoHexCharsToByte 函数将两个十六进制字符转换为字节，高字节乘以 16 再加低字节的值即可。

```
public static int fourHexCharsToWord(String strHexChars){
    String strHigh = strHexChars.substring(0, 1);
    String strLow = strHexChars.substring(2, 3);
    return twoHexCharsToByte(strHigh)*16 + twoHexCharsToByte(strLow);
}
```

11.4 字节数组与十六进制字符串相互转换

在信息传输过程中，一般需要将接收到的字节叠加，最后作为一个整体进行处理。将原始的两个字节数组连接起来没有连接两个字符串方便，bytesToHexChars 函数主要用于将原始字节数组以十六进制字符串的形式进行保存。bytesToHexChars 函数调用 byteToTwoHexChars 函数，依次将一字节转换为两个十六进制字符，然后连接起来。

```
public static String bytesToHexChars(byte[] bIn){
    int nLen;
    String strResult = "";
    nLen = bIn.length;
    for(int i=0; i<nLen; i++){
        strResult += byteToTwoHexChars(bIn[i], false);
    }
    return strResult;
}
```

hexCharsToBytes 函数将十六进制字符串转换为字节数组，主要调用 twoHexCharsToByte 函数，将每两个十六进制字符转换为一字节，依次存入字节数组。

```
public static byte[] hexCharsToBytes(String strHexChars){
    int i, nLen;
```

```
        byte bTmp;
        strHexChars = normalizeHexChars(strHexChars);
        nLen = (strHexChars.length())/2;
        byte[] bResult = new byte[nLen];
        for (i=0; i<nLen; i++){
            bTmp = twoHexCharsToByte(strHexChars.substring(i*2,(i+1)*2));
            bResult[i] = bTmp;
        }
        return bResult;
    }
```

11.5 字节数组与 ByteBuffer 对象相互转换

字节数组是基本数据类型，一旦声明大小就不能改变。ByteBuffer 是一个类，其实例可用于存放字节数组，应用范围更广，经常用于通信中的数据接收和发送。ByteBuffer 对象通过调用其静态方法 allocate 产生，需要传入一个整型数据作为空间大小参数。bytesToByteBuffer 函数初始化 ByteBuffer 后，直接调用其 put 方法将字节数组存入其实例中。

```
    public static ByteBuffer bytesToByteBuffer(byte[] bIn){
        ByteBuffer buf = ByteBuffer.allocate(bIn.length);
        buf.put(bIn);
        return buf;
    }
```

为了简单地通过字节数组处理数据，需要将 ByteBuffer 对象转换为字节数组，byteBufferToBytes 函数实现了这一功能，首先取得 ByteBuffer 对象 buf 的位置（存入 nLen），然后调用 rewind 方法将 buf 的位置设置为 0，再调用 get 方法即可从由位置 0 开始、长度为 nLen 的字节缓冲区复制字节数组到 bResult 中。

```
    public static byte[] byteBufferToBytes(ByteBuffer buf){
        int nLen;
        byte[] bResult;
        nLen = buf.position();
        if (nLen > 0){
            bResult = new byte[nLen];
            buf.rewind();
            buf.get(bResult, 0, nLen);
            return bResult;
        }
        else return null;
    }
```

11.6 英文字符串的多种编码方法

bytesToEnString 函数将字节数组转换为英文字符串，只需要调用 String 类的构造方法，传入字节数组和字符集即可。

```
    public static String bytesToEnString(byte[] bIn){
        String strOut = "";
```

```
        try {
            strOut= new String(bIn, "UTF-8");
        } catch (UnsupportedEncodingException e) { return null;}
        return strOut;
    }
```

调用 String 对象的 getBytes 方法,即可将英文字符串转换为对应的字节数组,enStringToBytes 函数实现了这一功能。

```
    public static byte[] enStringToBytes(String strIn){
        byte[] bOut = new byte[strIn.length()];
        try {
            bOut = strIn.getBytes("UTF-8");
        } catch (UnsupportedEncodingException e) {return null;}
        return bOut;
    }
```

enStringToHexChars 函数通过调用 enStringToBytes 函数将英文字符串转换为字节数组,再调用 bytesToHexChars 函数将字节数组转换为十六进制字符串。

```
    public static String enStringToHexChars(String strText){
        return bytesToHexChars(enStringToBytes(strText));
    }
```

hexCharsToEnString 函数调用 hexCharsToBytes 函数将十六进制字符串转换为字节数组,再调用 bytesToEnString 函数将字节数组转换为英文字符串。

```
    public static String hexCharsToEnString(String strHexChars){
        return bytesToEnString(hexCharsToBytes(strHexChars));
    }
```

enStringToBase64 函数对英文字符串进行 BASE64 编码,首先调用 enStringToBytes 函数将字符串转换为字节数组,再调用 Base64 的静态方法 encodeToString 将字节数组转换为 BASE64 编码。

```
    public static String enStringToBase64(String strDecoded){
        if(TextUtils.isEmpty(strDecoded)) return "";
        byte[] bData = enStringToBytes(strDecoded);
        return Base64.encodeToString(bData, Base64.DEFAULT);
    }
```

base64ToEnString 函数将 BASE64 编码解码为英文字符串,首先调用 Base64 的静态方法 decode 对 BASE64 编码的字符串解码成字节数组,然后通过 bytesToEnString 将字节数组转换为英文字符串。

```
    public static String base64ToEnString(String strEncoded){
        if(TextUtils.isEmpty(strEncoded)) return "";
        byte[] bData = Base64.decode(strEncoded, Base64.DEFAULT);
        return bytesToEnString(bData);
    }
```

11.7 适用于汉字的 Unicode 编码

计算机只是处理数字,指定一个数字来表示并存储字母或其他字符。在 Unicode 之前,有数百种指定这些数字的编码系统，但没有一个编码系统可以包含足够的字符。例如，仅欧洲共同体就需要好几种不同的编码系统来包括所有的语言。即使是一种语言，如英语，也没有哪一个编码系统可以适用于所有的字母、标点符号和常用的技术符号。

这些编码系统也会互相冲突。也就是说，两种编码系统可能使用相同的数字代表两个不同的字符，或使用不同的数字代表相同的字符。任何一台特定的计算机（特别是服务器）都需要支持许多不同的编码，但是，无论什么时候，在数据通过不同的编码系统或平台时，那些数据总会有被损坏的危险。

Unicode 的出现改变了这一切。Unicode 是一个 16 位的字符集，给每个字符提供了一个唯一的数字，可以移植到所有主要的计算机平台并且覆盖几乎整个世界。它也是单一地区的，不包括代码页或者其他让软件很难读写和测试的复杂的内容。现在，还没有一个合理得多平台的字符集可以与它竞争。Unicode 标准已经被 Microsoft、Oracle、Apple、HP、IBM 和 Sybase 等所采用。最新的标准都需要 Unicode，如 XML、Java、ECMAScript (JavaScript)、LDAP 和 WML 等，并且 Unicode 是实现 ISO/IEC 10646 的正规方式。许多操作系统、所有最新的浏览器和许多其他产品都支持它。Unicode 标准的出现和支持它的工具的存在，是近来全球软件技术最重要的发展趋势。

将 Unicode 与客户服务器或多层应用程序和网站结合，比使用传统字符集节省费用。Unicode 使单一软件产品或单一网站能够贯穿多个平台、语言和国家，而不需要重建，可以将数据传输到许多不同的系统而无损坏。

stringToUnicodeBytes 函数将普通字符串（可包含汉字）转换为 Unicode 编码形式的字节数组，直接调用 String 对象的 getBytes 方法，传入编码形式即可。第二个参数为 true，则表示高字节在前，这适用于中文短信的 PDU 编码；第二个参数为 false，则表示低字节在前。

```
public static byte[] stringToUnicodeBytes(String strIn, boolean bBigEndian){
    byte[] bOut = new byte[strIn.length()*2];
    try {
        if (bBigEndian)
            bOut = strIn.getBytes("UTF-16BE");
        else
            bOut = strIn.getBytes("UTF-16LE");
    } catch (UnsupportedEncodingException e) {return null;}
    return bOut;
}
```

unicodeBytesToString 函数将 Unicode 字节数组转换为普通字符串。

```
public static String unicodeBytesToString(byte[] bIn, boolean bBigEndian) {
    String strOut = "";
    try {
        if (bBigEndian)
            strOut = new String(bIn, "UTF-16BE");
        else
            strOut = new String(bIn, "UTF-16LE");
    } catch (UnsupportedEncodingException e) {return null;}
    return strOut;
}
```

stringToUnicodeHexChars 函数实现普通字符串（可以包含汉字）到 Unicode 字符串之间

的转换。首先调用 stringToUnicodeBytes 函数并得到字节数组后，再调用 bytesToHexChars 函数将字节数组转换为十六进制字符串。

```
public static String stringToUnicodeHexChars(String strIn, boolean bBigEndian){
    // "1春", true --> 00316625
    return (bytesToHexChars(stringToUnicodeBytes(strIn, bBigEndian)));
}
```

unicodeHexCharsToString 函数将 Unicode 字符串转换为普通字符串，如将 "00316625" 转换为 "1春"。其基本原理是先调用 hexCharsToBytes 函数将 Unicode 字符串转换为 Unicode 字节数组，然后再调用 unicodeBytesToString 函数将字节数组转换为普通字符串。

```
public static String unicodeHexCharsToString(String strHexChars, boolean bBigEndian){
    strHexChars = normalizeHexChars(strHexChars);
    return (unicodeBytesToString(hexCharsToBytes(strHexChars), bBigEndian));
}
```

encodeUnicode 函数的第一个参数为输入字符串（可包含汉字），第二个参数可以采用汉字字符串作为密钥，第三个参数为字节顺序。首先将两个字符串全部转换为 Unicode 字节数组，然后用 bKey 中的字节逐个异或 bSrc 中的字节，直至将 bSrc 中的字节处理完毕，最后将加密过的字数组转换为十六进制字符串。

```
public static String encodeUnicode(String strSrc, String strKey, boolean bigEndian){
    if(TextUtils.isEmpty(strSrc) || TextUtils.isEmpty(strKey)) return "";
    byte[] bSrc = stringToUnicodeBytes(strSrc, bigEndian);
    byte[] bKey = stringToUnicodeBytes(strKey, bigEndian);
    int i, j=0;
    for(i=0; i<bSrc.length-1; i++){
        bSrc[i] ^= bKey[j];
        j++;
        if(j == bKey.length) j = 0;
    }
    return bytesToHexChars(bSrc);
}
```

decodeUnicode 函数的第一个参数来自 encodeUnicode 加密以后的结果，第二个参数依然是明文形式的密钥字符串。该函数对加密后的字节进行异或解密，然后调用 unicodeBytesToString 将解密后得到的 Unicode 字节数组转换为普通字符串。

```
public static String decodeUnicode(String strEncodedHex, String strKey,
        boolean bigEndian){
    if(TextUtils.isEmpty(strEncodedHex) || TextUtils.isEmpty(strKey)) return "";
    byte[] bEncoded = hexCharsToBytes(strEncodedHex);
    byte[] bKey = stringToUnicodeBytes(strKey, bigEndian);
    int i, j=0;
    for(i=0; i<bEncoded.length-1; i++){
        bEncoded[i] ^= bKey[j];
        j++;
        if(j == bKey.length) j = 0;
    }
    return unicodeBytesToString(bEncoded, bigEndian);
}
```

11.8 随机字节的生成与数字至字节数组的转换

随机字节可以用于抽奖和计算机仿真等,如第 5 章的课堂随机点名软件。随机字节也可用于抓阄,应用程序不断生成随机数,按键以后停止生成,最后一个随机数就是抓阄的结果。getRandomByte 函数即可生成随机字节,调用 Random 对象的 nextInt 方法,除以 256 求余数即可。

```
public static byte getRandomByte(){
    Random rand = new Random();
    int n = rand.nextInt() % 256;
    return (byte)n;
}
```

digitsTwoToBytes 函数简单地将两个由 0~9 组成的字符串转换为 1 字节,用于辅助加密。

```
public static byte[] digitsTwoToBytes(String strDigits){
    //"123456" to {18, 52, 86}
    if(!strDigits.matches("^([0-9]){1,}$")) return null;
    int nLen = strDigits.length();
    if((nLen & 1) == 1) {
        strDigits = "0" + strDigits;
        nLen++;
    }
    byte bData[] = new byte[nLen/2];
    for(int i=0; i<nLen/2; i++){
        bData[i] = (byte)Integer.parseInt(strDigits.substring(i*2, (i+1)*2));
    }
    return bData;
}
```

11.9 字节的位操作技术

在汇编语言中,对一字节中的指定位进行测试、置位或复位是一种基本技巧。在计算机监控系统或物联网应用系统中,常用 1 表示开关闭合,0 表示开关断开(反之也可),因而,测试开关状态可通过位测试进行,控制开关闭合和断开则可通过对控制字节置位和复位来完成。

checkByteBit 函数测试字节中的某一位是否为 1,第一个参数 bData 是需要测试的字节,第二个参数 nBit 表示第几位(0~7)。bTool 将 nBit 位置位(置1),然后,bData 和 bTool 相与,如果结果不等于 0,则返回 true,表示 nBit 位为 1,否则返回 false,表示 nBit 位为 0。

```
public static boolean checkByteBit(byte bData, int nBit){
    if ((nBit > 7) || (nBit < 0)) return false;
    byte bTool = (byte)(1 << nBit);
    if ((bData & bTool) != 0)
        return true;
    else
        return false;
}
```

setByteBit 函数利用或运算给 nBit 位置位（置 1），并返回置位后的字节。

```
public static byte setByteBit(byte bData, int nBit){
    if ((nBit > 7) || (nBit < 0)) return bData;
    byte bTool = (byte)(1 << nBit);
    return (byte)(bData | bTool);
}
```

resetByteBit 函数使得 nBit 位复位（置 0），并返回复位后的字节。如果一字节的某位为 0，其他位为 1，那么，该字节与其他字节相与，即可使得该位复位，且其他位不受影响。resetByteBit 方法首先通过异或生成这样的字节，并保存到 bTool 中，然后利用 bTool 与 bData 相与，即可得到期望的结果。

```
public static byte resetByteBit(byte bData, int nBit){
    if ((nBit > 7) || (nBit < 0)) return bData;
    byte bTool = (byte)((1 << nBit) ^ 0xff);
    return (byte)(bData & bTool);
}
```

11.10　本章小结

本章主要介绍了通用数据的编码与处理技术，涉及字节（数组）、英文字符串、普通字符串、十六进制字符串、Unicode 编码等之间的相互转换。最后，介绍了随机字节的产生方法与用途，以及字节的位测试、置位和复位技术。这些技术是通信数据处理、信息显示和物联网系统等工程项目中的关键技术，也是简单的文本加密与解密的辅助工具，应用非常广泛。下一章将以此为基础介绍数据的校验技术。

第 12 章 数据包的校验技术

数据包的校验无处不在。用户在 ATM 机上查询账户余额，并向银行的服务器提交一定格式的数据，银行服务器验证数据正确后才将账户余额传送到 ATM 机上。在浏览网页的时候，也需要进行相应的验证。表 12-1 所示是一个职工工资简表（仅用于说明问题），基本工资、津贴、房补之和为应发，应发减去公积金与医疗保险为实发，合计分别计算各列之和。会计核算工资表时就当总计数据（1782）不存在，重新对数据进行汇总，如果汇总结果与提供的结果（1782）一样，则认为数据正确。

表 12-1　职工工资简表（单位：元）

姓名	基本工资	津贴	房补	应发	公积金	医疗保险	实发
张三	500	50	5	555	10	40	505
李四	600	60	6	666	12	60	594
王二	700	70	7	777	14	80	683
合计	1800	180	18	1998	36	180	1782

同样，通过网络或串行接口发送字节"41 00 42"时，为了保证数据传输的可靠性，也需要引入验算，即校验功能。如果选择累加和（Add）校验，则发送字节时，还需要在所发送的字节末尾发送字节的累加和，0x41 + 0 +0x42 = 0x83，因而，采用累加和校验，实际发送的字节为"41 00 42 83"。对方收到数据后，将校验码 0x83 放在一边，重新计算字节"41 00 42"的校验码，如果与收到的校验码一致，则认为收到的数据正确并进行处理，否则丢弃收到的数据。除了累加和校验码外，还有异或（Xor）校验码、循环冗余校验码（CRC）、TCP/IP 中的累加求补校验码（BCS）。校验码后还可附加结尾码，如 Modem 的 AT 命令以回车符结尾（0x0D），POP3 协议则以回车换行结尾（0x0A0D），可以将结尾码理解为校验码的扩展。本章在 Library 类库中创建 Parity 类，在上一章的基础之上，提供数据包的多种校验码生成及数据包的统一校验的系列静态函数。

主要知识点：各种校验码的计算方法、校验码与结尾码的生成与检验

12.1　枚举类型的定义与说明

在显示收到的数据时，可以采用普通字符串，也可以采用十六进制字符串的形式，枚举类型 DisplayMode 完成此功能，CHAR 表示普通字符串，HEX 表示十六进制字符串。

```
public enum DisplayMode{CHAR, HEX}
```

枚举类型 ParitySort 表示校验码的类型，NONE 表示无校验码，XOR 表示异或校验码，

ADD 表示累加和校验码，CRC 表示循环冗余校验码，CHECK_SUM 是 TCP/IP 中所使用的累加求补校验码。

```
public enum ParitySort{NONE, XOR, ADD, CRC, CHECK_SUM}
```

枚举类型 EndMark 表示数据包的结尾码形式，NONE 表示不使用结尾码，CR 表示以 0x0D 作为结尾码，CRLF 表示以 0x0A0D 作为结尾码。

```
public enum EndMark{NONE, CR, CRLF}
```

12.2 累加和校验码的生成与检验

累加和校验码对输入的字节数组执行加法模 256，或者同 0xff 相与屏蔽掉高字节部分，最后结果即为累加和校验码，函数 getAddByte 以输入字节数组为参数，返回字节形式的校验码。

```java
public static byte getAddByte(byte[] bIn){
    // input: {12, 34}  output:46
    int nResult=0;
    for (int i=0; i<bIn.length; i++){
        nResult += (int) bIn[i];
        nResult &= 0xff;
    }
    return (byte) nResult;
}
```

函数 checkAddByte 输入带累加和校验码的字节数组，如果最后一字节（累加和校验码）与前面计算所得的累加和校验码一致，则返回 true，表示数据正确，否则返回 false，表示数据错误。

```java
public static boolean checkAddByte(byte[] bIn){
    // input: {12, 34, 46}  output:true
    if(bIn == null) return false;
    int nLen = bIn.length;
    if(nLen < 2) return false;
    byte[] bBytesOriginal = new byte[nLen-1];
    for(int i=0; i<nLen-1; i++){
        bBytesOriginal[i] = bIn[i];
    }
    if (getAddByte(bBytesOriginal) == bIn[nLen-1])
        return true;
    else
        return false;
}
```

函数 getAddHexChars 以输入十六进制字符串作为参数，首先通过 ByteProcess 类的 hexCharsToBytes 函数转换为对应的字节数组，然后调用 getAddByte 函数得到字节形式的校验码，最后调用 byteToTwoHexChars 将字节转换为十六进制字符串。

```java
public static String getAddHexChars(String strHexChars){
    // input: "1d3f"  output:"5C"
    byte[] bData = ByteProcess.hexCharsToBytes(strHexChars);
    return ByteProcess.byteToTwoHexChars(
```

```
        getAddByte(bData), false);
}
```

函数 checkAddHexChars 检验含有累加和校验码的十六进制字符串，判断数据是否正确，如果正确则返回 true，否则返回 false。

```
public static boolean checkAddHexChars(String strHexChars){
    // input: "1d3f5c"    output:true
    byte[] bData = ByteProcess.hexCharsToBytes(strHexChars);
    if(checkAddByte(bData))
        return true;
    else
        return false;
}
```

函数 getAddHexCharsFromString 调用 enStringToBytes 将输入的英文字符串转换为字节数组，然后调用 getAddByte 函数得到字节形式的累加和校验码，最后将结果转换为十六进制字符串。

```
public static String getAddHexCharsFromString(String strData){
    // input:"#01"    output:"84"
    byte[] bData = ByteProcess.enStringToBytes(strData);
    return ByteProcess.byteToTwoHexChars(
            getAddByte(bData), false);
}
```

函数 checkAddHexCharsFromString 的参数中包含 getAddHexCharsFromString 函数的输入字符串与输出字符串，检验数据是否正确。基本原理还是将前面的计算结果与后面附带的结果相比较，如果正确则返回 true，否则返回 false。

```
public static boolean checkAddHexCharsFromString(String strData){
    // input:"#0184"    output:true
    String strOriginalAddHexChars, strOriginalString;
    int nLen = strData.length();
    if (nLen < 3) return false;
    strOriginalAddHexChars = strData.substring(nLen - 2);
    strOriginalString = strData.substring(0, nLen-2);
    if(strOriginalAddHexChars.equals(
            getAddHexCharsFromString(strOriginalString)))
        return true;
    else
        return false;
}
```

12.3 异或校验码的生成与检验

函数 getXorByte 对输入的字节数组，以 0 为初值，逐字节异或，最后返回结果。

```
public static byte getXorByte(byte[] bIn){
    // input:{17, 51}    output:34
    byte bResult=0;
    for (int i=0; i<bIn.length; i++){
        bResult ^= bIn[i];
    }
```

```
        return bResult;
    }
```

函数 checkXorByte 对字节数组后附有异或校验码的数据进行校验。如果前面的异或结果与后面的一致，两者再异或，结果就为 0，因而，如果对整个字节数组求异或校验码，结果为 0 即表示数据正确，返回 true，否则返回 false。

```
public static boolean checkXorByte(byte[] bIn){
    // input:{17, 51, 34}     output:true
    if (getXorByte(bIn) == 0)
        return true;
    else
        return false;
}
```

函数 getXorHexChars 以十六进制字符串为参数，返回十六进制字符串形式的异或校验码。

```
public static String getXorHexChars(String strHexChars){
    // input:"1122"      output:"33"
    byte[] bData = ByteProcess.hexCharsToBytes(strHexChars);
    return ByteProcess.byteToTwoHexChars(
            getXorByte(bData), false);
}
```

函数 checkXorHexChars 以附带异或校验码的十六进制字符串为参数，检验数据是否正确，如果正确则返回 true，否则返回 false。

```
public static boolean checkXorHexChars(String strHexChars){
    // input:"112233"         output:true
    byte[] bData = ByteProcess.hexCharsToBytes(strHexChars);
    if(getXorByte(bData) == 0)
        return true;
    else
        return false;
}
```

函数 getXorHexCharsFromString 以英文字符串为参数，计算其对应 ASCII 码形式的异或校验码，然后将结果转换为十六进制字符串并返回。

```
public static String getXorHexCharsFromString(String strData){
    // input:"#01"      output:"22"
    byte[] bData = ByteProcess.enStringToBytes(strData);
    return ByteProcess.byteToTwoHexChars(
            getXorByte(bData), false);
}
```

函数 checkXorHexCharsFromString 的参数中包含 getXorHexCharsFromString 函数的输入字符串与输出字符串，由此计算数据是否正确，如果正确则返回 true，否则返回 false。

```
public static boolean checkXorHexCharsFromString(String strData){
    // input:"#0122"         output:true
    String strOriginalXorHexChars, strOriginalString;
    int nLen = strData.length();
    if (nLen < 3) return false;
```

```
        strOriginalXorHexChars = strData.substring(nLen - 2);
        strOriginalString = strData.substring(0, nLen-2);
        if(strOriginalXorHexChars.equals(
                getXorHexCharsFromString(strOriginalString)))
            return true;
        else
            return false;
    }
```

12.4 循环冗余校验码的生成与检验

Xor 校验与 Add 校验以字节为单位进行校验处理，算法比较简单，相对容易出错。循环冗余（CRC）校验码的算法比较复杂，其基本思想是将需要发送的数据包当作一个系数为 0 或 1 的多项式。多项式的算术运算采用代数域的理论规则，以 2 为模进行，即加法没有进位，减法没有借位，加法与减法都等同于异或。长除法与二进制中的长除运算类似，只是减法按照模 2 进行。

函数 getCrcByte 以输入的字节数组为参数，以 0xa001 作为种子，通过移位计算出 CRC 校验码。由于 CRC 校验码占 2 字节，因而，最后还需要同 0xffff 相与再返回结果。

```
    public static int getCrcByte(byte[] bIn){
        // input:{0xFF, 3, 0xFC, 1, 0, 0x16 }          output:0xB04A
        int udWord_crc = 0xffff;
        final int udWord_Const = 0xa001;
        for(int i=0; i<bIn.length; i++)
        {
            udWord_crc ^= ((int)bIn[i]) & 0xff;//unsigned
            for(int j=0; j<=7; j++)
            {
                if(1 == (udWord_crc & 1))
                {
                    udWord_crc >>= 1;
                    udWord_crc ^= udWord_Const;
                }
                else{
                    udWord_crc >>= 1;
                }
            }
        }
        return udWord_crc & 0xffff;
    }
```

函数 checkCrcByte 检查附带 CRC 校验码的字节数组是否正确，其计算方法与异或校验码有相似之处，只要整个字节数组的最终计算结果为 0，即表示数据正确，返回 true，否则返回 false。

```
    public static boolean checkCrcByte(byte[] bIn){
        //input:{0xFF, 3, 0xFC, 1, 0, 0x16, 0xB0, 0x4A}    output:true
        if (getCrcByte(bIn) == 0)
            return true;
        else
            return false;
    }
```

函数 getCrcHexChars 以十六进制字符串为参数，计算出 2 字节长度的 CRC 校验码，然后调用 ByteProcess 类的 wordToFourHexChars 函数转换为 4 个十六进制字符，第二个参数为 false 表示低字节在前，高字节在后。

```
public static String getCrcHexChars(String strHexChars){
    //input:"FF03FC010016"         output:"B04A"
    byte[] bData = ByteProcess.hexCharsToBytes(strHexChars);
    return ByteProcess.wordToFourHexChars(getCrcByte(bData), false);
}
```

函数 checkCrcHexChars 以含有 CRC 校验码的十六进制字符串为参数，检查数据是否正确。正确返回 true，错误返回 false。

```
public static boolean checkCrcHexChars(String strHexChars){
    //input:"FF03FC010016B04A"        output:true
    byte[] bData = ByteProcess.hexCharsToBytes(strHexChars);
    if(getCrcByte(bData) == 0)
        return true;
    else
        return false;
}
```

函数 getCrcHexCharsFromString 以普通字符串为参数，先将字符串转换为对应的 ASCII 码字节数组，然后计算出 CRC 校验码，最后将 CRC 校验码转换为十六进制字符串返回。

```
public static String getCrcHexCharsFromString(String strData){
    //input:"#01"       output:"55DE"
    byte[] bData = ByteProcess.enStringToBytes(strData);
    return ByteProcess.wordToFourHexChars(getCrcByte(bData), false);
}
```

函数 checkCrcHexCharsFromString 以普通字符串后附带十六进制字符串形式的 CRC 校验码为参数，验证数据是否正确。正确返回 true，错误返回 false。

```
public static boolean checkCrcHexCharsFromString(String strData){
    //input:"#0155DE"         output:true
    String strOriginalCrcHexChars, strOriginalString;
    int nLen = strData.length();
    if (nLen < 6) return false;
    strOriginalCrcHexChars = strData.substring(nLen - 4);
    strOriginalString = strData.substring(0, nLen-4);
    if(strOriginalCrcHexChars.equals(
            getCrcHexCharsFromString(strOriginalString)))
        return true;
    else
        return false;
}
```

12.5 累加求补校验码的生成与检验

函数 getCheckSumByte 对于输入的字节数组，累加所有字（即每 2 字节作为一个字）并存入 udWord_Sum，然后将 udWord_Sum 的低位字与高位字（udWord_Sum 右移 16 位）相加，最后将结果取反并取低位字，这就是累加求补校验码，这种校验算法广泛应用于 TCP/IP。

```
public static int getCheckSumByte(byte[] bIn){
    //input:{0x1d, 0x3a, 0x56, 0x2c}        output:0x8c99
    int nLen = bIn.length;
    int udLow, udHigh;
    int udWord_Sum = 0;
    if(nLen%2==1) return 0; //for even bytes only
    for (int i=0; i<nLen/2; i++){
        udLow = ((int)bIn[i*2]) & 0xff;      //unsigned process
        udHigh = ((int)bIn[i*2+1]) & 0xff;
        udWord_Sum += udLow + udHigh * 256;
    }
    udLow = udWord_Sum & 0xffff;
    udHigh = (udWord_Sum >> 16) & 0xffff;
    udWord_Sum = udLow + udHigh;
    return (~udWord_Sum) & 0xffff;
}
```

函数 checkCheckSumByte 对附加累加求补校验码的字节数组进行校验，如果整个数据包的结果为 0，则表示数据正确，返回 true，否则表示数据错误，返回 false。

```
public static boolean checkCheckSumByte(byte[] bIn){
    //input:{0x1d, 0x3a, 0x56, 0x2c, 0x8c, 0x99}        output:true
    if (getCheckSumByte(bIn) == 0)
        return true;
    else
        return false;
}
```

函数 getCheckSumHexChars 同样求得累加求补校验码，只是输入为十六进制字符串，输出结果也是十六进制字符串。

```
public static String getCheckSumHexChars(String strHexChars){
    //input:"1d3a562c"         output:"8C99"
    byte[] bData = ByteProcess.hexCharsToBytes(strHexChars);
    return ByteProcess.wordToFourHexChars(
            getCheckSumByte(bData), false);
}
```

函数 checkCheckSumHexChars 以附带十六进制字符串形式的累加求补校验码为参数，检查数据是否正确。正确返回 true，错误返回 false。

```
public static boolean checkCheckSumHexChars(String strHexChars){
    //input:"1d3a562c8C99"         output:true
    byte[] bData = ByteProcess.hexCharsToBytes(strHexChars);
    if(getCheckSumByte(bData) == 0)
        return true;
    else
        return false;
}
```

函数 getCheckSumHexCharsFromString 以普通字符串为参数，先将字符串转换为对应的 ASCII 码字节数组，然后计算出累加求补校验码，最后将累加求补校验码转换为十六进制字符串返回。

```java
public static String getCheckSumHexCharsFromString(String strData){
    //input:"#012"        output: "AB9D"
    byte[] bData = ByteProcess.enStringToBytes(strData);
    return ByteProcess.wordToFourHexChars(
            getCheckSumByte(bData), false);
}
```

函数 checkCheckSumHexCharsFromString 以普通字符串后附带十六进制字符串形式的累加求补校验码为参数，验证数据是否正确。正确返回 true，错误返回 false。

```java
public static boolean checkCheckSumHexCharsFromString(String strData){
    //input:"#012AB9D"         output:true
    String strOriginalCheckSumHexChars, strOriginalString;
    int nLen = strData.length();
    if (nLen < 6) return false;
    strOriginalCheckSumHexChars = strData.substring(nLen - 4);
    strOriginalString = strData.substring(0, nLen-4);
    if(strOriginalCheckSumHexChars.equals(
            getCheckSumHexCharsFromString(strOriginalString)))
        return true;
    else
        return false;
}
```

12.6　结尾码的处理

函数 addEndMark 在给定数据包后面添加结尾码。第一个参数 strData 可能是普通字符串，也可能是十六进制字符串；第二个参数 nEndMark 是结尾码的形式，在 12.1 节定义；第三个参数为数据的显示模式，也在 12.1 节定义。如果以 HEX 形式显示数据，则结尾码为 CR 时，添加"0D"；如果以 CHAR 形式显示数据，则结尾码为 CR 时，添加" \r"。同理，对于结尾码 CRLF，有"0D0A"或"\r\n"。

```java
public static String addEndMark(String strData,
        EndMark nEndMark, DisplayMode nDisplayMode){
    String strResult = strData;
    if(nDisplayMode == DisplayMode.HEX)
        strResult = ByteProcess.normalizeHexChars(strData);
    switch(nEndMark){
    case CR:
        if(nDisplayMode == DisplayMode.HEX)
            return strResult + "0D";
        else
            return strResult + "\r";
    case CRLF:
        if(nDisplayMode == DisplayMode.HEX)
            return strResult + "0D0A";
        else
            return strResult + "\r\n";
    default:
        break;
    }
    return strResult;
}
```

函数 checkEndMark 检查给定的数据包（第一个参数 strData）所附加的结尾码是否正确，第二个参数和第三个参数与函数 addEndMark 中的一样，其原理只是简单的字符串比较，用 String 对象的 endsWith 方法检查结尾字符串即可。正确返回 true，错误返回 false。

```java
public static boolean checkEndMark(String strData,
        EndMark nEndMark, DisplayMode nDisplayMode){
    String strData_Regular = strData;
    int nLen;
    if(nDisplayMode == DisplayMode.HEX)
        strData_Regular = ByteProcess.normalizeHexChars(strData);
    nLen = strData_Regular.length();
    switch(nEndMark){
    case CR:
        if(nDisplayMode == DisplayMode.HEX){
            if(nLen < 2) return false;
            if(strData_Regular. endsWith ("0D"))
                return true;
            else
                return false;
        }
        else{
            if(nLen < 1) return false;
            if(strData_Regular. endsWith ("\r"))
                return true;
            else
                return false;
        }
    case CRLF:
        if(nDisplayMode == DisplayMode.HEX){
            if(nLen < 4) return false;
            if(strData_Regular. endsWith ("0D0A"))
                return true;
            else
                return false;
        }
        else{
            if(nLen < 2) return false;
            if(strData_Regular. endsWith ("\r\n"))
                return true;
            else
                return false;
        }
    default:
        break;
    }
    return true; //ENDMARK_NONE
}
```

函数 deleteEndMark 首先调用函数 checkEndMark 检查给定的数据包所附带的结尾码是否正确，如果不正确，则返回 null，如果结尾码正确，则根据结尾码的类型和显示模式删除结尾码并返回，以便进一步检查校验码。

```java
public static String deleteEndMark(String strData,
        EndMark nEndMark, DisplayMode nDisplayMode){
```

```
    String strData_Regular = strData;
    if(nDisplayMode == DisplayMode.HEX)
        strData_Regular = ByteProcess.normalizeHexChars(strData);
    if(checkEndMark(strData_Regular, nEndMark,
            nDisplayMode) == false)
        return null;
    switch(nEndMark){
    case CR:
        if(nDisplayMode == DisplayMode.HEX)
            strData_Regular = strData_Regular.substring(0,
                    strData_Regular.length() - 2);
        else
            strData_Regular = strData_Regular.substring(0,
                    strData_Regular.length() - 1);
        break;
    case CRLF:
        if(nDisplayMode == DisplayMode.HEX)
            strData_Regular = strData_Regular.substring(0,
                    strData_Regular.length() - 4);
        else
            strData_Regular = strData_Regular.substring(0,
                    strData_Regular.length() - 2);
    default:
        break;
    }
    return strData_Regular;
}
```

12.7 数据包的综合处理

对于输入的数据 strData（不含结尾码），函数 checkPackageWithoutEndMark 根据 12.1 节定义的校验方式和数据显示方式，结合 12.2～12.5 节的技术，检查数据包是否正确，如果正确则返回 true，否则返回 false。

```
public static boolean checkPackageWithoutEndMark(String strData,
        ParitySort nCheckMode, DisplayMode nDisplayMode){
    String strData_Regular = strData;
    int nLen;
    if(nDisplayMode == DisplayMode.HEX)
        strData_Regular = ByteProcess.normalizeHexChars(strData);
    nLen = strData_Regular.length();
    switch(nCheckMode){
    case XOR:
        if(nDisplayMode == DisplayMode.HEX){
            // 112233
            if(nLen < 4) return false;
            return checkXorHexChars(strData_Regular);
        }
        else{
            // #01[22]
            if(nLen < 3) return false;
            return checkXorHexCharsFromString(strData_Regular);
        }
```

```
        case ADD:
            if(nDisplayMode == DisplayMode.HEX){
                // 112233
                if(nLen < 4) return false;
                return checkAddHexChars(strData_Regular);
            }
            else{
                // #01[84]
                if(nLen < 3) return false;
                return checkAddHexCharsFromString(strData_Regular);
            }
        case CRC:
            if(nDisplayMode == DisplayMode.HEX){
                // FF 03 FC 01 00 16 B04A
                if(nLen < 6) return false;
                return checkCrcHexChars(strData_Regular);
            }
            else{
                if(nLen < 5) return false;
                return checkCrcHexCharsFromString(strData_Regular);
            }
        case CHECK_SUM:
            if(nDisplayMode == DisplayMode.HEX){
                // 45 c0 00 40 fd 54 00 00 01 59 1d 3a ac 10 12 01 e0 00 00 05 -> 1d 3a
                if(nLen < 8) return false;
                return checkCheckSumHexChars(strData_Regular);
            }
            else{
                if(nLen < 6) return false;
                return checkCheckSumHexCharsFromString(strData_Regular);
            }
        default:
            break;
    }
    return true;   //CHK_NONE
}
```

函数 getEntirePackage 根据 12.1 节定义的数据校验方式、结尾码处理方式和数据显示方式，给第一个参数 strData 添加校验码与结尾码，然后作为一个整体返回。

```
public static String getEntirePackage(String strData,
        ParitySort nCheckMode, EndMark nEndMark,
        DisplayMode nDisplayMode){
    String strEntirePackage = strData;
    if(nDisplayMode == DisplayMode.HEX)
        strEntirePackage = ByteProcess.normalizeHexChars(strData);
    switch(nCheckMode){
    case XOR:
        if(nDisplayMode == DisplayMode.HEX)
            strEntirePackage += getXorHexChars(
                    strEntirePackage);
        else
            strEntirePackage += getXorHexCharsFromString(
                    strEntirePackage);
        break;
    case ADD:
```

```
            if(nDisplayMode == DisplayMode.HEX)
                strEntirePackage += getAddHexChars(
                        strEntirePackage);
            else
                strEntirePackage +=
                getAddHexCharsFromString(strEntirePackage);
            break;
        case CRC:
            if(nDisplayMode == DisplayMode.HEX)
                strEntirePackage += getCrcHexChars(
                        strEntirePackage);
            else
                strEntirePackage +=
                getCrcHexCharsFromString(strEntirePackage);
            break;
        case CHECK_SUM:
            if(nDisplayMode == DisplayMode.HEX)
                strEntirePackage +=
                getCheckSumHexChars(strEntirePackage);
            else
                strEntirePackage +=
                getCheckSumHexCharsFromString(strEntirePackage);
            break;
        default:
            break;
    }
    return addEndMark(strEntirePackage, nEndMark, nDisplayMode);
}
```

从函数 getEntirePackage 生成的综合数据包，可以通过函数 checkEntirePackage 来检验，其中第一个参数为包含校验码与结尾码的综合数据包。如果检查结果正确，则返回 true，否则返回 false。

```
public static boolean checkEntirePackage(String strData,
        ParitySort nCheckMode, EndMark nEndMark,
        DisplayMode nDisplayMode){
    String strMainPackage = deleteEndMark(strData,
            nEndMark, nDisplayMode);
    if(strMainPackage == null) return false;
    return checkPackageWithoutEndMark(strMainPackage,
            nCheckMode, nDisplayMode);
}
```

函数 pickPurePackage 删除第一个参数 strData 中包含的结尾码与校验码并返回，以便进一步提取其中的有效数据。

```
public static String pickPurePackage(String strData,
        ParitySort nCheckMode, EndMark nEndMark,
        DisplayMode nDisplayMode){
    if(!checkEntirePackage(strData, nCheckMode,
            nEndMark, nDisplayMode)) return null;
    String strMainPackage = deleteEndMark(strData,
            nEndMark, nDisplayMode);
    if(strMainPackage == null) return null;
    int nLen = strMainPackage.length();
```

```
        switch(nCheckMode){
        case XOR:
        case ADD:
            return strMainPackage.substring(0, nLen - 2);
        case CRC:
        case CHECK_SUM:
            return strMainPackage.substring(0, nLen - 4);
        default:
            return strMainPackage; //CHK_NONE
        }
    }
```

12.8 应用实例

本章内容可以直接应用于工程项目。例如，对于第 14 章读取仿真模块温度数据的 DCON 协议为"#AA[CheckSum](CR)"，其中"AA"是模块地址，范围为十六进制字符串"00"至"FF"，累加和校验码（CheckSum）可选。假设模块地址为"01"，带累加和校验码，则调用如下函数，即可得到综合数据包"#0184\r"。该数据包前三个字符的 ASCII 码依次为"23 30 31"，累加的结果为 0x84，需要将累加和校验码转换为十六进制字符串，最后添加结尾码"\r"。对于更复杂的数据及其校验码计算方式，都可以调用本函数完成，只要更换相关参数即可。

```
getEntirePackage("#01", ParitySort.ADD, EndMark.CR, DisplayMode.CHAR);
```

利用第 11 章和第 12 章的技术，只需关注基本的通信协议格式，而不必关注附带校验码和结尾码的数据包如何生成，也不必担心收到的附带校验码和结尾码的数据包是否正确，只要调用 getEntirePackage 函数生成综合数据包并调用 checkEntirePackage 函数检查综合数据包，即可解决诸多问题。

12.9 本章小结

本章主要介绍了累加和、异或、循环冗余和累加求补校验码的生成算法及校验算法，以及结尾码的处理方法，然后在此基础之上完成了数据包的综合生成方法与校验方法，并在上一节给出了数据包的综合生成实例。本章与上一章的技术广泛应用于物联网监控系统，第 13～15 章正是对这些技术的典型应用。

第13章 通用 TCP 客户机与服务器测试软件

移动互联网正进入高速普及期，成功的产品和服务模式不断向其他产业领域延伸渗透，其中最重要的就是数据的传输与处理。本章结合第 11 章的数据编码与处理技术及第 12 章的数据包校验技术，设计一个通用客户机与服务器类，保障数据传输的正确性和可靠性。在这些基础之上，设计一个通用 TCP 客户机软件，可以对服务器软件进行测试；同时设计一个通用 TCP 服务器软件，可以配合客户机软件的测试。用手机作为通用 TCP 客户机或服务器，具有便携的特性，可以对现场设备进行测试。

主要知识点：多线程和消息机制、IP 地址获取、Socket 对象的应用、客户机与服务器的实现、AlertDialog 组件的使用

13.1 主要功能和技术特点

通用 TCP 客户机测试软件与对应的服务器软件有相同的数据处理功能，使用同一个类进行数据的接收与发送，只是初始连接方式不同。为了避免引起混淆，这里开始仅介绍 TCP 客户机测试软件，在 13.8 节重点介绍 TCP 服务器的关键代码与性能。

通用 TCP 客户机测试软件（安装后的 App 名为 TCP Client，下文简称 TCP Client）基于 TCP。在数据处理方面，TCP Client 可以自动添加校验码和结尾码，选择以普通字符串或者十六进制字节形式显示，在主窗口自动记录数据收发及发生的时间（精确到毫秒），并且可以根据需要将某时间片内的数据自动合并为一个数据包。下面是 TCP Client 的一些典型功能和技术特点：

- 自动添加异或、累加和、CRC 和累加求补校验码以及 CR 和 CRLF 结尾码，以普通字符串或十六进制字节形式显示收发数据，同时显示时间戳；
- 用来测试以"基于 TCP 服务器模式"运行的"设备 / 模块 / 系统"（为方便行文，下文统称"智能设备"），获取通信参数，为系统开发做前期准备；
- 只要与智能设备处于同一个网段，可随时随地对智能设备进行测试，具有便携性。

13.2 软件操作

TCP Client 运行后的软件界面如图 13-1 所示，左侧上面的单行编辑框（EditText）显示当前接收到的数据或需要发送的数据。下面的多行文本框（TextView）显示历史收发的数据，时间戳中的 TickCount 是开机以来的毫秒数，其后是当前时间；以"<"为前导符的为接收到的数据，以">"为前导符的为发送出去的数据。

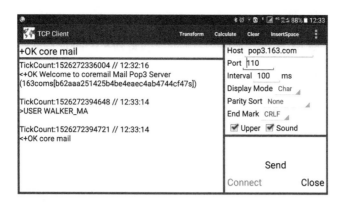

图 13-1　TCP Client 软件界面

右侧上半部分为参数设置，Host 为远程服务器的名称或 IP 地址，Port 为服务器开放的端口号，Interval 为数据间隔的时间长度，这里为 100ms，在此间隔之内到达的数据都从属于一个数据包；Display Mode 为显示方式，分为 Char（文本）和 Hex（十六进制）两种形式；Parity Sort 为数据包的校验码方式，分为 None（无校验码）、Add（累加和校验码）、Xor（异或校验码）、CRC（循环冗余校验码）和 CheckSum（累加求补校验码）；如果核选 Upper，则发送出去的数据自动转换为大写形式，图 13-1 左侧下半部分中的 "USER WALKER_MA" 全部是大写，就是因为核选了 Upper；核选 Sound 则当收到数据时，发出提醒的声音，表示数据到。

菜单项【Transform】将编辑框中的普通字符串转换为十六进制字符串（Display Mode 为 Char 时），在下面的文本框中显示；或者将编辑框中的十六进制字符串转换为普通字符串（Display Mode 为 Hex 时），在下面的文本框中显示。菜单项【Calculate】将编辑框中的数据添加校验码和结尾码后，在下面的文本框中显示。菜单项【Clear】用来清除编辑框和文本框中的数据。菜单项【InsertSpace】用来将编辑框中的数据，每两个字符添加一个空格，为了便于阅读字节。

在溢出菜单项中，【DeleteSpace】用来删除编辑框中字符串所含的空格，菜单项【POP3】用来将远程服务器设置为网易 POP3 服务器，将数据格式设置为 POP3 协议形式。菜单项【Interval】用来测量客户机与远程服务器之间的时间间隔，只有在 CHAR 显示模式、无校验码和结尾码非空且与服务器处于连接状态时，该命令才有效。菜单项【About】显示版权信息。

右侧下半部分为命令按钮区域，其中【Connect】用来连接远程服务器，【Close】用来关闭当前连接，【Send】用来发送数据。无论是发生连接、关闭还是收发数据错误，都将以浮动消息的方式进行提示。

13.3　界面布局

TCP Client 软件的项目名称为 TcpTest，在其界面布局中，大框架采用线性布局 LinearLayout，左侧和右侧都采用相对布局 RelativeLayout，左侧的相对布局宽度权重 layout_weight 为 2，右侧的相对布局宽度权重 layout_weight 为 1。左侧上面为 EditText 标签，用来显示当前收到的数据，或者编辑需要发送的数据；下面为收发的历史数据。右侧的相对布局又分为两个子相对布局 RelativeLayout，上半部分为参数设置，下半部分为命令按钮。

```xml
<LinearLayout xmlns:android="http://schemas.android.com/apk/res/android"
    xmlns:tools="http://schemas.android.com/tools"
    android:layout_width="match_parent"
    android:layout_height="match_parent"
    android:background="@android:color/holo_green_dark"
    android:baselineAligned="false"
    android:orientation="horizontal"
    android:paddingBottom="@dimen/activity_vertical_margin"
    android:paddingLeft="@dimen/activity_horizontal_margin"
    android:paddingRight="@dimen/activity_horizontal_margin"
    android:paddingTop="@dimen/activity_vertical_margin"
    tools:context="com.walkerma.tcptest.MainActivity" >
    <RelativeLayout
        android:id="@+id/rl_left"
        android:layout_width="0dp"
        android:layout_height="wrap_content"
        android:layout_weight="2" >
        <EditText
            android:id="@+id/edCurrent"
            android:layout_width="match_parent"
            android:layout_height="wrap_content"
            android:layout_alignParentLeft="true"
            android:layout_alignParentTop="true"
            android:background="@color/white"
            android:hint="@string/current_data"
            android:inputType="text"
            android:textAppearance="?android:attr/textAppearanceLarge" />
        <TextView
            android:id="@+id/txtHistory"
            android:layout_width="match_parent"
            android:layout_height="match_parent"
            android:layout_alignParentLeft="true"
            android:layout_below="@+id/edCurrent"
            android:layout_marginTop="@dimen/activity_vertical_margin"
            android:background="@color/white"
            android:hint="@string/content_history"
            android:maxLines="20"
            android:textAppearance="?android:attr/textAppearanceMedium" />
    </RelativeLayout>
    <RelativeLayout
        android:id="@+id/rl_right"
        android:layout_width="0dp"
        android:layout_height="wrap_content"
        android:layout_marginLeft="@dimen/activity_horizontal_margin"
        android:layout_weight="1" >
        <RelativeLayout
            android:id="@+id/rl_settings"
            android:layout_width="match_parent"
            android:layout_height="wrap_content"
            android:layout_alignParentLeft="true"
            android:layout_alignParentTop="true"
            android:background="@color/white"
```

```xml
        android:paddingLeft="@dimen/activity_horizontal_margin" >
    <TextView
        android:id="@+id/txtHost"
        android:layout_width="wrap_content"
        android:layout_height="wrap_content"
        android:layout_alignParentLeft="true"
        android:layout_alignParentTop="true"
        android:text="@string/host"
        android:textAppearance="?android:attr/textAppearanceMedium" />
    <EditText
        android:id="@+id/edHost"
        android:layout_width="wrap_content"
        android:layout_height="wrap_content"
        android:layout_alignBaseline="@+id/txtHost"
        android:layout_toRightOf="@+id/txtHost"
        android:hint="@string/iphint"
        android:inputType="text"
        android:maxLength="15"
        android:textAppearance="?android:attr/textAppearanceMedium" />
    <TextView
        android:id="@+id/txtPort"
        android:layout_width="wrap_content"
        android:layout_height="wrap_content"
        android:layout_alignParentLeft="true"
        android:layout_below="@+id/txtHost"
        android:layout_marginTop="@dimen/activity_vertical_margin"
        android:text="@string/port"
        android:textAppearance="?android:attr/textAppearanceMedium" />
    <EditText
        android:id="@+id/edPort"
        android:layout_width="wrap_content"
        android:layout_height="wrap_content"
        android:layout_alignBaseline="@+id/txtPort"
        android:layout_toRightOf="@+id/txtPort"
        android:hint="@string/porthint"
        android:inputType="number"
        android:maxLength="5"
        android:textAppearance="?android:attr/textAppearanceMedium" />
    <TextView
        android:id="@+id/txtInterval"
        android:layout_width="wrap_content"
        android:layout_height="wrap_content"
        android:layout_alignParentLeft="true"
        android:layout_below="@+id/txtPort"
        android:layout_marginTop="@dimen/activity_vertical_margin"
        android:text="@string/interval"
        android:textAppearance="?android:attr/textAppearanceMedium" />
    <EditText
        android:id="@+id/edInterval"
        android:layout_width="wrap_content"
        android:layout_height="wrap_content"
        android:layout_alignBaseline="@+id/txtInterval"
        android:layout_toRightOf="@+id/txtInterval"
        android:hint="@string/intervalhint"
        android:inputType="number"
        android:maxLength="4"
```

```xml
        android:textAppearance="?android:attr/textAppearanceMedium" />
    <TextView
        android:id="@+id/txtMs"
        android:layout_width="wrap_content"
        android:layout_height="wrap_content"
        android:layout_alignBaseline="@+id/txtInterval"
        android:layout_toRightOf="@+id/edInterval"
        android:text="@string/ms"
        android:textAppearance="?android:attr/textAppearanceMedium" />
    <TextView
        android:id="@+id/txtMode"
        android:layout_width="wrap_content"
        android:layout_height="wrap_content"
        android:layout_alignParentLeft="true"
        android:layout_below="@+id/txtInterval"
        android:layout_marginTop="@dimen/activity_vertical_margin"
        android:text="@string/mode"
        android:textAppearance="?android:attr/textAppearanceMedium" />
    <Spinner
        android:id="@+id/spMode"
        android:layout_width="wrap_content"
        android:layout_height="wrap_content"
        android:layout_alignBaseline="@+id/txtMode"
        android:layout_toRightOf="@+id/txtMode"
        android:entries="@array/items_Mode"
        android:gravity="center_horizontal" />
    <TextView
        android:id="@+id/txtParity"
        android:layout_width="wrap_content"
        android:layout_height="wrap_content"
        android:layout_alignParentLeft="true"
        android:layout_below="@+id/txtMode"
        android:layout_marginTop="@dimen/activity_vertical_margin"
        android:text="@string/parity"
        android:textAppearance="?android:attr/textAppearanceMedium" />
    <Spinner
        android:id="@+id/spParity"
        android:layout_width="wrap_content"
        android:layout_height="wrap_content"
        android:layout_alignBaseline="@+id/txtParity"
        android:layout_toRightOf="@+id/txtParity"
        android:entries="@array/items_Parity"
        android:gravity="center_horizontal" />
    <TextView
        android:id="@+id/txtEndMark"
        android:layout_width="wrap_content"
        android:layout_height="wrap_content"
        android:layout_alignParentLeft="true"
        android:layout_below="@+id/txtParity"
        android:layout_marginTop="@dimen/activity_vertical_margin"
        android:text="@string/endmark"
        android:textAppearance="?android:attr/textAppearanceMedium" />
    <Spinner
        android:id="@+id/spEndMark"
        android:layout_width="wrap_content"
        android:layout_height="wrap_content"
```

```xml
            android:layout_alignBaseline="@+id/txtEndMark"
            android:layout_toRightOf="@+id/txtEndMark"
            android:entries="@array/items_EndMark"
            android:gravity="center_horizontal" />
    <CheckBox
            android:id="@+id/chkUpper"
            android:layout_width="wrap_content"
            android:layout_height="wrap_content"
            android:layout_alignParentLeft="true"
            android:layout_below="@+id/txtEndMark"
            android:layout_marginTop="@dimen/activity_vertical_margin"
            android:checked="false"
            android:onClick="onClickCheckBox"
            android:text="@string/upper"
            android:textAppearance="?android:attr/textAppearanceMedium" />
    <CheckBox
            android:id="@+id/chkSound"
            android:layout_width="wrap_content"
            android:layout_height="wrap_content"
            android:layout_alignBaseline="@+id/chkUpper"
            android:layout_marginLeft="@dimen/activity_horizontal_margin"
            android:layout_toRightOf="@+id/chkUpper"
            android:checked="false"
            android:onClick="onClickCheckBox"
            android:text="@string/sound"
            android:textAppearance="?android:attr/textAppearanceMedium" />
</RelativeLayout>
<RelativeLayout
        android:id="@+id/rl_right_bottom"
        android:layout_width="match_parent"
        android:layout_height="match_parent"
        android:layout_alignParentLeft="true"
        android:layout_below="@+id/rl_settings"
        android:layout_marginTop="@dimen/activity_vertical_margin"
        android:background="@color/white" >
    <Button
            android:id="@+id/btSend"
            android:layout_width="wrap_content"
            android:layout_height="wrap_content"
            android:layout_centerInParent="true"
            android:background="@null"
            android:onClick="onClick"
            android:text="@string/send"
            android:textAppearance="?android:attr/textAppearanceLarge" />
    <Button
            android:id="@+id/btConnect"
            android:layout_width="wrap_content"
            android:layout_height="wrap_content"
            android:layout_alignBaseline="@+id/btClose"
            android:layout_below="@+id/btSend"
            android:layout_alignParentBottom="true"
            android:layout_alignParentLeft="true"
            android:layout_marginLeft="@dimen/activity_horizontal_margin"
            android:background="@null"
            android:onClick="onClick"
            android:text="@string/connect"
```

```
                    android:textAppearance="?android:attr/textAppearanceLarge" />
                <Button
                    android:id="@+id/btClose"
                    android:layout_width="wrap_content"
                    android:layout_height="wrap_content"
                    android:layout_alignParentBottom="true"
                    android:layout_alignParentRight="true"
                    android:background="@null"
                    android:onClick="onClick"
                    android:text="@string/close"
                    android:textAppearance="?android:attr/textAppearanceLarge" />
            </RelativeLayout>
        </RelativeLayout>
</LinearLayout>
```

13.4 配置文件

由于需要查询网络状态，进行网络通信，因而需要有访问网络状态的权限及网络通信的权限。收到数据需要播放声音，因而需要调整音量的权限，确保收到数据时能够以最大音量通知用户。在对智能设备测试过程中，需要保持手机屏幕常亮，因而应设置 WAKE_LOCK 权限。

```
<?xml version="1.0" encoding="utf-8"?>
<manifest xmlns:android="http://schemas.android.com/apk/res/android"
    package="com.walkerma.tcptest"
    android:versionCode="1"
    android:versionName="1.0" >
    <uses-sdk
        android:minSdkVersion="14"
        android:targetSdkVersion="21" />
    <uses-permission android:name=
        "android.permission.INTERNET" />
    <uses-permission android:name=
        "android.permission.ACCESS_NETWORK_STATE" />
    <uses-permission android:name=
        "android.permission.WAKE_LOCK" />
    <uses-permission android:name=
        "android.permission.ACCESS_WIFI_STATE" />
    <uses-permission android:name=
        "android.permission.MODIFY_AUDIO_SETTINGS"/>
    <application
        android:allowBackup="true"
        android:icon="@drawable/ic_launcher"
        android:label="@string/app_name"
        android:theme="@style/AppTheme" >
        <activity
            android:name=".MainActivity"
            android:label="@string/app_name"
            android:screenOrientation="landscape" >
            <intent-filter>
                <action android:name="android.intent.action.MAIN" />
                <category android:name="android.intent.category.LAUNCHER" />
            </intent-filter>
```

```
            </activity>
        </application>
</manifest>
```

13.5 网络处理类

网络处理类是 Library 类库的成员，主要用来获取网络状态及本地 IP 地址。静态函数 isNetworkAvailable 用来判断当前设备是否处于联网状态，可能通过 WiFi，也可能通过移动数据；静态函数 isWifiEnabled 用来判断当前设备是否打开 WiFi（但不一定处于联网状态）；静态函数 isWifiConnected 用来判断 WiFi 是否处于联网状态，即当前设备处于联网状态，而且是通过 WiFi 联网，则返回 true，否则返回 false；静态函数 isMobileConnected 用来判断移动数据是否处于联网状态。

静态函数 getLocalWifiIP 用于获取通过 WiFi 联网的本地 IP 地址，而 getLocalGprsIP 用于获取通过移动数据联网的本地 IP 地址，这两个函数都比较耗时，因而，获取当前联网的本地 IP 地址时，通过多线程静态类 ThreadIP 来实现，首先调用函数 isWifiConnected 以确认联网方式，然后分别调用 WiFi 或移动数据的相关函数以获取 IP 地址。在初始化 ThreadIP 对象之前，需要调用静态函数 isNetworkAvailable，确保当前设备处于联网状态。ThreadIP 类主要用于通用 TCP 服务器中，以便将本地 IP 地址告知远程客户机。

静态函数 setMobileDataState 用于设置移动数据的状态，该函数要求 Root 权限，否则调用无效；静态函数 getMobileDataState 则用于获取移动数据的状态；静态函数 isMobileDataEnabled 用于判断移动数据是否处于打开状态（不一定联网）。

网络处理类需要 ACCESS_NETWORK_STATE、CHANGE_NETWORK_STATE、ACCESS_WIFI_STATE、CHANGE_WIFI_STATE 和 WRITE_SETTINGS 等权限，如果只需要 WiFi 或移动数据中的一种，则选择相关权限即可。

```
package com.walkerma.library;
import java.lang.reflect.Method;
import java.net.InetAddress;
import java.net.NetworkInterface;
import java.net.SocketException;
import java.util.Enumeration;
import java.util.Locale;
import android.content.Context;
import android.net.ConnectivityManager;
import android.net.NetworkInfo;
import android.net.wifi.WifiInfo;
import android.net.wifi.WifiManager;
import android.os.Handler;
import android.telephony.TelephonyManager;
public class NetworkProcess {
    public final static int MSG_IP = 30123;
    public static boolean isNetworkAvailable(Context context) {
        ConnectivityManager connectivityManager = (ConnectivityManager)
            context.getSystemService(Context.CONNECTIVITY_SERVICE);
        NetworkInfo activeNetworkInfo = connectivityManager.getActiveNetworkInfo();
        if(activeNetworkInfo==null) return false;
```

```java
            return activeNetworkInfo.isConnected();
    }
    public static boolean isWifiEnabled(Context context){
        WifiManager wifiManager = (WifiManager)
                context.getSystemService(Context.WIFI_SERVICE);
        return wifiManager.isWifiEnabled();
    }
    public static boolean isWifiConnected(Context context){
        ConnectivityManager connectivityManager = (ConnectivityManager)
                context.getSystemService(Context.CONNECTIVITY_SERVICE);
        NetworkInfo activeNetworkInfo = connectivityManager.getActiveNetworkInfo();
        if(activeNetworkInfo==null) return false;
        if(!activeNetworkInfo.isConnected()) return false;
        if(activeNetworkInfo.getType()==ConnectivityManager.TYPE_WIFI)
            return true;
        else
            return false;
    }
    public static boolean isMobileConnected(Context context) {
        ConnectivityManager connectivityManager = (ConnectivityManager)
                context.getSystemService(Context.CONNECTIVITY_SERVICE);
        NetworkInfo activeNetworkInfo = connectivityManager.getActiveNetworkInfo();
        if(activeNetworkInfo==null) return false;
        if(!activeNetworkInfo.isConnected()) return false;
        if(activeNetworkInfo.getType()==ConnectivityManager.TYPE_MOBILE)
            return true;
        else
            return false;
    }
    public static String getLocalWifiIP(Context context){
        WifiManager wm = (WifiManager) context
                .getSystemService(android.content.Context.WIFI_SERVICE );
        WifiInfo wfInfo = wm.getConnectionInfo();
        int ipAddress = wfInfo.getIpAddress();
        return String.format(Locale.ENGLISH, "%d.%d.%d.%d",
                (ipAddress & 0xff),
                (ipAddress >> 8 & 0xff),
                (ipAddress >> 16 & 0xff),
                (ipAddress >> 24 & 0xff));
    }
    public static String getLocalGprsIP(){
        try {
            List<NetworkInterface> interfaces = Collections.list(
                    NetworkInterface.getNetworkInterfaces());
            for (NetworkInterface intf : interfaces) {
                List<InetAddress> addrs = Collections.list(
                        intf.getInetAddresses());
                for (InetAddress addr : addrs) {
                    if (!addr.isLoopbackAddress()) {
                        String sAddr = addr.getHostAddress();
                        //boolean isIPv4 = InetAddressUtils.isIPv4Address(sAddr);
                        boolean isIPv4 = sAddr.indexOf(':')<0;
                        if (useIPv4) {
                            if (isIPv4)
                                return sAddr;
                        } else {
```

```java
                            if (!isIPv4) {
                                int delim = sAddr.indexOf('%');
                                return delim<0 ? sAddr.toUpperCase(Locale.US) :
                                    sAddr.substring(0, delim).
                                    toUpperCase(Locale.US);
                            }
                        }
                    }
                }
            }
        } catch (Exception ex) { }
        return "";
    }
    public static class ThreadIP extends Thread{
        private Handler mHandler;
        private Context context;
        public ThreadIP(Handler mHandler, Context context) {
            this.mHandler = mHandler;
            this.context = context;
        }
        public void run(){
            String strIP;
            if (isWifiConnected(context))
                strIP = getLocalWifiIP(context); // TYPE_WIFI
            else
                strIP = getLocalGprsIP(); // TYPE_MOBILE
            mHandler.obtainMessage(MSG_IP, 0, 0, strIP).sendToTarget();
        }
    }
    public static void setMobileDataState(Context context, boolean mobileDataEnabled){
        try{
            TelephonyManager telephonyService = (TelephonyManager)
                    context.getSystemService(Context.TELEPHONY_SERVICE);
            Method setMobileDataEnabledMethod = telephonyService.getClass().
                    getDeclaredMethod("setDataEnabled", boolean.class);
            if (null != setMobileDataEnabledMethod)
                setMobileDataEnabledMethod.invoke(
                        telephonyService, mobileDataEnabled);
        }
        catch (Exception e){
            e.printStackTrace();
        }
    }
    public static boolean getMobileDataState(Context context){
        try{
            TelephonyManager telephonyService = (TelephonyManager)
                    context.getSystemService(Context.TELEPHONY_SERVICE);
            Method getMobileDataEnabledMethod = telephonyService.
                    getClass().getDeclaredMethod("getDataEnabled");
            if (null != getMobileDataEnabledMethod){
                boolean mobileDataEnabled = (Boolean)
                        getMobileDataEnabledMethod.invoke(telephonyService);
                return mobileDataEnabled;
            }
        }
        catch (Exception e){
```

```
            e.printStackTrace();
        }
        return false;
    }
    public static Boolean isMobileDataEnabled(Context context){
        Object connectivityService = context.getSystemService(
                Context.CONNECTIVITY_SERVICE);
        ConnectivityManager cm = (ConnectivityManager) connectivityService;
        try {
            Class<?> c = Class.forName(cm.getClass().getName());
            Method m = c.getDeclaredMethod("getMobileDataEnabled");
            m.setAccessible(true);
            return (Boolean)m.invoke(cm);
        } catch (Exception e) {
            e.printStackTrace();
            return null;
        }
    }
}
```

13.6 通用 TCP 客户机与服务器类

TCP 客户机和服务器编程只是在获取 Socket 的方式上不同，两者均共享基本的读取和发送函数及大部分消息，基本原理都是通过 Handler 对象发送消息和传递数据给主线程，因而将两者合为一个类 TcpClientServer，可以方便软件维护。

13.6.1 各种声明的说明

TcpClientServer 类中所定义的消息如表 13-1 所示，无论是工作在客户机还是服务器模式，如果软件收到"Interval?!\r"，则立即返回"OK!\r"，同时向主线程发送消息 MSG_ReplyOK，该方法从接收数据到返回数据，中间没有耽误任何时间（使用快速接收算法，具体原理在 13.6.4 节介绍），因而可以用来测试客户机与服务器之间的时间间隔，为软件研发提供参数设置支持。

表 13-1　TcpClientServer 类中的自定义消息说明

名称	说明
MSG_Connected	连接成功
MSG_ConnectError	连接失败
MSG_Closed	主动关闭连接，释放占用资源
MSG_DataArrived	收到数据
MSG_TransmitError	传输错误，一般是对方关闭连接或出现故障引起
MSG_ReadError	读取数据错误
MSG_ReplyOK	响应"OK!"，用于测量两个端点之间的时间间隔

变量 nDelay 用于设置数据包之间的间隔时间，在此间隔时间之内到达的数据从属于一个数据包，将作为一个整体来处理。变量 nDelay 的最小值为 DEFAULT_MIN_DELAY，最

大值为 DEFAULT_MAX_DELAY。

枚举类型 DataSort 中定义了数据的 4 种形式，TEXT 表示普通字符串文本，BYTES 表示字节数据，TEXT_CR 表示以回车符结尾的字符串文本，TEXT_CRLF 表示以回车换行为结尾符的字符串文本。当前数据类型用变量 nDataSort 表示，默认为 TEXT_CR。

Handler 对象 mHandler 由主线程在初始化 TcpClientServer 对象的时候传入，用于向主线程传递消息和数据；String 对象 strServer 是远程服务器的名称或 IP 地址；int 类型变量 nPort 为端口地址，如果以客户机方式工作，则 nPort 为远程服务器的端口号，如果以服务器方式工作，则 nPort 为本地侦听的端口号。

多线程对象 threadConnect 用于连接远程服务器或者等待远程客户机连接，多线程对象 threadRead 则用于在后台读取远程服务器或客户机发来的数据。如果 nDataSort 为字符串文本类型，则采用 BufferedReader 对象 inReader 读取文本数据，并采用 BufferedWriter 对象 outWriter 发送数据。如果 nDataSort 为字节类型，则采用 BufferedInputStream 对象 bis 读取字节数据，并采用 BufferedOutputStream 对象 bos 发送数据。

13.6.2 构造函数

TCP 客户机的构造函数需要使用三个参数，即 Handler 对象 mHandler、String 类型的服务器名称或 IP 地址 strServer、int 类型的端口号 nPort，另外还需要指定工作方式 nWorkMode 为 WorkMode.CLIENT，TcpClientServer 类正是根据此工作方式来确定不同的 Socket 获取方式。

TCP 服务器的构造函数只需要两个参数，即 Handler 对象 mHandler 和 int 类型的端口号 nPort，需要指定工作方式 nWorkMode 为 WorkMode.SERVER。

13.6.3 获取 Socket 对象与多线程的启动

通过构造函数初始化 TcpClientServer 对象后，并不能马上侦听本地端口或连接远程服务器。在连接建立之前，需要调用 setDataSort 函数设置 nDataSort。对于 CLIENT 工作方式，调用 connect 函数，启动多线程 threadConnect 对象与远程服务器建立连接；在 SERVER 工作方式下，调用 listen 函数侦听本地端口，其实也是调用 connect 函数，如此做主要是为了符合编程习惯。在多线程 threadConnect 对象中调用内部函数 getSocket 来获取 Socket 对象 socket，在 CLIENT 工作方式下，socket 对象直接通过 Socket 类实例化得到；在 SERVER 工作方式下，通过调用 ServerSocket 对象的 accept 方法来获取 socket 对象。

在得到 socket 以后，即可根据数据类型 nDataSort 来初始化接收与发送数据的流对象，接着启动多线程对象 threadRead，在后台调用内部函数 readData 接收数据，并通过 mHandler 对象以消息形式将数据传递到主线程。

13.6.4 数据接收与发送

数据接收函数全部是私有函数，因为通过 mHandler 发送数据到达消息 MSG_DataArrived 时附有接收到的数据对象。在 readData 函数中，如果接收字节类型的数据，则调用 readByteBuffer 函数，否则调用 readText 函数。这里以后者为例详细展开数据接收过程，

StringBuffer 对象 sb 通过 append 方法将收到的字符累加，通过 toString 方法转换为字符串并保存于 String 对象 strData 中，在规定的时间间隔之内，通过 while 循环不断重复这一工作。

在 while 循环中，调用 isValidOfTextEnd 函数检查字符串数据的结尾，如果是附带回车符或回车换行符的文本数据，则立即返回接收到的数据，此即高效接收算法。如果收到了以"Interval?!"开头的查询时间间隔的文本数据，则立即发送响应数据并用 MSG_ReplyOK 消息通知主线程。

发送数据函数都是公有函数，sendText 用来发送文本数据，sendBytes 用来发送字节数组形式的数据。无论是使用 outWriter 还是 bos 对象发送数据，都要通过缓冲区进行，为了强迫输出流（或缓冲的流）立即发送数据，即使此时缓冲区还没有填满，也要调用 flush 方法。

13.6.5　TcpClientServer 源代码

从以上分析可知，TcpClientServer 类主要通过初始化对象、设置数据格式，然后与远程服务器连接或侦听本地端口等待远程客户机的连接，连接成功后在后台接收数据，收到的数据通过消息形式传递到主线程，发送数据则直接在主线程中调用公有数据发送函数完成。任务完成后，需要调用公有函数 close 关闭连接，释放占用的资源。

```java
package com.walkerma.library;
import java.io.BufferedInputStream;
import java.io.BufferedOutputStream;
import java.io.BufferedReader;
import java.io.BufferedWriter;
import java.io.IOException;
import java.io.InputStreamReader;
import java.io.OutputStreamWriter;
import java.net.InetAddress;
import java.net.InetSocketAddress;
import java.net.ServerSocket;
import java.net.Socket;
import java.net.UnknownHostException;
import java.nio.ByteBuffer;
import java.util.Locale;
import android.os.Handler;
public class TcpClientServer {
    public static final int MSG_Connected = 10001;
    public static final int MSG_ConnectError = 10002;
    public static final int MSG_Closed = 10003;
    public static final int MSG_DataArrived = 10004;
    public static final int MSG_TransmitError = 10005;
    public static final int MSG_ReadError = 10006;
    public static final int MSG_ReplyOK = 10007;
    public static final int CONNECT_TIMEOUT = 3000;     //ms
    private final int DEFAULT_MIN_DELAY = 50;
    private final int DEFAULT_MAX_DELAY = 10000;
    private int nDelay;
    public enum DataSort{TEXT, BYTES, TEXT_CR, TEXT_CRLF}
    //default data format
```

```java
private DataSort nDataSort=DataSort.TEXT_CR;
private enum WorkMode{CLIENT, SERVER}
private WorkMode nWorkMode = WorkMode.CLIENT;
private Handler mHandler;
private String strServer;
private int nPort;
private ThreadConnect threadConnect;
private ThreadRead threadRead;
private ServerSocket serverSocket=null;
private Socket socket=null;
private boolean bStopConnect=false;
private BufferedReader inReader;     //for text
private BufferedWriter outWriter;
private BufferedInputStream bis;     // for bytes
private BufferedOutputStream bos;
private boolean bConnected;
private void getSocket() {
    if(nWorkMode==WorkMode.CLIENT){
        socket = new Socket();
        try {
            socket.connect(new InetSocketAddress(
                    strServer, nPort), CONNECT_TIMEOUT);
        } catch (IOException e) {
            e.printStackTrace();
        }
    }
    else{
        try {
            serverSocket = new ServerSocket(nPort);
            socket = serverSocket.accept();
            serverSocket.close();
            serverSocket = null;
        } catch (IOException e) {
            e.printStackTrace();
        }
    }
}
public final class ThreadConnect extends Thread{
    public void run() {
        bConnected = false;
        try {
            getSocket();
            if(socket==null) {
                sendResetMsg(MSG_ConnectError);
                return;
            }
            if(nDataSort == DataSort.BYTES){
                bis = new BufferedInputStream(socket.getInputStream());
                bos = new BufferedOutputStream(socket.getOutputStream());
            }
            else{
                inReader = new BufferedReader(new InputStreamReader(
                        socket.getInputStream()));
                outWriter = new BufferedWriter(new OutputStreamWriter(
                        socket.getOutputStream()));
            }
```

```java
                    bConnected = true;
                } catch (UnknownHostException e) {
                    e.printStackTrace();
                } catch (IOException e) {
                    e.printStackTrace();
                }
                if (bConnected){
                    mHandler.obtainMessage(MSG_Connected,
                            0, 0,null).sendToTarget();
                    threadRead = new ThreadRead();
                    threadRead.start();
                }
                else{
                    sendResetMsg(MSG_ConnectError);
                }
            }
        }
    }
    public final class ThreadRead extends Thread{
        boolean bDataArrived;
        public void run() {
            while(true){
                if(bConnected != true) return;
                if(bStopConnect) return;
                try {
                    bDataArrived = false;
                    if(nDataSort==DataSort.BYTES){
                        if(bis.available()>0) bDataArrived = true;
                    }
                    else{
                        if(inReader.ready()) bDataArrived = true;
                    }
                    if(bDataArrived){
                        Object oData = readData();
                        if(bConnected != true) return; //be necessary!
                        mHandler.obtainMessage(MSG_DataArrived, 0, 0,
                                oData).sendToTarget();
                    }
                } catch (IOException e) {
                    e.printStackTrace();
                    sendResetMsg(MSG_ReadError);
                }
            }
        }
    }
    public TcpClientServer(Handler mHandler, String strServer, int nPort ) {
        this.mHandler = mHandler;
        this.strServer = strServer;
        this.nPort = nPort;
        nWorkMode = WorkMode.CLIENT;
    }
    public TcpClientServer(Handler mHandler, int nPort) {
        this.mHandler = mHandler;
        this.nPort = nPort;
        nWorkMode = WorkMode.SERVER;
    }
    public String getRemoteIP(){
```

```java
        if(socket==null) return "";
        InetAddress ia = socket.getInetAddress();
        byte[] ipArray = ia.getAddress();
        return String.format(Locale.US, "%d.%d.%d.%d",
                ((int)ipArray[0])&0xff, ((int)ipArray[1])&0xff,
                ((int)ipArray[2])&0xff, ((int)ipArray[3])&0xff);
    }
    public void setDataSort(DataSort nDataSort){
        if(bConnected) return;
        this.nDataSort = nDataSort;
    }
    public void setReadDelay(int nDelay){
        if((nDelay<DEFAULT_MIN_DELAY)||
                (nDelay>DEFAULT_MAX_DELAY))
            this.nDelay = DEFAULT_MIN_DELAY;
        else
            this.nDelay = nDelay;
    }
    public void connect(){
        threadConnect = new ThreadConnect();
        threadConnect.start();
    }
    public void listen(){
        connect();
    }
    private Object readData(){
        if(nDataSort == DataSort.BYTES)
            return readByteBuffer();
        else
            return readText();
    }
    private void sendResetMsg(int msg){
        bConnected = false;
        bStopConnect = true;
        mHandler.obtainMessage(msg,
                0, 0, null).sendToTarget();
    }
    private String readText(){
        StringBuffer sb = new StringBuffer();
        String strData="";
        long lStart = System.currentTimeMillis();
        while(System.currentTimeMillis()-lStart < nDelay){
            try{
                while(inReader.ready()){
                    sb.append((char)(inReader.read()));
                    strData = sb.toString();
                    if(isValidOfTextEnd(strData)) return strData;
                    lStart = System.currentTimeMillis();
                }
            }catch (IOException e) {
                e.printStackTrace();
                sendResetMsg(MSG_ReadError);
            }
        }
        return strData;
    }
```

```java
    private boolean isValidOfTextEnd(String strData){
        switch(nDataSort){
        case TEXT_CR:
            if(strData.startsWith("Interval?!\r")){
                sendText("OK!"+"\r");
                mHandler.obtainMessage(MSG_ReplyOK,
                        0, 0,null).sendToTarget();
            }
            if(strData.endsWith("\r")) return true;
            break;
        case TEXT_CRLF:
            if(strData.startsWith("Interval?!\r\n")){
                sendText("OK!"+"\r\n");
                mHandler.obtainMessage(MSG_ReplyOK,
                        0, 0,null).sendToTarget();
            }
            if(strData.endsWith("\r\n")) return true;
            break;
        default:
            break;
        }
        return false;
    }
    private ByteBuffer readByteBuffer(){
        ByteBuffer buf = ByteBuffer.allocate(1024);
        long lStart = System.currentTimeMillis();
        while(System.currentTimeMillis()-lStart < nDelay){
            try{
                while(bis.available()>0){
                    byte[] bIn = new byte[bis.available()];
                    bis.read(bIn, 0, bis.available());
                    buf.put(bIn);
                    lStart = System.currentTimeMillis();
                }
            }catch (IOException e) {
                e.printStackTrace();
                sendResetMsg(MSG_ReadError);
            }
        }
        return buf;
    }
    public void sendText(String strText){
        try {
            outWriter.write(strText);
            outWriter.flush();
        } catch (IOException e) {
            e.printStackTrace();
            sendResetMsg(MSG_TransmitError);
        }
    }
    public void sendBytes(byte[] bOut){
        try {
            bos.write(bOut, 0, bOut.length);
            bos.flush();
        } catch (IOException e) {
            e.printStackTrace();
```

```
                sendResetMsg(MSG_TransmitError);
            }
        }
        public void close(){
            try {
                if(nDataSort == DataSort.BYTES){
                    if(bis!=null) bis.close();
                    if(bos!=null) bos.close();
                }
                else{
                    if(inReader!=null) inReader.close();
                    if(outWriter!=null) outWriter.close();
                }
                if(socket!=null)
                    if(!socket.isClosed()) socket.close();
                if(serverSocket!=null)
                    if(!serverSocket.isClosed()) serverSocket.close();
                sendResetMsg(MSG_Closed);
            } catch (IOException e) {
                e.printStackTrace();
            }
        }
    }
```

13.7 窗体源代码

TCP Client 软件需要引用 Library 类库，调用以上两节的类和其他相关类。String 常量 INTERVAL_TEST 中定义了测量客户机与服务器之间时间间隔的字符串。布尔变量 bUpper 用来控制是否需要大写发送的字符串，bSound 用来控制接收到数据时是否进行声音提醒。整型变量 nInterval 定义接收的数据间隔，在此间隔之内的数据从属于一个数据包，默认为 50 毫秒，在整型常量 MIN_INTERVAL 中定义，也可在软件运行时在图 13-1 中进行设置。字符串常量 DEFAULT_HOST 中定义了一个默认远程服务器 pop3.163.com，用来测试 WiFi 是否连接正常。默认远程服务器端口在整型常量 DEFAULT_PORT 中定义，方便快速操作。ArrayList<String> 对象 listHistory 用来保存收发的历史数据，仅保留 5（在常量 MAX_LIST_HISTORY 中定义）条数据，如果超过 5 条，则清除最先发送或接收到的第一条数据。

对于 TCP Client 软件所使用的连接远程服务器的参数以及数据处理参数，通过在 onDestroy 方法中调用 saveData 函数而保存到内部存储器；而在软件启动初始化的时候，调用相应的 restoreData 函数恢复参数，并在界面显示。最核心的初始化工作在 onClick 方法的 btConnect 选项中，在此完成客户机对象 client 的初始化工作，设置延迟和数据类型，然后调用 connect 函数连接远程服务器。发送数据比较简单，在 btSend 选项中调用 sendData 函数，在其中根据数据处理参数添加校验码和结尾码，然后再调用 client 对象的 sendText 或 sendBytes 函数将数据发送出去。

客户机对象 client 向主线程传递消息和数据时，触发 Handler 对象 mHandler 的 handleMessage 方法，在其中调用 processMessage 函数对消息和数据进行综合处理，对于数据到达消息 MSG_DataArrived，调用 processData 函数，检查数据是否正确，并调用 displayHistory 函数在历史区域显示收到的数据。对于连接成功消息 MSG_Connected 和连接关闭消息 MSG_Closed 都要调用 setButtonStatus，用来更新命令按钮和菜单项【Interval】的状态。

```java
package com.walkerma.tcptest;
import java.nio.ByteBuffer;
import java.text.SimpleDateFormat;
import java.util.ArrayList;
import java.util.Calendar;
import java.util.Date;
import java.util.Locale;
import android.app.Activity;
import android.app.AlertDialog;
import android.content.Context;
import android.media.AudioManager;
import android.media.MediaPlayer;
import android.os.Bundle;
import android.os.Handler;
import android.os.Message;
import android.text.TextUtils;
import android.view.KeyEvent;
import android.view.Menu;
import android.view.MenuItem;
import android.view.View;
import android.view.WindowManager;
import android.widget.AdapterView;
import android.widget.Button;
import android.widget.CheckBox;
import android.widget.EditText;
import android.widget.Spinner;
import android.widget.TextView;
import android.widget.Toast;
import android.widget.AdapterView.OnItemSelectedListener;
import com.walkerma.library.GeneralProcess;
import com.walkerma.library.NetworkProcess;
import com.walkerma.library.FileProcess;
import com.walkerma.library.TcpClientServer;
import com.walkerma.library.ByteProcess;
import com.walkerma.library.Parity;
import com.walkerma.library.Parity.DisplayMode;
import com.walkerma.library.Parity.ParitySort;
import com.walkerma.library.Parity.EndMark;
import com.walkerma.library.TcpClientServer.DataSort;
public class MainActivity extends Activity {
    private EditText edCurrent, edHost, edPort, edInterval;
    private TextView txtHistory;
    private Spinner spMode, spParity, spEndMark;
    private CheckBox chkUpper, chkSound;
    private Button btConnect, btClose, btSend;
    private final String INTERVAL_TEST = "Interval?!";
    private MenuItem menuInterval;
    private boolean bMenuCreated=false;
    private DisplayMode nDisplayMode=DisplayMode.CHAR;
    private ParitySort nParitySort=ParitySort.NONE;
    private EndMark nEndMark=EndMark.NONE;
```

```java
private boolean bUpper=true, bSound=false;
private final int MIN_INTERVAL = 50;
private int nInterval; //ms
private ArrayList<String> listHistory = new ArrayList<String>();
private final int MAX_LIST_HISTORY = 5;
private final String DEFAULT_HOST = "pop3.163.com";
private String strHost;
private final int DEFAULT_PORT = 1024;
private int nPort;
private TcpClientServer client;
private boolean bConnected = false;
private boolean bConnectSent=false;
private boolean bCloseSent=false;
private Handler mHandler;
private AudioManager am;
private int nMode;
private int nCurrentVolume;
private boolean bSpeakerStatus;
private MediaPlayer mPlayer;
private boolean bCanExit = false;
private final int MSG_Delay = 10016;
private void setSoundMode(){
    am = (AudioManager)getSystemService(Context.AUDIO_SERVICE);
    if(am == null) return;
    nMode = am.getMode();
    am.setMode(AudioManager.MODE_IN_CALL);
    nCurrentVolume = am.getStreamVolume(AudioManager.STREAM_MUSIC);
    am.setStreamVolume(AudioManager.STREAM_MUSIC,
            am.getStreamMaxVolume(AudioManager.STREAM_MUSIC),0);
    bSpeakerStatus = am.isSpeakerphoneOn();
    am.setSpeakerphoneOn(true);
}
private void restoreSoundMode(){
    if(am == null) return;
    am.setMode(nMode);
    am.setStreamVolume(AudioManager.STREAM_MUSIC, nCurrentVolume, 0);
    am.setSpeakerphoneOn(bSpeakerStatus);
}
class IncomingHandlerCallback implements Handler.Callback{
    @Override
    public boolean handleMessage(Message msg) {
        processMessage(msg);
        return true;
    }
}
private void processData(Message msg){
    String strData, strFormat;
    if(nDisplayMode == DisplayMode.CHAR){
        strData = (String)msg.obj;
        strFormat = strData;
    }
    else{
        strData = ByteProcess.bytesToHexChars(
                ByteProcess.byteBufferToBytes((ByteBuffer)msg.obj));
        strFormat = ByteProcess.insertSpaceToHexChars(strData);
    }
```

```java
            edCurrent.setText(strFormat);
            displayHistory(strData, "<");
            if(!Parity.checkEntirePackage(strData, nParitySort,
                    nEndMark, nDisplayMode))
                Toast.makeText(this, "Package Error.",
                        Toast.LENGTH_SHORT).show();
    }
    private void processMessage(Message msg){
        switch(msg.what){
        case MSG_Delay:
            bCanExit = false;
            return;
        case TcpClientServer.MSG_DataArrived:
            processData(msg);
            if(bSound) playSound();
            return;
        case TcpClientServer.MSG_ReplyOK:
            displayHistory("Auto replay: " +
                    Parity.getEntirePackage("OK!",
                            ParitySort.NONE, nEndMark,
                            DisplayMode.CHAR), ">");
            return;
        case TcpClientServer.MSG_Connected:
            bConnected = true;
            Toast.makeText(this, "Connected.",
                    Toast.LENGTH_SHORT).show();
            listHistory.clear();
            break;
        case TcpClientServer.MSG_ReadError:
            bConnected = false;
            if(bCloseSent) break;
            Toast.makeText(this, "Read error.",
                    Toast.LENGTH_SHORT).show();
            break;
        case TcpClientServer.MSG_ConnectError:
            bConnected = false;
            Toast.makeText(this, "Connect error.",
                    Toast.LENGTH_SHORT).show();
            break;
        case TcpClientServer.MSG_TransmitError:
            bConnected = false;
            Toast.makeText(this, "Transmit error.",
                    Toast.LENGTH_SHORT).show();
            break;
        case TcpClientServer.MSG_Closed:
            bConnected = false;
            if(bConnectSent) break;
            Toast.makeText(this, "Closed.",
                    Toast.LENGTH_SHORT).show();
        }
        setButtonStatus(bConnected);
    }
    private void setButtonStatus(boolean bConnected){
        if(bConnected){
            btSend.setEnabled(true);
            btClose.setEnabled(true);
```

```java
            btConnect.setEnabled(false);
            spMode.setEnabled(false);
            spParity.setEnabled(false);
            spEndMark.setEnabled(false);
            if(bMenuCreated && (nDisplayMode==DisplayMode.CHAR) &&
                    (nParitySort==ParitySort.NONE) &&
                    (nEndMark!=EndMark.NONE))
                menuInterval.setEnabled(true);
        }
        else{
            btConnect.setEnabled(true);
            btSend.setEnabled(false);
            btClose.setEnabled(false);
            spMode.setEnabled(true);
            spParity.setEnabled(true);
            spEndMark.setEnabled(true);
            if(bMenuCreated) menuInterval.setEnabled(false);
        }
}
private void playSound(){
    try {
        mPlayer = MediaPlayer.create(this, R.raw.notify);
        if(mPlayer.getCurrentPosition()>0)
            mPlayer.seekTo(0);
        mPlayer.start();
    } catch (IllegalStateException e) {
        // TODO Auto-generated catch block
        e.printStackTrace();
    }
}
@Override
protected void onCreate(Bundle savedInstanceState) {
    super.onCreate(savedInstanceState);
    setContentView(R.layout.activity_main);
    edCurrent = (EditText)findViewById(R.id.edCurrent);
    txtHistory = (TextView)findViewById(R.id.txtHistory);
    edHost = (EditText)findViewById(R.id.edHost);
    edPort = (EditText)findViewById(R.id.edPort);
    edInterval = (EditText)findViewById(R.id.edInterval);
    chkUpper = (CheckBox)findViewById(R.id.chkUpper);
    chkSound = (CheckBox)findViewById(R.id.chkSound);
    btConnect = (Button)findViewById(R.id.btConnect);
    btClose = (Button)findViewById(R.id.btClose);
    btSend = (Button)findViewById(R.id.btSend);
    initDisplayMode();
    initParitySort();
    initEndMark();
    getWindow().addFlags(
            WindowManager.LayoutParams.FLAG_KEEP_SCREEN_ON);
    restoreData();
    setButtonStatus(bConnected);
    mHandler = new Handler(new IncomingHandlerCallback());
}
private void restoreData(){
    String strData = FileProcess.readInternalData(this,
            getString(R.string.app_name));
```

```java
            if(TextUtils.isEmpty(strData)) return;
            String[] itemArray = strData.split("/");
            if(!(itemArray[0]=="#")) edHost.setText(itemArray[0]);
            edPort.setText(itemArray[1]);
            nPort = Integer.parseInt(itemArray[1]);
            edInterval.setText(itemArray[2]);
            nInterval = Integer.parseInt(itemArray[2]);
            nDisplayMode = getDisplayMode(Integer.parseInt(itemArray[3]));
            spMode.setSelection(nDisplayMode.ordinal());
            nParitySort = getParitySort(Integer.parseInt(itemArray[4]));
            spParity.setSelection(nParitySort.ordinal());
            nEndMark = getEndMark(Integer.parseInt(itemArray[5]));
            spEndMark.setSelection(nEndMark.ordinal());
            bUpper = Boolean.parseBoolean(itemArray[6]);
            chkUpper.setChecked(bUpper);
            bSound = Boolean.parseBoolean(itemArray[7]);
            chkSound.setChecked(bSound);
            setSoundMode();
        }
        private void saveData(){
            getEditTextData();
            String strData = TextUtils.isEmpty(strHost) ? "#": strHost;
            strData += "/" + Integer.toString(nPort);
            strData += "/" + Integer.toString(nInterval);
            strData += "/" + Integer.toString(nDisplayMode.ordinal());
            strData += "/" + Integer.toString(nParitySort.ordinal());
            strData += "/" + Integer.toString(nEndMark.ordinal());
            strData += "/" + Boolean.toString(bUpper);
            strData += "/" + Boolean.toString(bSound);
            FileProcess.writeInternalData(this,
                    getString(R.string.app_name), strData);
        }
        private void initDisplayMode(){
            spMode = (Spinner)findViewById(R.id.spMode);
            spMode.setOnItemSelectedListener(new OnItemSelectedListener(){
                @Override
                public void onItemSelected(AdapterView<?> parent, View view,
                        int position, long id) {
                    nDisplayMode = getDisplayMode(position);
                }
                @Override
                public void onNothingSelected(AdapterView<?> parent) {
                    //TODO
                }});
        }
        private DisplayMode getDisplayMode(int nIndex){
            if(nIndex == 0)
                return DisplayMode.CHAR;
            else
                return DisplayMode.HEX;
        }
        private void initParitySort(){
            spParity = (Spinner)findViewById(R.id.spParity);
            spParity.setOnItemSelectedListener(new OnItemSelectedListener(){
                @Override
                public void onItemSelected(AdapterView<?> parent, View view,
```

```java
                        int position, long id) {
                    nParitySort = getParitySort(position);
                }
                @Override
                public void onNothingSelected(AdapterView<?> parent) {
                    // TODO Auto-generated method stub
                }});
    }
    private ParitySort getParitySort(int nIndex){
        switch(nIndex){
        case 1:
            return ParitySort.XOR;
        case 2:
            return ParitySort.ADD;
        case 3:
            return ParitySort.CRC;
        case 4:
            return ParitySort.CHECK_SUM;
        default:
            return ParitySort.NONE;
        }
    }
    private void initEndMark(){
        spEndMark = (Spinner)findViewById(R.id.spEndMark);
        spEndMark.setOnItemSelectedListener(new OnItemSelectedListener(){
            @Override
            public void onItemSelected(AdapterView<?> parent, View view,
                    int position, long id) {
                nEndMark = getEndMark(position);
            }
            @Override
            public void onNothingSelected(AdapterView<?> parent) {
                // TODO Auto-generated method stub
            }});
    }
    private EndMark getEndMark(int nIndex){
        switch(nIndex){
        case 1:
            return EndMark.CR;
        case 2:
            return EndMark.CRLF;
        default:
            return EndMark.NONE;
        }
    }
    @Override
    public boolean onCreateOptionsMenu(Menu menu) {
        // Inflate the menu; this adds items to the action bar if it is present.
        getMenuInflater().inflate(R.menu.main, menu);
        menuInterval = menu.findItem(R.id.action_interval);
        menuInterval.setEnabled(false);
        bMenuCreated = true;
        return true;
    }
    @Override
    public boolean onOptionsItemSelected(MenuItem item) {
```

```java
int id = item.getItemId();
String strCurrent = edCurrent.getText().toString();
String strRet;
switch(id){
case R.id.action_transform:
    if(TextUtils.isEmpty(strCurrent)) return true;
    if(nDisplayMode==DisplayMode.HEX)
        strRet = ByteProcess.hexCharsToEnString(strCurrent);
    else{
        strRet = ByteProcess.enStringToHexChars(strCurrent);
        strRet = ByteProcess.insertSpaceToHexChars(strRet);
    }
    txtHistory.setText(strRet);
    return true;
case R.id.action_calculate:
    if(nParitySort==ParitySort.NONE &&
    nEndMark==EndMark.NONE) {
        Toast.makeText(this, "No any PairtySort or EndMark.",
                Toast.LENGTH_SHORT).show();
        return true;
    }
    if(TextUtils.isEmpty(strCurrent)) return true;
    strRet = Parity.getEntirePackage(strCurrent,
            nParitySort, nEndMark, nDisplayMode);
    txtHistory.setText(strRet);
    return true;
case R.id.action_clear:
    edCurrent.setText("");
    txtHistory.setText("");
    return true;
case R.id.action_insert:
    if(TextUtils.isEmpty(strCurrent)) return true;
    strCurrent = ByteProcess.deleteSpaceFromHexChars(strCurrent);
    edCurrent.setText(ByteProcess.insertSpaceToHexChars(strCurrent));
    return true;
case R.id.action_delete:
    if(TextUtils.isEmpty(strCurrent)) return true;
    edCurrent.setText(ByteProcess.deleteSpaceFromHexChars(strCurrent));
    return true;
case R.id.action_pop3:
    edHost.setText(DEFAULT_HOST);
    edPort.setText("110");
    edInterval.setText("100");
    spMode.setSelection(0);
    nDisplayMode = DisplayMode.CHAR;
    spParity.setSelection(0);
    nParitySort = ParitySort.NONE;
    spEndMark.setSelection(2);
    nEndMark = EndMark.CRLF;
    chkSound.setChecked(true);
    bSound = true;
    setSoundMode();
    return true;
case R.id.action_interval:
    sendData(INTERVAL_TEST);
    return true;
```

```java
            case R.id.action_about:
                GeneralProcess.fireAboutDialog(this,
                        getString(R.string.app_name),
                        getString(R.string.strCopyrights));
                return true;
        }
        return super.onOptionsItemSelected(item);
    }
    @Override
    public boolean onKeyDown(int keyCode, KeyEvent event) {
        // TODO Auto-generated method stub
        switch(keyCode){
        case KeyEvent.KEYCODE_BACK:
            if(!bCanExit){
                bCanExit = true;
                Toast.makeText(getApplicationContext(),
                        "Press again to quit.", Toast.LENGTH_SHORT).show();
                // if not pressed within 1 seconds then will be setted(canExit) as false
                mHandler.sendEmptyMessageDelayed(MSG_Delay, 1000);
                return true;
            }
            else break;
        }
        return super.onKeyDown(keyCode, event);
    }
    @Override
    protected void onDestroy() {
        mHandler.removeCallbacksAndMessages(null);
        if(client!=null) client.close();
        if(mPlayer!=null) mPlayer.release();
        if(bSound) restoreSoundMode();
        saveData();
        getWindow().clearFlags(
                WindowManager.LayoutParams.FLAG_KEEP_SCREEN_ON);
        super.onDestroy();
    }
    public void onClickCheckBox(View v){
        int id = v.getId();
        switch(id){
        case R.id.chkUpper:
            if(chkUpper.isChecked())
                bUpper = true;
            else
                bUpper = false;
            break;
        case R.id.chkSound:
            if(chkSound.isChecked()){
                bSound = true;
                setSoundMode();
            }
            else{
                bSound = false;
                restoreSoundMode();
            }
            break;
        }
```

```java
    }
    private void getEditTextData(){
        String strData;
        strHost = edHost.getText().toString();
        if(TextUtils.isEmpty(strHost)){
            strHost = DEFAULT_HOST;
            edHost.setText(strHost);
        }
        strData = edPort.getText().toString();
        if(TextUtils.isEmpty(strData)){
            nPort = DEFAULT_PORT;
            edPort.setText(Integer.toString(nPort));
        }
        else
            nPort = Integer.parseInt(strData);
        strData = edInterval.getText().toString();
        if(TextUtils.isEmpty(strData)){
            nInterval = MIN_INTERVAL;
            edInterval.setText(Integer.toString(nInterval));
        }
        else
            nInterval = Integer.parseInt(strData);
    }
    private void alertForNetwork(){
        AlertDialog.Builder builder = new AlertDialog.Builder(this);
        builder.setIcon(R.drawable.ic_action_warning);
        builder.setTitle("Wi-Fi/GPRS: ");
        builder.setMessage("Wi-Fi/GPRS is disconnected.");
        builder.setPositiveButton("OK", null);
        builder.setCancelable(false);
        AlertDialog dialog = builder.create();
        dialog.show();
    }
    public void onClick(View v){
        String strData;
        int id = v.getId();
        switch(id){
        case R.id.btConnect:
            bConnectSent = true;
            bCloseSent = false;
            getEditTextData();
            if(!NetworkProcess.isNetworkAvailable(this)){
                alertForNetwork();
                return;
            }
            Toast.makeText(this, "Connecting...",
                    Toast.LENGTH_SHORT).show();
            if(client!=null) client.close();
            client = new TcpClientServer(mHandler, strHost, nPort);
            client.setReadDelay(nInterval);
            if(nDisplayMode == DisplayMode.CHAR){
                switch(nEndMark){
                case CR:
                    client.setDataSort(DataSort.TEXT_CR);
                    break;
                case CRLF:
```

```java
                        client.setDataSort(DataSort.TEXT_CRLF);
                        break;
                    default:
                        client.setDataSort(DataSort.TEXT);
                        break;
                }
            }
            else
                client.setDataSort(DataSort.BYTES);
            client.connect();
            break;
        case R.id.btSend:
            strData = edCurrent.getText().toString();
            if(TextUtils.isEmpty(strData)){
                Toast.makeText(this, "Data empty.", Toast.LENGTH_SHORT).show();
                return;
            }
            sendData(strData);
            break;
        case R.id.btClose:
            bCloseSent = true;
            bConnectSent = false;
            if(client!=null){
                client.close();
                client = null;
            }
            break;
    }
}
private void sendData(String strData){
    if(bUpper) strData = strData.toUpperCase(Locale.ENGLISH);
    strData = Parity.getEntirePackage(strData,
            nParitySort, nEndMark, nDisplayMode);
    if(nDisplayMode==DisplayMode.CHAR)
        client.sendText(strData);
    else
        client.sendBytes(ByteProcess.hexCharsToBytes(strData));
    displayHistory(strData, ">");
}
private void displayHistory(String strData, String mark){
    String strFormat = strData;
    String strStamp = getTimeStamp() + "\r\n";
    if(nDisplayMode==DisplayMode.HEX)
        strFormat = ByteProcess.insertSpaceToHexChars(strData);
    listHistory.add(strStamp + mark + strFormat + "\r\n");
    while(listHistory.size()>MAX_LIST_HISTORY) listHistory.remove(0);
    strStamp = "";
    for(int i=0; i<listHistory.size(); i++)
        strStamp += listHistory.get(i);
    txtHistory.setText(strStamp);
}
private String getCurrentTime(String format){
    //format = "H:mm:ss", 9:59:59
    Calendar cl = Calendar.getInstance();
    Date dt = cl.getTime();
    SimpleDateFormat formatter = new SimpleDateFormat(format, Locale.US);
```

```
        return formatter.format(dt);
    }
    private String getTimeInMs(){
        Calendar cl = Calendar.getInstance();
        return Long.toString(cl.getTimeInMillis());
    }
    private String getTimeStamp(){
        //TickCount:2062050543 // Time:17:22:09
        return "TickCount:" + getTimeInMs() + " // " +
        getCurrentTime("H:mm:ss");
    }
}
```

13.8　TCP 服务器的关键代码

通用 TCP 服务器测试软件（安装后的 App 名为 TCP Server，下文简称 TCP Server）的项目名称为 TCP_Server，在初始化 TcpClientServer 类的对象时，只需要提供 Handler 对象 mHandler 和端口号，前者用于向主线程提交数据和发送消息，后者用于侦听本地端口。在侦听本地端口的命令按钮的 onClick 方法中，也只需要使用 mHandler 和端口号。另外，TCP Server 需要获取本地 IP 地址，这需要启用 13.5 节中的静态类 ThreadIP 对象，以便告知客户机进行连接。

13.9　本章小结

本章利用第 11 章的数据编码与处理技术及第 12 章的数据包校验技术，结合本章的网络处理类和通用 TCP 客户机与服务器类，实现了通用 TCP 客户机测试软件和通用 TCP 服务器测试软件，可以自动生成多种校验码与结尾码，也可以对多种数据进行检验，借助这两个软件可以方便地测试两者之间的时间间隔，为软件编程提供合理的时间参数。本章的技术可以广泛应用于物联网的数据检测与监控，随后的两章内容正是对本章的 TCP 客户机与服务器的具体化应用。

第 14 章　I-7013D 模块仿真软件

物联网广泛应用于众多领域，数据采集与输出控制模块是物联网系统中直接与被监控对象关联的不可或缺的输入输出模块，学习和研究这些模块对物联网系统的开发、测试与教学等都具有重要意义。教育部倡导提高教学质量，加强实践技能的培养，更好地为社会主义建设服务。但是，一方面企业招不到具有工程项目经验的物联网系统研发人才，另一方面学校由于设备投入不足也难以培养物联网人才。

I-7013D 模块是台湾泓格公司的产品，基于 RS-485 工业控制总线，可以采集一路温度模拟量。作者前期完成了 Windows 版"I-7013D 模块仿真软件"（软件著作权号：2017SR374721），采用该模块的实物产品图片作为背景，模拟实际的工程环境和信号指示，并且全部采用该产品提供的协议进行操作。为了进一步培养学生的网络编程能力，本章研发 Android 环境下的 TCP 版"I-7013D 模块仿真软件"。

主要知识点：TCP 服务器的具体实现、AlertDialog 组件的使用

14.1　主要功能和技术特点

I-7013D 模块仿真软件（安装后的 App 名为 i-7013D Simulated，下文简称"仿真模块"）采用 TCP 服务器模式工作，提供 IP 地址和端口号供远程客户机连接。其主要特点为：

- 与实物模块一样，采集一路温度量数据。
- 温度可以以三种方式进行仿真：手动方式，温度可以单向升高（设置 37.19℃为上限）或单向下降（设置 20℃为下限）；自动方式，温度从下限逐渐升高到上限，然后从上限逐渐下降到下限，如此周而复始；锁定方式，当仿真模块运行于手动或自动方式时，锁定方式可以停止数据变化，这样便于观察温度读取效果。
- 为了方便检测软件可视化显示温度趋势，仿真模块以正弦波形式产生温度。
- 温度查询协议与实物模块的一致且显示于主界面，协议采用字符格式，累加和校验码可选，以回车符结尾。
- 仿真模块只要与检测软件处于同一网络，均可提供便携的 TCP 服务器连接，除手机外无须增加任何设备。

14.2　软件操作

仿真模块运行后的软件界面如图 14-1 所示，底部的图片来自 Windows 版仿真模块的示意图，由于 TCP 版无须设置波特率相关工作参数，因而，仅实现温度读取及响应协议。点

击【Listen】将显示本地 IP 地址,可以设置服务器侦听的端口号 Port(这里为 1024),仿真模块的 ID 范围在 00～FF 之间。

可以用上一章的通用 TCP 客户机测试软件连接此 IP 地址和 Port,连接成功后,第一行的电源指示灯由灰色变成红色。当工作模式为自动(Auto)时,温度在下限和上限之间变化,当温度上升时,温度数值以红色显示,反之以蓝色显示。当工作模式为手动(Manual)时,选择温度左侧的复选框,温度上升,当升至上限时停止变化;清除复选框,则温度下降,当降至下限时停止变化。无论是在 Auto 还是 Manual 模式下,如果切换至锁定模式(Lock),则温度立即停止变化,此时可以查看监测软件所读取的温度数据是否正确。选择 CheckSum 用来确定是否添加累加和校验码,【Stop】按钮用于关闭连接,【Quit】按钮用于退出应用软件,溢出菜单中的【About】菜单项用于显示版权信息。

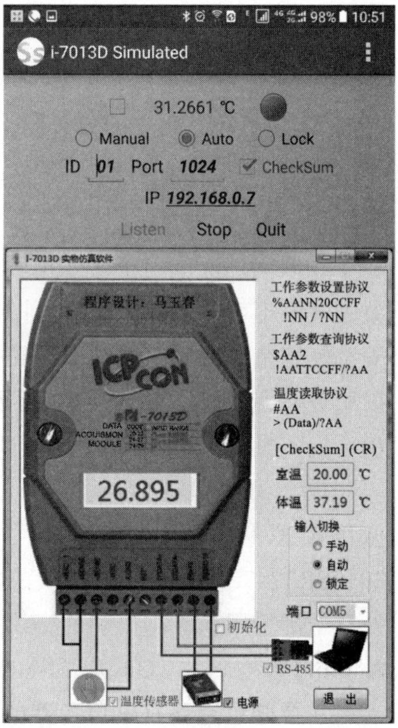

14.3 界面布局

仿真模块软件的项目名称为 I7013D_Simulated,界面布局的大框架采用线性布局,从上到下排列大组件;内部分别采用线性布局或相对布局等,从左到右排列最终组件。第一行最左侧的复选框 chkUp 实现温度传感器的功能,选择时温度升高,反之温度下降。

图 14-1 I-7013D 模块仿真软件

```
<LinearLayout xmlns:android="http://schemas.android.com/apk/res/android"
    android:layout_width="match_parent"
    android:layout_height="match_parent"
    android:gravity="top"
    android:orientation="vertical" >
    <LinearLayout
        android:layout_width="wrap_content"
        android:layout_height="wrap_content"
        android:layout_gravity="center"
        android:layout_marginTop="@dimen/activity_horizontal_margin"
        android:orientation="horizontal" >
        <CheckBox
            android:id="@+id/chkUp"
            android:layout_width="wrap_content"
            android:layout_height="wrap_content"
            android:layout_marginEnd="@dimen/activity_horizontal_margin"
            android:checked="false"
            android:onClick="onClick_CheckUp"
            android:textAppearance="?android:attr/textAppearanceMedium" />
        <TextView
            android:id="@+id/txtTemp"
            style="@style/TextStyle"
            android:layout_marginLeft="@dimen/user6_margin"
```

```xml
            android:layout_marginRight="@dimen/user6_margin"
            android:text="@string/strOriginalTemp" />
    <TextView
        android:id="@+id/txtUnit"
        style="@style/TextStyle"
        android:text="@string/strUnit" />
    <ImageView
        android:id="@+id/imgPower"
        android:layout_width="50dp"
        android:layout_height="32dp"
        android:layout_marginLeft="@dimen/activity_horizontal_margin"
        android:contentDescription="@null"
        android:scaleType="fitCenter"
        android:src="@drawable/grey" />
</LinearLayout>
<RadioGroup
    android:id="@+id/radioGroup1"
    android:layout_width="wrap_content"
    android:layout_height="wrap_content"
    android:layout_gravity="center"
    android:orientation="horizontal" >
    <RadioButton
        android:id="@+id/rdManual"
        android:layout_width="108dp"
        android:layout_height="wrap_content"
        android:checked="true"
        android:onClick="onClick_WorkMode"
        android:text="@string/strManual" />
    <RadioButton
        android:id="@+id/rdAuto"
        android:layout_width="84dp"
        android:layout_height="wrap_content"
        android:onClick="onClick_WorkMode"
        android:text="@string/strAuto" />
    <RadioButton
        android:id="@+id/rdLock"
        android:layout_width="84dp"
        android:layout_height="wrap_content"
        android:onClick="onClick_WorkMode"
        android:text="@string/strLock" />
</RadioGroup>
<RelativeLayout
    android:layout_width="wrap_content"
    android:layout_height="wrap_content"
    android:layout_gravity="center" >
    <TextView
        android:id="@+id/txtLabel_ID"
        style="@style/TextStyle"
        android:text="@string/strID" />
    <EditText
        android:id="@+id/edID"
        style="@style/TextStyle.Edit"
        android:layout_alignBaseline="@+id/txtLabel_ID"
        android:layout_marginLeft="@dimen/user4_margin"
        android:layout_toRightOf="@+id/txtLabel_ID"
        android:hint="@string/idhint"
```

```xml
            android:inputType="text"
            android:maxLength="2" />
        <TextView
            android:id="@+id/txtLabelPort"
            style="@style/TextStyle"
            android:layout_alignBaseline="@+id/txtLabel_ID"
            android:layout_gravity="center_vertical"
            android:layout_marginLeft="@dimen/user4_margin"
            android:layout_toRightOf="@+id/edID"
            android:text="@string/strPort" />
        <EditText
            android:id="@+id/edPort"
            style="@style/TextStyle.Edit"
            android:layout_alignBaseline="@+id/txtLabel_ID"
            android:layout_marginLeft="@dimen/user4_margin"
            android:layout_toRightOf="@+id/txtLabelPort"
            android:hint="@string/idhint"
            android:inputType="number"
            android:maxLength="4" />
        <CheckBox
            android:id="@+id/chkSum"
            android:layout_width="wrap_content"
            android:layout_height="wrap_content"
            android:layout_alignBaseline="@+id/txtLabel_ID"
            android:layout_marginLeft="@dimen/user4_margin"
            android:layout_toRightOf="@+id/edPort"
            android:text="@string/strCheckSum" />
    </RelativeLayout>
    <LinearLayout
        android:layout_width="wrap_content"
        android:layout_height="wrap_content"
        android:layout_gravity="center"
        android:orientation="horizontal" >
        <TextView
            android:id="@+id/txtLabelIP"
            style="@style/TextStyle"
            android:text="@string/strIP" />
        <TextView
            android:id="@+id/txtIP"
            style="@style/TextStyle"
            android:layout_marginLeft="6dp"
            android:textStyle="bold|italic"
            android:text="@null" />
    </LinearLayout>
    <LinearLayout
        android:layout_width="wrap_content"
        android:layout_height="40dp"
        android:layout_gravity="center"
        android:layout_marginLeft="@dimen/activity_horizontal_margin"
        android:layout_marginRight="@dimen/activity_horizontal_margin"
        android:orientation="horizontal" >
        <Button
            android:id="@+id/btListen"
            style="?android:attr/buttonBarButtonStyle"
            android:layout_width="90dp"
            android:layout_height="wrap_content"
```

```xml
            android:background="?android:attr/selectableItemBackground"
            android:onClick="onClick_Button"
            android:text="@string/strListen" />
        <Button
            android:id="@+id/btStop"
            style="?android:attr/buttonBarButtonStyle"
            android:layout_width="60dp"
            android:layout_height="wrap_content"
            android:background="?android:attr/selectableItemBackground"
            android:onClick="onClick_Button"
            android:text="@string/strStop" />
        <Button
            android:id="@+id/btQuit"
            style="?android:attr/buttonBarButtonStyle"
            android:layout_width="60dp"
            android:layout_height="wrap_content"
            android:background="?android:attr/selectableItemBackground"
            android:onClick="onClick_Button"
            android:text="@string/strQuit" />
    </LinearLayout>
    <ImageView
        android:id="@+id/imageView1"
        android:layout_width="match_parent"
        android:layout_height="match_parent"
        android:contentDescription="@null"
        android:scaleType="fitCenter"
        android:src="@drawable/i7013d" />
</LinearLayout>
```

14.4 配置文件

仿真模块基于 TCP 通信，因而需要网络相关的权限。在配合监测软件的工作过程中，需要随时查看仿真模块的当前温度及工作状态，因而，需要保持手机屏幕常亮，应设置 WAKE_LOCK 权限。

```xml
<?xml version="1.0" encoding="utf-8"?>
<!DOCTYPE xml>
<manifest xmlns:android="http://schemas.android.com/apk/res/android"
    package="com.walkerma.i7013d_simulated"
    android:versionCode="1"
    android:versionName="1.0" >
    <uses-sdk
        android:minSdkVersion="14"
        android:targetSdkVersion="21" />
    <uses-permission android:name=
        "android.permission.INTERNET" />
    <uses-permission android:name=
        "android.permission.ACCESS_NETWORK_STATE" />
    <uses-permission android:name=
        "android.permission.WAKE_LOCK" />
    <uses-permission android:name=
        "android.permission.ACCESS_WIFI_STATE" />
    <application
        android:allowBackup="true"
```

```xml
            android:icon="@drawable/ic_launcher"
            android:label="@string/app_name"
            android:theme="@style/AppTheme" >
            <activity
                android:name=".MainActivity"
                android:label="@string/app_name"
                android:screenOrientation="portrait" >
                <intent-filter>
                    <action android:name="android.intent.action.MAIN" />
                    <category android:name="android.intent.category.LAUNCHER" />
                </intent-filter>
            </activity>
        </application>
</manifest>
```

14.5 窗体源代码

仿真模块软件需要引用 Library 类库，调用第 13 章的网络类和第 11 章的数据编码与处理技术，以及第 12 章的数据包校验技术。常量 TEMP_TOP 为上限温度值，TEMP_BOTTOM 为下限温度值，dT_New 中保存最新温度，dT_Old 为最近的历史温度，如果 dT_New 中的温度大于 dT_Old 的温度，就用红色字体显示温度，否则用蓝色字体显示温度。校验码类型变量 nParitySort 为 ParitySort.ADD，如果有校验码就添加，没有校验码则使用 ParitySort.NONE。工作模式通过枚举类型 WorkMode 定义，默认为 WorkMode.MANUAL（手动）。常量 DELAY_READ_WIFI 存放使用 WiFi 连接时的接收数据延迟（单位为毫秒），即在该时间之内的数据被当作一个数据包来处理，常量 DELAY_READ_GPRS 中存放使用 GPRS 移动数据连接时的接收数据延迟，实际使用的接收数据延迟存于变量 nDelayRead 中，将根据连接类型自动选择延迟常量。常量 DELAY_INTERVAL 用来定制温度产生的频率，这里为 15 毫秒，即每隔这一时间段，温度就上升或下降一个单位。

仿真模块以正弦函数的形式产生温度，常量 PI 中定义了 π 的值，在两个 PI 范围之内，温度形成一个周期，在一个周期之内分为 STEPS_TOTAL 个节拍，一个节拍所占的弧度用常量 PI_STEP 表示。产生温度的函数在 getCurrentTemp 中定义，当温度上升时，当前温度用公式（1）表示，当温度下降时，当前温度用公式（2）表示。

$$dT\_New = TEMP\_BOTTOM + (TEMP\_TOP - TEMP\_BOTTOM) * (1 + Math.sin(PI*3/2 + PI\_STEP*nSteps))/2; \quad (1)$$

$$dT\_New = TEMP\_BOTTOM + (TEMP\_TOP - TEMP\_BOTTOM) * (1 + Math.sin(PI/2 + PI\_STEP*nSteps))/2; \quad (2)$$

退出仿真模块时调用 saveSettings 函数保存网络参数和仿真模块的参数，在仿真模块初始化时，在 onCreate 方法中调用 restoreSettings 函数恢复保存的参数。当点击【Listen】按钮时，调用 listenProcess 函数，获取本地 IP 地址，初始化 TcpClientServer 类，并设置延迟值，最后侦听端口。当有客户机连接成功时，在 handleMessage 方法中调用 processMessage 函数，对于 MSG_Connected 消息，每隔 DELAY_INTERVAL 毫秒执行一次 runnable 对象中的 run 方法，即调用 getCurrentTemp 在函数取得当前温度，通过消息 MSG_T_CHANGED

将温度值传递到主线程。在函数 processMessage 中对于 MSG_T_CHANGED 消息，如果温度升高，则将温度显示组件 txtTemp 的文本颜色设置为红色，否则设置为蓝色。

以上温度的变化只是在仿真模块内部显示，外界并不知道。当外部的温度监测软件发送温度读取命令"#AA[CheckSum](CR)"（详细说明见 12.8 节）时，函数 processMessage 就会得到消息 MSG_DataArrived，在这里调用 checkEntirePackage 函数检查数据包是否正确，如果正确，则将温度数据写入字符串 strReply 中，再根据协议要求添加前导字符">"、校验码和结尾码发送给监测软件。

```java
package com.walkerma.i7013d_simulated;
import java.util.Locale;
import com.walkerma.library.GeneralProcess;
import com.walkerma.library.NetworkProcess;
import com.walkerma.library.NetworkProcess.ThreadIP;
import com.walkerma.library.Parity;
import com.walkerma.library.Parity.DisplayMode;
import com.walkerma.library.Parity.EndMark;
import com.walkerma.library.Parity.ParitySort;
import com.walkerma.library.TcpClientServer;
import android.app.Activity;
import android.app.AlertDialog;
import android.content.Context;
import android.content.SharedPreferences;
import android.content.SharedPreferences.Editor;
import android.graphics.Color;
import android.graphics.Paint;
import android.os.Bundle;
import android.os.Handler;
import android.os.Message;
import android.view.KeyEvent;
import android.view.Menu;
import android.view.MenuItem;
import android.view.View;
import android.view.WindowManager;
import android.widget.Button;
import android.widget.CheckBox;
import android.widget.EditText;
import android.widget.ImageView;
import android.widget.RadioButton;
import android.widget.TextView;
import android.widget.Toast;
public class MainActivity extends Activity {
    private Context context;
    private ImageView imgPower;
    private CheckBox chkUp, chkSum;
    private EditText edID, edPort;
    private final double TEMP_TOP = 37.19;
    private final double TEMP_BOTTOM = 20.0;
    private double dT_New, dT_Old=TEMP_TOP;
    private ParitySort nParitySort = ParitySort.ADD;
    private TextView txtTemp, txtIP;
    private Button btListen, btStop;
    private RadioButton rdManual;
    private enum WorkMode{MANUAL, AUTO, LOCK};
```

```java
    private WorkMode nWorkMode = WorkMode.MANUAL;
    private TcpClientServer server=null;
    private int nPort;
    private Handler mHandler;
    private final int DELAY_READ_WIFI = 1200;
    private final int DELAY_READ_GPRS = 1500;
    private int nDelayRead = DELAY_READ_WIFI;
    private final int DELAY_INTERVAL = 15; //for temperature
    private boolean bCanExit = false;
    private final int MSG_T_CHANGED = 1010;
    private final int MSG_ExitDelay = 1011;
    private final double PI = 3.14159;
    private final int STEPS_TOTAL = 2048;
    private final int STEPS_HALF = STEPS_TOTAL/2;
    private final double PI_STEP = PI*2/STEPS_TOTAL;
    private int nSteps;
    Runnable runnable = new Runnable() {
        @Override
        public void run() {
            String strTemp = String.format(Locale.ENGLISH,
                    "%.4f", getCurrentTemp());
            mHandler.obtainMessage(MSG_T_CHANGED, 0, 0,
                    strTemp).sendToTarget();
            mHandler.postDelayed(this, DELAY_INTERVAL);
        }
    };
    private double getCurrentTemp(){
        if (!((nWorkMode == WorkMode.LOCK) ||
                ((nWorkMode == WorkMode.MANUAL)
                        && (nSteps == STEPS_HALF)))
                        && (nSteps < STEPS_HALF))
            nSteps++;
        if(chkUp.isChecked()){
            //steps=STEPS_HALF, dt_new=TEMP_TOP
            dT_New = TEMP_BOTTOM + (TEMP_TOP - TEMP_BOTTOM) *
                    (1 + Math.sin(PI*3/2 + PI_STEP*nSteps))/2;
        }
        else{
            //steps=0, dT_New=TEMP_TOP
            dT_New = TEMP_BOTTOM + (TEMP_TOP - TEMP_BOTTOM) *
                    (1 + Math.sin(PI/2 + PI_STEP*nSteps))/2;
        }
        if((nSteps == STEPS_HALF) && (nWorkMode == WorkMode.AUTO)){
            nSteps = 0;
            if(chkUp.isChecked())
                chkUp.setChecked(false);
            else
                chkUp.setChecked(true);
        }
        return dT_New;
    }
    @Override
    protected void onCreate(Bundle savedInstanceState) {
        super.onCreate(savedInstanceState);
        setContentView(R.layout.activity_main);
        context = this;
```

```java
        getWindow().addFlags(
                WindowManager.LayoutParams.FLAG_KEEP_SCREEN_ON);
        initView();
        restoreSettings();
        setStatusOff();
        mHandler = new Handler(new IncomingHandlerCallback());
    }
    class IncomingHandlerCallback implements Handler.Callback{
        @Override
        public boolean handleMessage(Message msg) {
            processMessage(msg);
            return true;
        }
    }
    private void processMessage(Message msg){
        switch(msg.what){
        case MSG_T_CHANGED:
            txtTemp.setText((String)msg.obj);
            if(dT_New>dT_Old)
                txtTemp.setTextColor(Color.RED);
            if(dT_New<dT_Old)
                txtTemp.setTextColor(Color.BLUE);
            dT_Old = dT_New;
            break;
        case MSG_ExitDelay:
            bCanExit = false;
            return;
        case NetworkProcess.MSG_IP:
            txtIP.setText((String)msg.obj);
            txtIP.setPaintFlags(txtIP.getPaintFlags() |
                    Paint.UNDERLINE_TEXT_FLAG);
            if(NetworkProcess.isWifiConnected(context))
                nDelayRead = DELAY_READ_WIFI;
            else
                nDelayRead = DELAY_READ_GPRS;
            break;
        case TcpClientServer.MSG_Connected:
            setStatusOn();
            mHandler.postDelayed(runnable, DELAY_INTERVAL);
            break;
        case TcpClientServer.MSG_DataArrived:
            String strData = (String)msg.obj;
            boolean bValid = Parity.checkEntirePackage(
                    strData, nParitySort,
                    EndMark.CR, DisplayMode.CHAR);
            if(bValid == false){
                Toast.makeText(this, "Data Format Error!",
                        Toast.LENGTH_SHORT).show();
                break;
            }
            if(strData.startsWith("#"+edID.getText().toString())==false){
                Toast.makeText(this, "Leading char or ID Error!",
                        Toast.LENGTH_SHORT).show();
                break;
            }
            String strReply = ">+0" + txtTemp.getText().toString();
```

```java
                strReply = Parity.getEntirePackage(strReply, nParitySort,
                        EndMark.CR, DisplayMode.CHAR);
                server.sendText(strReply);
                break;
            case TcpClientServer.MSG_Closed:
            case TcpClientServer.MSG_ReadError:
            case TcpClientServer.MSG_TransmitError:
                setStatusOff();
                break;
        }
    }
    private void initView(){
        imgPower = (ImageView)findViewById(R.id.imgPower);
        chkUp = (CheckBox)findViewById(R.id.chkUp);
        chkSum = (CheckBox)findViewById(R.id.chkSum);
        txtTemp = (TextView)findViewById(R.id.txtTemp);
        edID = (EditText)findViewById(R.id.edID);
        txtIP = (TextView)findViewById(R.id.txtIP);
        edPort = (EditText)findViewById(R.id.edPort);
        btListen = (Button)findViewById(R.id.btListen);
        btStop = (Button)findViewById(R.id.btStop);
        rdManual = (RadioButton)findViewById(R.id.rdManual);
    }
    private void restoreSettings(){
        String strAppName = getString(R.string.app_name);
        SharedPreferences sp = getSharedPreferences(strAppName, 0);
        edID.setText(sp.getString("ID", "01"));
        chkSum.setChecked(sp.getBoolean("CheckSum", true));
        edPort.setText(sp.getString("Port", "1024"));
        nPort = Integer.parseInt(edPort.getText().toString());
    }
    private void saveSettings(){
        String strID = edID.getText().toString();
        String strPort = edPort.getText().toString();
        String strAppName = getString(R.string.app_name);
        SharedPreferences sp = getSharedPreferences(strAppName, 0);
        Editor editor = sp.edit();
        editor.putString("Port", strPort);
        editor.putString("ID", strID);
        editor.putBoolean("CheckSum", chkSum.isChecked());
            editor.commit();
    }
    private void setStatusOn(){
        imgPower.setImageResource(R.drawable.red);
        btListen.setEnabled(false);
        btStop.setEnabled(true);
        chkSum.setEnabled(false);
            nSteps = STEPS_HALF;      //=>primariy T=TEMP_BOTTOM
    }
    private void setStatusOff(){
        imgPower.setImageResource(R.drawable.grey);
        btListen.setEnabled(true);
        btStop.setEnabled(false);
        chkSum.setEnabled(true);
        txtIP.setText("###.###.###.###");
        txtTemp.setText("00.0000");
        txtTemp.setTextColor(Color.BLACK);
```

```java
        rdManual.setChecked(true);
    }
    public void onClick_CheckUp(View v){
        nSteps = STEPS_HALF - nSteps;
    }
    public void onClick_WorkMode(View v){
        RadioButton rb = (RadioButton)v;
        chkUp.setEnabled(false);
        switch(rb.getId()){
        case R.id.rdManual:
            nWorkMode = WorkMode.MANUAL;
            chkUp.setEnabled(true);
            break;
        case R.id.rdAuto:
            nWorkMode = WorkMode.AUTO;
            break;
        case R.id.rdLock:
            nWorkMode = WorkMode.LOCK;
            break;
        }
    }
    private void alertForNetwork(){
        AlertDialog.Builder builder = new AlertDialog.Builder(this);
        builder.setIcon(R.drawable.ic_action_warning);
        builder.setTitle("Wi-Fi/GPRS: ");
        builder.setMessage("Wi-Fi/GPRS is disconnected.");
        builder.setPositiveButton("OK", null);
        builder.setCancelable(false);
        AlertDialog dialog = builder.create();
        dialog.show();
    }
    public void onClick_Button(View v){
        int id = v.getId();
        switch(id){
        case R.id.btListen:
            listenProcess();
            Toast.makeText(this, "Listening...",
                    Toast.LENGTH_SHORT).show();
            break;
        case R.id.btStop:
            mHandler.removeCallbacks(runnable);
            server.close();
            server = null;
            break;
        case R.id.btQuit:
            saveSettings();
            finish();
            break;
        }
    }
    private void listenProcess(){
        if(!NetworkProcess.isNetworkAvailable(this)){
            alertForNetwork();
            return;
        }
        nParitySort = chkSum.isChecked()? ParitySort.ADD:ParitySort.NONE;
        ThreadIP threadIP = new ThreadIP(mHandler, context);
```

```java
            threadIP.start();
            if(server!=null) server.close();       // 02/24/2018
            server = new TcpClientServer(mHandler, nPort);
            server.setReadDelay(nDelayRead);
            server.listen();
    }
    @Override
    public boolean onCreateOptionsMenu(Menu menu) {
            getMenuInflater().inflate(R.menu.main, menu);
             return true;
    }
    @Override
    public boolean onOptionsItemSelected(MenuItem item) {
            if(item.getItemId()==R.id.menu_about)
                GeneralProcess.fireAboutDialog(this,
                        getString(R.string.app_name),
                        getString(R.string.strCopyrights));
            return super.onOptionsItemSelected(item);
    }
    @Override
    public boolean onKeyDown(int keyCode, KeyEvent event) {
        // TODO Auto-generated method stub
        switch(keyCode){
        case KeyEvent.KEYCODE_BACK:
            if(!bCanExit){
                bCanExit = true;
                Toast.makeText(getApplicationContext(),
                        "Press again to quit.", Toast.LENGTH_SHORT).show();
                // if not pressed within 1 seconds then will be setted(canExit) as false
                mHandler.sendEmptyMessageDelayed(MSG_ExitDelay, 1000);
                return true;
            }
            else break;
        }
        return super.onKeyDown(keyCode, event);
    }
    @Override
    protected void onDestroy() {
        // TODO Auto-generated method stub
        mHandler.removeCallbacksAndMessages(null);
        if(server!=null) server.close();       // 02/24/2018
        getWindow().clearFlags(
                WindowManager.LayoutParams.FLAG_KEEP_SCREEN_ON);
        super.onDestroy();
    }
}
```

14.6 本章小结

本章将基于 RS-485 工业控制总线的 I-7013D 模块仿真为 TCP 服务器形式，这是上一章通用 TCP 服务器的一个特例，这样可以为学生学习 TCP 客户机编程提供一个对等实验的便携平台。而且，让温度以正弦函数形式产生，可以提高监测软件的可视化效果。下一章的 TCP 版温度监测软件将实时读取仿真模块的温度，并以实时数据线显示。

第 15 章　I-7013D 模块监测软件

物联网通过信息传感设备,按照约定的协议,将任何物品与互联网连接,进行信息交换和通信,以实现智能化识别、定位、跟踪、监控和管理。第 14 章完成了 TCP 版"I-7013D 模块仿真软件",可以正弦曲线形式产生实时温度,本章的 TCP 版"I-7013D 模块监测软件"能够实时监测仿真模块的温度,两者构成一个简单直观的基于物联网的计算机监控系统。

主要知识点:TCP 客户机的具体实现、定时功能的实现、短信发送、播放音频、绘图、横屏和竖屏切换

15.1　主要功能和技术特点

I-7013D 模块监测软件(安装后的 App 名为 Supervisor for i-7013D,下文简称"监测软件")采用 TCP 客户机模式工作,可以根据 IP 地址和端口号连接到远程服务器。其主要特点为:

- 实时显示当前温度数值及温度趋势线;
- 温度上升采用红色文本显示,趋势线采用红线,温度采样点采用蓝点;
- 温度下降采用蓝色文本显示,趋势线采用蓝线,温度采样点采用红点;
- 远程服务器的 IP 地址和端口地址、仿真模块的 ID 号和校验码在同一界面设置;
- 超过警戒温度播放报警声音,并可向指定手机发送报警短信;
- 手机横屏与竖屏均实现了持久化处理,温度趋势线显示不中断;
- 实时提示通信异常,软件退出时显示监测软件与仿真模块之间通信的时间间隔。

15.2　软件操作

监测软件运行后的界面如图 15-1 所示,这是与仿真模块连接成功的状态,温度趋势线显示区域最右侧的点为当前温度,温度值显示于第一行左侧,当右侧新的温度点加入时,如果屏幕已满,则最左侧的温度点被移出。点击【Stop】按钮关闭与仿真模块的连接,并显示"CommInterval = 53ms",即监测软件与仿真模块之间的通信时间间隔为 53 毫秒(不同距离有不同的时间),监测软件发出查询命令 53 毫秒后即可收到温度数据;点击【Quit】退出监测软件。

对于图 15-1 中的初始状态(即未与仿真模块连接,下同),点击【Setup】进入图 15-2 所示的参数设置界面,在这里可以设置远程服务器的 IP 地址和侦听 Port、仿真模块的 ID 及是否携带累加和校验码。

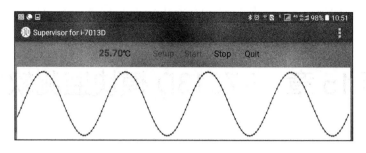

图 15-1　I-7013D 模块监测软件

在图 15-2 中，如果选择 SMS，则可以设置报警电话号码，即温度达到 36.5℃时，向设定的目标手机发送报警短信，短信内容如图 15-3 所示，由于发送短信和接收短信都是同一手机，所以，这里有两条短信。温度超限会同时播放报警声音。

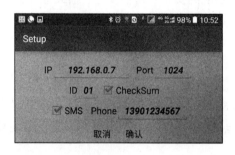

图 15-2　监测软件参数设置

图 15-3　报警短信

在图 15-1 中的初始状态下，只有设置完参数，才能点击【Start】与仿真模块建立连接，否则将显示"Please [Setup] at first."；如果没有打开网络，则点击【Start】时将显示"Wi-Fi/GPRS is disconnected."；如果仿真模块没有开始侦听或监测软件所设置的服务器参数与仿真模块的实际参数不一致，则点击【Start】时将显示"Please check the IP and Port of remote server, then listen."；正常情况下，仿真模块开始侦听，点击【Start】，则依次显示"Starting..."和"Connected"，监测软件开始按照图 15-1 运行。当监测软件处于图 15-1 所示的状态时，如果仿真模块出现故障或主动关闭连接，则监测软件发送的查询命令将不会得到响应，此时将显示"Data Missing..."和"Transmit error."，并主动关闭连接。

图 15-1 为横屏状态，如果让手机竖屏，数据不会丢失，温度趋势线将连续显示，但界面布局将发生改变，第一行的文本和按钮将显示为两行，这里不再截图说明。

15.3　配置文件

监测软件的项目名称为 I7013D_TCP，使用网络通信，因而需要有访问网络状态的权限及网络通信的权限。温度超过警戒线时需要播放声音，因而需要调整音量的权限，确保以最大音量通知用户；同时，还需要向目标手机发送报警信息，因而需要发送短信的权限。在对仿真模块的温度测试过程中，需要保持手机屏幕常亮，因而也需要 WAKE_LOCK 权限。另外，还需要 READ_PHONE_STATE 权限，以便读取本机电话号码。

在手机横屏或竖屏变化时，要确保温度趋势线连续显示，这需要设置主窗体 Main Activity 的 configChanges 属性，以便在主窗体的 onConfigurationChanged 事件处理方法中对主窗体重新布局和重绘温度趋势线。

```xml
<?xml version="1.0" encoding="utf-8"?>
<!DOCTYPE xml>
<manifest xmlns:android="http://schemas.android.com/apk/res/android"
    package="com.walkerma.i7013d_tcp"
    android:versionCode="1"
    android:versionName="1.0" >
    <uses-sdk
        android:minSdkVersion="14"
        android:targetSdkVersion="21" />
    <uses-permission android:name=
        "android.permission.INTERNET"/>
    <uses-permission android:name=
        "android.permission.READ_PHONE_STATE"/>
    <uses-permission android:name=
        "android.permission.SEND_SMS"/>
    <uses-permission android:name=
        "android.permission.ACCESS_NETWORK_STATE"/>
    <uses-permission android:name=
        "android.permission.WAKE_LOCK"/>
    <uses-permission android:name=
        "android.permission.ACCESS_WIFI_STATE"/>
    <uses-permission android:name=
        "android.permission.MODIFY_AUDIO_SETTINGS"/>
    <application
        android:allowBackup="true"
        android:icon="@drawable/ic_launcher"
        android:label="@string/app_name"
        android:theme="@style/AppTheme" >
        <activity
            android:name=".MainActivity"
            android:configChanges="orientation|screenSize"
            android:label="@string/app_name">
            <intent-filter>
                <action android:name="android.intent.action.MAIN" />
                <category android:name="android.intent.category.LAUNCHER" />
            </intent-filter>
        </activity>
        <activity
            android:name=".SetupActivity"
            android:label="Setup" >
        </activity>
    </application>
</manifest>
```

15.4 参数设置窗体

为了使监测软件层次更加清晰，创建参数设置窗体 SetupActivity，对监测软件的参数进行统一设置。本节内容包括参数设置窗体的界面布局说明及源代码介绍。

15.4.1 界面布局

参数设置窗体的界面布局文件名为 activity_setup.xml，其中的大框架采用相对布局，内部采用线性布局且各组件居中显示。为了在第一层的 EditText 组件 editIP 中顺利输入 IP 地址，需要设置其 digits 属性的范围包括数字和小数点，inputType 属性则为 number 或 numberDecimal。在图 15-2 中，软件界面为英文，但是两个按钮的文本却为中文，以【取消】为例，这是因为按钮的 text 属性使用了 "@android:string/cancel"，即 Android 系统定义的 cancel 字符串，因为手机的语言环境为中文，所以就自动设置为"取消"了。

```xml
<RelativeLayout xmlns:android="http://schemas.android.com/apk/res/android"
    xmlns:tools="http://schemas.android.com/tools"
    android:layout_width="match_parent"
    android:layout_height="match_parent"
    android:layout_gravity="center" >
    <LinearLayout
        android:id="@+id/layout_IP"
        android:layout_width="wrap_content"
        android:layout_height="wrap_content"
        android:layout_alignParentTop="true"
        android:layout_centerHorizontal="true"
        android:orientation="horizontal"
        android:paddingTop="@dimen/activity_horizontal_margin" >
        <TextView
            android:id="@+id/txtIP"
            style="@style/TextStyle"
            android:text="@string/strIP" />
        <EditText
            android:id="@+id/editIP"
            style="@style/TextStyle.Edit"
            android:layout_width="wrap_content"
            android:digits="0123456789."
            android:hint="@string/hintip"
            android:inputType="number|numberDecimal"
            android:maxLength="15"/>
        <TextView
            android:id="@+id/txtPort"
            style="@style/TextStyle"
            android:layout_marginLeft="10dp"
            android:text="@string/strPort" />
        <EditText
            android:id="@+id/editPort"
            style="@style/TextStyle.Edit"
            android:layout_width="wrap_content"
            android:inputType="number"
            android:maxLength="5"
            android:hint="@string/hintport"
            android:text="@null" />
    </LinearLayout>
    <LinearLayout
        android:id="@+id/layout_ID"
        android:layout_width="wrap_content"
        android:layout_height="wrap_content"
        android:layout_below="@+id/layout_IP"
```

```xml
        android:layout_centerHorizontal="true"
        android:orientation="horizontal" >
    <TextView
        android:id="@+id/txtID"
        style="@style/TextStyle"
        android:text="@string/strID" />
    <EditText
        android:id="@+id/editID"
        style="@style/TextStyle.Edit"
        android:layout_width="wrap_content"
        android:maxLength="2" />
    <CheckBox
        android:id="@+id/chkSum"
        android:layout_width="wrap_content"
        android:layout_height="wrap_content"
        android:layout_marginEnd="@dimen/activity_horizontal_margin"
        android:checked="true"
        android:text="@string/strCheckSum"
        android:textAppearance="?android:attr/textAppearanceMedium" />
</LinearLayout>
<LinearLayout
    android:id="@+id/layout_Phone"
    android:layout_width="wrap_content"
    android:layout_height="wrap_content"
    android:layout_below="@+id/layout_ID"
    android:layout_centerHorizontal="true"
    android:orientation="horizontal"
    android:paddingLeft="@dimen/activity_horizontal_margin"
    android:paddingRight="@dimen/activity_horizontal_margin" >
    <CheckBox
        android:id="@+id/chkSMS"
        android:layout_width="wrap_content"
        android:layout_height="wrap_content"
        android:onClick="onClick_CheckSMS"
        android:text="@string/strSMS"
        android:textAppearance="?android:attr/textAppearanceMedium" />
    <TextView
        android:id="@+id/textView1"
        style="@style/TextStyle"
        android:paddingLeft="@dimen/activity_horizontal_margin"
        android:text="@string/strPhone" />
    <EditText
        android:id="@+id/editPhone"
        style="@style/TextStyle.Edit"
        android:maxLength="11"
        android:hint="@string/hintphone"
        android:inputType="phone" />
</LinearLayout>
<LinearLayout
    android:layout_width="wrap_content"
    android:layout_height="wrap_content"
    android:layout_below="@+id/layout_Phone"
    android:layout_centerHorizontal="true"
    android:orientation="horizontal" >
    <Button
        android:id="@+id/btCancel"
```

```xml
            style="?android:attr/buttonBarButtonStyle"
            android:layout_width="wrap_content"
            android:layout_height="wrap_content"
            android:onClick="onClick"
            android:background="?android:attr/selectableItemBackground"
            android:text="@android:string/cancel" />
        <Button
            android:id="@+id/btOK"
            style="?android:attr/buttonBarButtonStyle"
            android:layout_width="wrap_content"
            android:layout_height="wrap_content"
            android:onClick="onClick"
            android:background="?android:attr/selectableItemBackground"
            android:text="@android:string/ok" />
    </LinearLayout>
</RelativeLayout>
```

15.4.2 源代码

采用 4.2 节的技术，主窗体 MainActivity 调用 SetupActivity 获取当前工作参数，定义的公有静态整型常量 RESULT_CODE 表示返回的结果代码。在 onCreate 方法中对各组件进行初始化，然后调用 restoreSettings 函数读取保存的参数并在各组件中显示。在 onClick 方法中，处理用户点击 btOK 的功能，即调用 saveSettings 函数检查并保存当前参数，并返回到 MainActivity。

函数 getLocalPhone 用于获取本机电话号码。标准的 SDK 不支持两张 SIM 卡，但是，即使只有一张 SIM 卡，也不能保证通过该函数可获取手机号码，这是因为移动网络是通过 SIM 卡的 IMSI 号码来识别用户的，所以有的移动运营商并没有将手机号码写入 SIM 卡。

在 saveSettings 函数中，调用 checkIP 函数检查 IP 地址是否符合规范，调用 checkPort 函数检查端口号是否在规定的范围之内，调用 checkID 函数检查仿真模块的 ID 是否为两个十六进制字符。如果 IP 地址错误，则调用 generalAlert 函数，以对话框的形式显示错误信息。如果所有参数格式均正确，则写入手机。

```java
package com.walkerma.i7013d_tcp;
import android.app.ActionBar;
import android.app.Activity;
import android.app.AlertDialog;
import android.content.Context;
import android.content.Intent;
import android.content.SharedPreferences;
import android.content.SharedPreferences.Editor;
import android.os.Bundle;
import android.telephony.TelephonyManager;
import android.text.TextUtils;
import android.view.View;
import android.widget.CheckBox;
import android.widget.EditText;
public class SetupActivity extends Activity {
    private EditText editIP, editPort, editID, editPhone;
    private CheckBox chkSum, chkSMS;
    public final static int RESULT_CODE = 1;
    @Override
```

```java
protected void onCreate(Bundle savedInstanceState) {
    super.onCreate(savedInstanceState);
    setContentView(R.layout.activity_setup);
    ActionBar actionBar = getActionBar();
    actionBar.setDisplayShowHomeEnabled(false);
    actionBar.setDisplayShowTitleEnabled(true);
    editIP = (EditText)findViewById(R.id.editIP);
    editPort = (EditText)findViewById(R.id.editPort);
    editID = (EditText)findViewById(R.id.editID);
    editPhone = (EditText)findViewById(R.id.editPhone);
    chkSum = (CheckBox)findViewById(R.id.chkSum);
    chkSMS = (CheckBox)findViewById(R.id.chkSMS);
    chkSMS.setChecked(getIntent().getExtras()
            .getBoolean("CheckSMS"));
    editPhone.setEnabled(chkSMS.isChecked());
    restoreSettings();
}
public void onClick(View v){
    int id = v.getId();
    switch(id){
    case R.id.btOK:
        if (0 != saveSettings()) return;
        Bundle bundle = new Bundle();
        bundle.putString("IP", editIP.getText().toString());
        bundle.putInt("Port", Integer.parseInt(
                editPort.getText().toString()));
        bundle.putString("ID", editID.getText().toString());
        bundle.putBoolean("CheckSum", chkSum.isChecked());
        bundle.putBoolean("CheckSMS", chkSMS.isChecked());
        bundle.putString("Phone", editPhone.getText().toString());
        Intent it = new Intent();
        it.putExtras(bundle);
        setResult(RESULT_CODE, it);
        break;
    case R.id.btCancel:
        break;
    }
    finish();
}
private String getLocalPhone(){
    TelephonyManager tm = (TelephonyManager)
            getSystemService(Context.TELEPHONY_SERVICE);
    return tm.getLine1Number();
}
public void onClick_CheckSMS(View v){
    chkSMS = (CheckBox)v;
    editPhone.setEnabled(chkSMS.isChecked());
    if(chkSMS.isChecked()==false) return;
    String strPhone = editPhone.getText().toString();
    if(TextUtils.isEmpty(strPhone))
        editPhone.setText(getLocalPhone());
}
private void restoreSettings(){
    String strAppName = getString(R.string.app_name);
    SharedPreferences sp = getSharedPreferences(strAppName, 0);
    String strIP = sp.getString("IP", "");
```

```java
        String strPort = sp.getString("Port", "1024");
        String strID = sp.getString("ID", "01");
        Boolean bSum = sp.getBoolean("CheckSum", true);
        String strPhone = sp.getString("Phone", "");     //Uncared of chkSMS
        editIP.setText(strIP);
        editPort.setText(strPort);
        editID.setText(strID);
        chkSum.setChecked(bSum);
        editPhone.setText(strPhone);
    }
    private boolean checkIP(String IP){
        String strIP_Head = "^(([01]?\\d\\d?|2[0-4]\\d|25[0-5])\\.){3}";
        String strIP_End = "([01]?\\d\\d?|2[0-4]\\d|25[0-5])$";
        return IP.matches(strIP_Head + strIP_End);
    }
    private boolean checkPort(String Port){
        int nPort = Integer.parseInt(Port);
        if (nPort<1024 || nPort>65535)
            return false;
        else
            return true;
    }
    private boolean checkID(String ID){
        return ID.matches("^([0-9A-F]){2}$");
    }
    public void generalAlert(int nResource,
            String strTitle, String strMessage){
        AlertDialog.Builder builder = new
                AlertDialog.Builder(SetupActivity.this);
        builder.setIcon(nResource);
        builder.setCancelable(false);
        builder.setTitle(strTitle);
        builder.setMessage(strMessage);
        builder.setPositiveButton("OK", null);
        AlertDialog dialog = builder.create();
        dialog.show();
    }
    private int saveSettings(){
        String strIP = editIP.getText().toString();
        if (checkIP(strIP)==false) {
            generalAlert(R.drawable.ic_action_warning,"IP Warning: ",
                    "Please input regular IP.");
            return -1;
        }
        String strPort = editPort.getText().toString();
        if(checkPort(strPort)==false){
            generalAlert(R.drawable.ic_action_warning,"Port Warning: ",
                    "The Port is between 1024 and 65535.");
            return -1;
        }
        String strID = editID.getText().toString();
        if(checkID(strID) == false) {
            generalAlert(R.drawable.ic_action_warning,"ID Warning: ",
                    "Please input regular ID([0-9A-F]){2}).");
            return -1;
        }
```

```
            String strPhone = editPhone.getText().toString();
            String strAppName = getString(R.string.app_name);
            SharedPreferences sp = getSharedPreferences(strAppName, 0);
            Editor editor = sp.edit();
            editor.putString("IP", strIP);
            editor.putString("Port", strPort);
            editor.putString("ID", strID);
            editor.putBoolean("CheckSum", chkSum.isChecked());
            editor.putString("Phone", strPhone);        //chk_SMS is not saved.
            editor.commit();
            return 0;
        }
    }
```

15.5 主窗体

监测软件运行后首先进入主窗体（MainActivity），DrawView 是主窗体中的一个自定义组件，用来显示实时温度的趋势线。主窗体的布局文件名为 activity_main.xml，允许横屏和竖屏显示，因而，在资源目录 res 下创建 layout-land 文件夹以存放横屏布局文件，创建 layout-port 文件夹以存放竖屏布局文件，两者都采用相同的布局文件名 activity_main.xml，监测软件在运行过程中会根据手机所处状态自动选择布局文件。

15.5.1 实时温度显示组件

自定义组件 DrawView 用来绘制实时温度趋势线，Paint 对象 bkPaint 用来绘制白色背景，linePaint 对象用来绘制温度数据线，ptPaint 对象用来绘制温度数据点：在上升途中 linePaint 采用红色，ptPaint 采用蓝色；在下降途中 linePaint 采用蓝色，ptPaint 采用红色。ArrayList<Float> 对象 pList 中存放温度序列。组件高度用 nHeight 表示，宽度用 nWidth 表示，在宽度范围内的温度点用 nMaxPoints 表示，竖屏为 MAX_POINTS（80）个点，横屏为 160 个点。在构造函数中，主要完成 Paint 对象和 pList 对象的初始化工作。TEMP_TOP 为最大温度值，TEMP_BOTTOM 为最小温度值。

DrawView 对象通过公有函数 clearData 清除 pList 对象中的数据，通过 setTemperatureY 函数调用私有函数 addDataToArray 将当前温度添加到 pList 序列中，如果温度序列大小超过 nMaxPoints，则将最前面的温度删除。通过公有函数 setPointList 对内部 pList 赋值，通过 getPointList 函数获取内部 pList 的值。

温度序列 pList 中的值通过 getPoint 函数转换成以左下角为坐标原点的坐标，存放到 Bundle 对象的键值对中，最高温度的纵坐标转换为"0.1* 组件高度"，最低温度的纵坐标转换为"0.9* 组件高度"。DrawView 对象通过调用继承的方法 invalidate 触发 onDraw 函数，获取组件的高度和宽度，并根据横屏或竖屏状态获得当前的 nMaxPoints 值，随后调用私有函数 drawBackGround 绘制背景，调用私有函数 drawLine 绘制数据线（并在其中绘制数据点）。

```
package com.walkerma.i7013d_tcp;
import java.util.ArrayList;
import android.content.Context;
```

```java
import android.content.res.Configuration;
import android.graphics.Canvas;
import android.graphics.Color;
import android.graphics.Paint;
import android.os.Bundle;
import android.util.AttributeSet;
import android.view.View;
public class DrawView extends View {
    private Paint bkPaint;
    private Paint linePaint;
    private Paint ptPaint;
    private ArrayList<Float> pList;
    private float nHeight;
    private float nWidth;
    private final int MAX_POINTS = 80;
    private int nMaxPoints = MAX_POINTS;
    private final int WIDTH_LINE = 4;
    private final float TEMP_TOP = 37.19f;
    private final float TEMP_BOTTOM = 20f;
    public DrawView(Context context, AttributeSet attrs) {
        super(context, attrs);
        // TODO Auto-generated constructor stub
        bkPaint=new Paint();                            //不要放入onDraw中
        linePaint = new Paint();
        linePaint.setStyle(Paint.Style.STROKE);
        linePaint.setAntiAlias(false);
        linePaint.setStrokeWidth(WIDTH_LINE);
        ptPaint= new Paint();
        pList = new ArrayList<Float>(nMaxPoints + 1);   //add, get, size, remove
    }
    public void setPointList(ArrayList<Float> pList){
        this.pList = pList;
    }
    public ArrayList<Float> getPointList(){
        return pList;
    }
    @Override
    protected void onDraw(Canvas canvas) {
        // TODO Auto-generated method stub
        nHeight = this.getHeight();                     //不要放入Constructor中
        nWidth = this.getWidth();
        if (this.getResources().getConfiguration().orientation ==
                Configuration.ORIENTATION_LANDSCAPE)
            nMaxPoints = MAX_POINTS * 2;
        else
            nMaxPoints = MAX_POINTS;
        drawBackGround(canvas, bkPaint);
        drawLine(canvas, pList);
        super.onDraw(canvas);
    }
    private void drawBackGround(Canvas canvas, Paint paint){
        bkPaint.setColor(Color.WHITE);                  //在绘图之前先设置白底
        canvas.drawRect(0, 0, nWidth, nHeight, bkPaint);
    }
    private Bundle getPoint(int nLocation, float fd){
        Bundle p = new Bundle();
```

```
            p.putFloat("x",(nWidth / nMaxPoints) * nLocation);
            p.putFloat("y",nHeight *
                    (1.125f- (fd-TEMP_BOTTOM)/(TEMP_TOP-TEMP_BOTTOM))
                    * 0.8f);
            return p;
        }
        private void drawLine(Canvas canvas, ArrayList<Float> fList){
            int nLen = fList.size();
            if(nLen < 3) return;
            Bundle p;
            float x0, y0, x1, y1;
            for(int i=1; i<nLen; i++){
                p = getPoint(i-1, fList.get(i-1));
                x0 = p.getFloat("x");
                y0 = p.getFloat("y");
                p = getPoint(i, fList.get(i));
                x1 = p.getFloat("x");
                y1 = p.getFloat("y");
                if(y1 < y0) {
                    linePaint.setColor(Color.RED);
                    ptPaint.setColor(Color.BLUE);
                }
                if(y1 > y0) {
                    linePaint.setColor(Color.BLUE);
                    ptPaint.setColor(Color.RED);
                }
                canvas.drawLine(x0, y0, x1, y1, linePaint);
                ptPaint.setStrokeWidth(WIDTH_LINE*2); //draw last point
                canvas.drawPoint(x1, y1, ptPaint);
            }
            //Log.d("drawLine", fList.toString());
        }
        private void addDataToArray(float fd){
            pList.add(fd);
            //当横屏变为竖屏时，删掉多余的点
            while (pList.size() > nMaxPoints) pList.remove(0);
        }
        public void setTemperatureY(float t){
            addDataToArray(t);
        }
        public void clearData(){
            pList.clear();
        }
    }
}
```

15.5.2 portrait 布局

在 portrait 布局方式中，大框架采用线性布局，从上到下分为三个子布局。第一个子布局也为线性布局，依次排放 txtValue 显示温度，txtUnit 显示温度单位，btSetup 则是一个参数设置按钮；第二个子布局也为线性布局，依次排放 btStart、btStop 和 btQuit 三个按钮；最后一个布局为自定义组件 DrawView 对象 vRealTime。

```
<LinearLayout xmlns:android="http://schemas.android.com/apk/res/android"
    xmlns:tools="http://schemas.android.com/tools"
    android:layout_width="match_parent"
```

```xml
    android:layout_height="match_parent"
    android:orientation="vertical" >
    <!-- android:keepScreenOn="true" -->
    <LinearLayout
        android:layout_width="wrap_content"
        android:layout_height="wrap_content"
        android:layout_gravity="center"
        android:orientation="horizontal" >
        <TextView
            android:id="@+id/txtValue"
            android:layout_width="wrap_content"
            android:layout_height="wrap_content"
            android:gravity="center"
            android:text="@string/strOriginalValue"
            android:textSize="20sp"
            android:textStyle="bold" />
        <TextView
            android:id="@+id/txtUnit"
            android:layout_width="wrap_content"
            android:layout_height="wrap_content"
            android:text="@string/strUnit"
            android:textAppearance="?android:attr/textAppearanceMedium" />
        <Button
            android:id="@+id/btSetup"
            android:layout_width="wrap_content"
            android:layout_height="wrap_content"
            android:layout_marginLeft="40dp"
            android:background="?android:attr/selectableItemBackground"
            android:onClick="onClick"
            android:text="@string/strSetup" />
    </LinearLayout>
    <LinearLayout
        android:layout_width="wrap_content"
        android:layout_height="wrap_content"
        android:layout_gravity="center"
        android:orientation="horizontal" >
        <Button
            android:id="@+id/btStart"
            style="?android:attr/buttonBarButtonStyle"
            android:layout_width="wrap_content"
            android:layout_height="wrap_content"
            android:background="?android:attr/selectableItemBackground"
            android:onClick="onClick"
            android:text="@string/strStart" />
        <Button
            android:id="@+id/btStop"
            style="?android:attr/buttonBarButtonStyle"
            android:layout_width="wrap_content"
            android:layout_height="wrap_content"
            android:background="?android:attr/selectableItemBackground"
            android:onClick="onClick"
            android:text="@string/strStop" />
        <Button
            android:id="@+id/btQuit"
            style="?android:attr/buttonBarButtonStyle"
            android:layout_width="wrap_content"
```

```xml
            android:layout_height="wrap_content"
            android:background="?android:attr/selectableItemBackground"
            android:onClick="onClick"
            android:text="@string/strQuit" />
    </LinearLayout>
    <com.walkerma.i7013d_tcp.DrawView
        android:id="@+id/vRealTime"
        android:layout_width="match_parent"
        android:layout_height="match_parent"
        android:layout_margin="5dp" />
</LinearLayout>
```

15.5.3 landscape 布局

在 landscape 布局方式中，由于横屏带来横向空间的扩大，将 portrait 布局中的前两个子布局合并到一个子布局中，即文本温度显示和按钮的设置全部位于一行，其他所有组件及其属性均不变，实际效果如图 15-1 所示（landscape 布局文件略）。

15.5.4 源代码

主窗体中主要实现 TCP 客户机功能，对仿真模块进行定时查询以便获取实时温度数据，以及对返回数据进行处理。关于调整手机音量播放报警声音、利用消息机制处理后退键及发送报警短信的技术，在第 3 章和第 4 章都有涉及，这里不再赘述。

在主窗体的全局变量中声明了各组件对象和报警温度 ALARM_T。布尔变量 bAlarmed 表示报警是否发出，如果已经发出（true），则必须等温度低于报警温度才能复位（false）。布尔变量 chkSum 表示是否添加累加和校验码，chkSMS 表示是否发送报警短信。字符串变量 strID 中存放仿真模块的 ID 号（十六进制字符串形式），strData 中存放查询命令，在与仿真模块连接成功的时候于 processMessage 函数中自动生成。

函数 setButtonStatus 根据布尔变量 bConnected 来设置按钮的状态；函数 prepareForStart 主要检查校验方法，初始化校验方式变量 nParitySort，同时初始化客户机对象 client，首次将接收延迟设置为最大值 DELAY_MAX；函数 clearForStop 主要用于关闭连接后释放资源并复原软件界面。

如果向仿真模块发送查询命令的间隔时间过长，则实时效果较差；如果间隔时间过短，没有收到仿真模块的回复就接着发送下一个查询命令，则容易引起通信冲突。这里采用自适应算法，通过测量时间间隔来确定查询间隔时间。

一旦与仿真模块连接成功，则在 processMessage 函数的 MSG_Connected 消息中启动测试通信时间。其中，布尔变量 bTested 表示是否完成测量（初始值为 false），lStart 是测量开始时间点，生成完整的查询命令 strData 后，向仿真模块发送，接着启动多线程对象 threadIntervalCommTest 进行处理。

多线程类 ThreadIntervalCommTest 的 run 方法执行一个 while 死循环，整型变量 nInterval_Comm 中存放监测软件与仿真模块之间的实际时间间隔，如果该值大于常量 DELAY_MAX，则发送消息 MSG_ParametersError，即通信参数错误，比如校验码或结尾码不一致，导致仿真模块无响应。如果监测软件收到仿真模块返回的数据，则在 processMessage 函数的 MSG_DataArrived 消息中设置 bTested 为 true，这时，if(bTested) 条件成立，nInterval_Comm 与常量 DELAY_INQUIRY_ADD（一个经验值，这里为 800 毫秒）之和即为比较稳定可靠的查询间隔时间 nDelayInquiry，不会引起通信冲突。接着，在 client 对象中调用 setReadDelay 函数，将接收延迟 nDelayRead 设置为最小值 DELAY_READ_ADD，利用消息机制调用 Handler 对象 mHandler 的 postDelayed 方法，间隔 nDelayInquiry 毫秒后执行 runnable 接口中规定的内容，调用 vRealTime 对象中的 clearData 函数，为绘制温度趋势线做准备，最后通过 break 语句跳出 while 循环。

在 runnable 接口中只有一个 run 方法，主要用于定时发送查询命令。布尔变量 bDataMissing 表示数据丢失，在调用 client 对象的 sendText 函数发送数据前，先设置该值为 true，即先假定发出的数据没有响应。如果收到仿真模块的响应，则在 processMessage 函数的 MSG_DataArrived 消息中将 bDataMissing 设置为 false。在 run 方法中，如果数据丢失，则提示"Data Missing ..."，并认为接收延迟 nDelayRead 较小，尝试加上 DELAY_READ_ADD 进行调整。最后定时发送查询命令。

在 processMessage 函数的 MSG_DataArrived 消息中处理接收到的数据，函数 pickValidData 调用第 12 章介绍的 pickPurePackage 函数，删除结尾码和校验码从而获取有效数据 strPureData，再调用 substring 方法删除前导字符">"，最后解析为 double 类型的温度数据，并在文本框中显示。checkForAlarm 函数用于检查温度是否超过警戒值，以决定是否播放提醒声音和发送报警短信。最后调用自定义组件对象 vRealTime 的 setTemperatureY 函数，添加当前温度，并调用 invalidate 方法刷新组件，完成图形绘制。

当手机转屏时，自动执行 onConfigurationChanged 方法，首先调用 vRealTime 对象的 getPointList 函数，获取保存的温度序列，然后调用 setContentView 方法重新加载布局文件（系统自动选择 portrait 或 landscape 文件），初始化组件，如果温度序列 dataRestored 非空，则调用 vRealTime 对象的 setPointList 函数恢复温度序列，最后调用 setButtonStatus 函数重新设置按钮的状态。

```
package com.walkerma.i7013d_tcp;
import java.util.ArrayList;
import com.walkerma.library.GeneralProcess;
import com.walkerma.library.NetworkProcess;
import com.walkerma.library.Parity;
import com.walkerma.library.Parity.DisplayMode;
import com.walkerma.library.Parity.EndMark;
import com.walkerma.library.Parity.ParitySort;
import com.walkerma.library.TcpClientServer;
import android.app.Activity;
import android.app.AlertDialog;
import android.app.PendingIntent;
import android.content.Context;
import android.content.DialogInterface;
import android.content.Intent;
import android.content.IntentFilter;
import android.content.res.Configuration;
import android.graphics.Color;
import android.media.AudioManager;
import android.media.MediaPlayer;
import android.os.Bundle;
```

```java
import android.os.Handler;
import android.os.Message;
import android.telephony.SmsManager;
import android.widget.Button;
import android.widget.TextView;
import android.widget.Toast;
import android.view.*;
public class MainActivity extends Activity {
    private Context context;
    private boolean bDebug = false;
    /*------------------->SMS<-------------------*/
    private SmsSentBroadcastReceiver sentReceiver;
    private SmsDeliveredBroadcastReceiver deliveredReceiver;
    private Intent sentIntent;
    private PendingIntent sentPI;
    private Intent deliveryIntent;
    private PendingIntent deliveredPI;
    private final String SENT = "sent";
    private final String DELIVERED = "delivered";
    private AudioManager am;
    private int nMode;
    private int nCurrentVolume;
    private boolean bSpeakerStatus;
    private MediaPlayer mPlayer;
    private TextView txtValue;
    private Button btSetup, btStart, btStop;
    private DrawView vRealTime;
    private final static int REQUEST_CODE = 1;
    private final static double ALARM_T = 36.5;
    private boolean bAlarmed = false;
    private boolean bStartSent=false;
    private boolean bStopSent=false;
    private String strIP, strID, strData, strPhone;
    private ParitySort     nParitySort;
    private Boolean chkSum=false, chkSMS;
    private int nPort;
    private TcpClientServer client;
    private Handler mHandler;
    private ThreadIntervalCommTest threadIntervalCommTest;
    private static final int MSG_ParametersError = 11006;
    private final int MSG_ExitDelay = 11007;
    private boolean bCanExit=false;
    private boolean bTested;                //通信参数测试
    private int nInterval_Comm;             //来回的时间间隔
    private final int DELAY_INQUIRY_ADD = 800;
    private int nDelayInquiry;              // nInterval_Comm + DELAY_INQUIRY_ADD
    private final int DELAY_MAX = 8000;    //默认的最大延迟毫秒数
    private final int DELAY_READ_ADD = 100;
    private int nDelayRead;
    private long lStart;
    private double t_Old, t_New;
    private boolean bDataMissing = false;
    private boolean bConnected = false;    //    portrait和landscape布局切换
    Runnable runnable = new Runnable() {
        @Override
        public void run() {
```

```java
            if(bDataMissing){
                Toast.makeText(context, "Data Missing ...",
                        Toast.LENGTH_SHORT).show();
                if((nDelayRead+DELAY_READ_ADD)*2 < nDelayInquiry){
                    nDelayRead += DELAY_READ_ADD;
                    //自动调整
                    client.setReadDelay(nDelayRead);
                }
            }
            client.sendText(strData);
            bDataMissing = true;
            mHandler.postDelayed(this, nDelayInquiry);
        }
    };
    public  final class ThreadIntervalCommTest extends Thread{
        public void run() {
            while(true){
                nInterval_Comm = (int)(System.currentTimeMillis()-lStart);
                if(nInterval_Comm >= DELAY_MAX) {
                    mHandler.obtainMessage(MSG_ParametersError,
                            0, 0,null).sendToTarget();
                    break;
                }
                if(bTested) {
                    nDelayRead = DELAY_READ_ADD;
                    client.setReadDelay(nDelayRead); //优化
                    nDelayInquiry = nInterval_Comm + DELAY_INQUIRY_ADD;
                    mHandler.postDelayed(runnable, nDelayInquiry);
                    vRealTime.clearData();
                    break;
                }
            }
        }
    }
    private void setSoundMode(){
        am = (AudioManager)getSystemService(Context.AUDIO_SERVICE);
        if(am == null) return;
        nMode = am.getMode();
        am.setMode(AudioManager.MODE_IN_CALL);
        nCurrentVolume = am.getStreamVolume(AudioManager.STREAM_MUSIC);
        am.setStreamVolume(AudioManager.STREAM_MUSIC,
                am.getStreamMaxVolume(AudioManager.STREAM_MUSIC ),0);
        bSpeakerStatus = am.isSpeakerphoneOn();
        am.setSpeakerphoneOn(true);
    }
    private void restoreSoundMode(){
        if(am == null) return;
        am.setMode(nMode);
        am.setStreamVolume(AudioManager.STREAM_MUSIC, nCurrentVolume, 0);
        am.setSpeakerphoneOn(bSpeakerStatus);
    }
    @Override
    protected void onCreate(Bundle savedInstanceState) {
        super.onCreate(savedInstanceState);
        setContentView(R.layout.activity_main);
        context = this;
```

```java
        getWindow().addFlags(
                WindowManager.LayoutParams.FLAG_KEEP_SCREEN_ON);
        mPlayer = MediaPlayer.create(this, R.raw.notify);
        //mPlayer.setVolume(0.9f, 0.9f);
        setSoundMode();
        initializeWidgets();
        initSMS();
        mHandler = new Handler(new IncomingHandlerCallback());
    }
    @Override
    public void onConfigurationChanged(Configuration newConfig) {
        ArrayList<Float> dataRestored = vRealTime.getPointList();
        setContentView(R.layout.activity_main);
        initializeWidgets();
        if (dataRestored != null) vRealTime.setPointList(dataRestored);
        setButtonStatus();
        super.onConfigurationChanged(newConfig);
    }
    class IncomingHandlerCallback implements Handler.Callback{
        @Override
        public boolean handleMessage(Message msg) {
            processMessage(msg);
            return true;
        }
    }
    private void initializeWidgets(){
        vRealTime = (DrawView)findViewById(R.id.vRealTime);
        txtValue = (TextView)findViewById(R.id.txtValue);
        txtValue.setTextColor(Color.BLACK);
        btSetup = (Button)findViewById(R.id.btSetup);
        btStart = (Button)findViewById(R.id.btStart);
        btStop = (Button)findViewById(R.id.btStop);
        setButtonStatus();
    }
    private void setButtonStatus(){
        if(bConnected){
            btSetup.setEnabled(false);
            btStart.setEnabled(false);
            btStop.setEnabled(true);
        }
        else{
            btSetup.setEnabled(true);
            btStart.setEnabled(true);
            btStop.setEnabled(false);
        }
    }
    public void onClick(View v){
        switch(v.getId()){
        case R.id.btSetup:
            Intent it = new Intent();
            it.putExtra("CheckSMS", chkSMS); //初始chkSMS = false
            it.setClass(this, SetupActivity.class);
            startActivityForResult(it, REQUEST_CODE);
            break;
        case R.id.btStart:
            bStartSent = true;
```

```java
                bStopSent = false;
                prepareForStart();
                Toast.makeText(this, "Starting ...",
                        Toast.LENGTH_SHORT).show();
                break;
            case R.id.btStop:
                bStopSent = true;
                bStartSent = false;
                clearForStop();
                break;
            case R.id.btQuit:
                finish();
                break;
        }
    }
    private void prepareForStart(){
        // boolean must be initialized at first.
        nParitySort = chkSum?ParitySort.ADD:ParitySort.NONE;
        if(strIP==null){alertForStart(); return;}
        if(!bDebug){
            // AVD不支持WiFi
            if(!NetworkProcess.isNetworkAvailable(this)){
                alertForNetwork();
                return;
            }
        }
        if(client!=null) client.close();
        client = new TcpClientServer(mHandler, strIP, nPort);
        client.setReadDelay(DELAY_MAX); //Temporary
        client.connect();
    }
    private void clearForStop(){
        mHandler.removeCallbacks(runnable);
        client.close();
        txtValue.setText(R.string.strOriginalValue);
        txtValue.setTextColor(Color.BLACK);
    }
    @Override
    protected void onActivityResult(int requestCode,
            int resultCode, Intent data) {
        if (requestCode == REQUEST_CODE){
            if (resultCode == SetupActivity.RESULT_CODE){
                Bundle bundle = new Bundle();
                bundle = data.getExtras();
                strIP = bundle.getString("IP");
                nPort = bundle.getInt("Port");
                strID = bundle.getString("ID");
                chkSum = bundle.getBoolean("CheckSum");
                chkSMS = bundle.getBoolean("CheckSMS");
                strPhone = bundle.getString("Phone");
            }
        }
        super.onActivityResult(requestCode, resultCode, data);
    }
    /*------------------->SMS<-------------------*/
    private void initSMS(){
```

```java
        //public static PendingIntent getBroadcast (Context context,
        //                int requestCode, Intent intent, int flags)
        sentIntent = new Intent(SENT);
        sentPI = PendingIntent.getBroadcast(
                getApplicationContext(), 0, sentIntent,
                PendingIntent.FLAG_UPDATE_CURRENT);

        sentReceiver = new SmsSentBroadcastReceiver();
        registerReceiver(sentReceiver, new IntentFilter(SENT));

        deliveryIntent = new Intent(DELIVERED);
        deliveredPI = PendingIntent.getBroadcast(
                getApplicationContext(), 0, deliveryIntent,
                PendingIntent.FLAG_UPDATE_CURRENT);
        deliveredReceiver = new SmsDeliveredBroadcastReceiver();
        registerReceiver(deliveredReceiver, new IntentFilter(DELIVERED));
    }
    private void processMessage(Message msg){
        switch(msg.what){
        case MSG_ExitDelay:
            bCanExit = false;
            return;
        case TcpClientServer.MSG_Connected:
            bConnected = true;
            bStartSent = false;
            Toast.makeText(this, "Connected",
                    Toast.LENGTH_SHORT).show();
            strData = Parity.getEntirePackage("#"+strID,
                    nParitySort, EndMark.CR, DisplayMode.CHAR);
            bTested = false;
            lStart = System.currentTimeMillis();
            client.sendText(strData);
            threadIntervalCommTest = new ThreadIntervalCommTest();
            threadIntervalCommTest.start(); // once only
            break;
        case TcpClientServer.MSG_DataArrived:
            if (!bTested) bTested = true;
            t_New = pickValidData((String)msg.obj);
            txtValue.setText(String.format("%.2f", t_New));
            checkForAlarm(t_New);
            if (t_New > t_Old)
                txtValue.setTextColor(Color.RED);
            else
                txtValue.setTextColor(Color.BLUE);
            checkForAlarm(t_New);
            t_Old = t_New;
            vRealTime.setTemperatureY((float)t_New);
            vRealTime.invalidate();
            bDataMissing = false;
            break;
        case MSG_ParametersError:
            Toast.makeText(this,
                    "i-7013D parameters error ...",
                    Toast.LENGTH_SHORT).show();
            break;
        case TcpClientServer.MSG_ConnectError:
```

```java
                bConnected = false;
                alertForConnectError();
                clearForStop(); //07/25/15
                break;
            case TcpClientServer.MSG_TransmitError:
                bConnected = false;
                Toast.makeText(this, "Transmit error.",
                        Toast.LENGTH_SHORT).show();
                clearForStop(); //07/25/15
                break;
            case TcpClientServer.MSG_Closed:
                bConnected = false;
                if(bStartSent) break;
                Toast.makeText(this, "Closed.",
                        Toast.LENGTH_SHORT).show();
                //显示通信参数
                if(bTested) {
                    Toast.makeText(this, "CommInterval = " +
                            Integer.toString(nInterval_Comm) + " ms",
                            Toast.LENGTH_SHORT).show();
                }
                break;
            case TcpClientServer.MSG_ReadError:
                bConnected = false;
                if(bStopSent) break;
                Toast.makeText(this, "Read error.",
                        Toast.LENGTH_SHORT).show();
                break;
        }
        setButtonStatus();
    }
    private double pickValidData(String strData){
        String strPureData = Parity.pickPurePackage(strData, nParitySort,
                EndMark.CR, DisplayMode.CHAR);
        if(strPureData == null)
            return 0;
        else
            return Double.parseDouble(strPureData.substring(1));
    }
    private void checkForAlarm(double t){
        if((t>ALARM_T) && (bAlarmed == false)){
            bAlarmed = true;
            try {
                if(mPlayer.getCurrentPosition()>0)
                    mPlayer.seekTo(0);
                mPlayer.start();
            } catch (IllegalStateException e) {
                // TODO Auto-generated catch block
                e.printStackTrace();
            }
            if(chkSMS) sendSms(t);
        }
        if(t<ALARM_T-1) bAlarmed = false;
    }
    /*------------------->SMS<-------------------*/
    private void sendSms(double t){
```

```java
        if(strPhone == null) return;
        String strMessage = "Dangerous: " + txtValue.getText().toString() +
                " > " + String.format("%.2f", ALARM_T) + ". 【马玉春】";
        SmsManager sm = SmsManager.getDefault();
        sm.sendTextMessage(strPhone, null, strMessage,
                sentPI, deliveredPI);
    }
    private void alertForStart(){
        AlertDialog.Builder builder = new AlertDialog.Builder(this);
        builder.setIcon(R.drawable.ic_action_info);
        builder.setTitle("Start Info: ");
        builder.setMessage("Please [Setup] at first.");
        builder.setPositiveButton("OK", null);
        builder.setCancelable(false);
        AlertDialog dialog = builder.create();
        dialog.show();
    }
    private void alertForNetwork(){
        AlertDialog.Builder builder = new AlertDialog.Builder(this);
        builder.setIcon(R.drawable.ic_action_warning);
        builder.setTitle("Wi-Fi/GPRS: ");
        builder.setMessage("Wi-Fi/GPRS is disconnected.");
        builder.setPositiveButton("OK", null);
        builder.setCancelable(false);
        AlertDialog dialog = builder.create();
        dialog.show();
    }
    private void alertForConnectError(){
        AlertDialog.Builder builder = new AlertDialog.Builder(this);
        builder.setIcon(R.drawable.ic_action_warning);
        builder.setCancelable(false);
        builder.setTitle("Connect Error: ");
        builder.setMessage("Please check the IP and Port of remote server, then listen.");
        builder.setPositiveButton("OK", new DialogInterface.OnClickListener() {
            @Override
            public void onClick(DialogInterface dialog, int which) {
                // TODO Auto-generated method stub
                ;
            }
        });
        AlertDialog dialog = builder.create();
        dialog.show();
    }
    @Override
    public boolean onCreateOptionsMenu(Menu menu) {
        // Inflate the menu; this adds items to the action bar if it is present.
        getMenuInflater().inflate(R.menu.main, menu);
        //menu.add(groupId, itemId, order, title);
        return true;
    }
    @Override
    public boolean onOptionsItemSelected(MenuItem item) {
        if(item.getItemId()==R.id.menu_about)
            GeneralProcess.fireAboutDialog(this,
                    getString(R.string.app_name),
                    getString(R.string.strCopyrights));
```

```
            return super.onOptionsItemSelected(item);
    }
    @Override
    public boolean onKeyDown(int keyCode, KeyEvent event) {
        // TODO Auto-generated method stub
        switch(keyCode){
        case KeyEvent.KEYCODE_BACK:
            if(!bCanExit){
                bCanExit = true;
                Toast.makeText(getApplicationContext(),
                        "Press again to quit.", Toast.LENGTH_SHORT).show();
                //假如在1秒以内没有再次按下"返回"键,那么将再次设置canExit为false
                mHandler.sendEmptyMessageDelayed(MSG_ExitDelay, 1000);
                return true;
            }
            else break;
        }
        return super.onKeyDown(keyCode, event);
    }
    @Override
    protected void onDestroy() {
        // TODO Auto-generated method stub
        mHandler.removeCallbacksAndMessages(null);
        if(client!=null) client.close();
        mPlayer.release();
        /*------------------->SMS<-------------------*/
        unregisterReceiver(sentReceiver);
        unregisterReceiver(deliveredReceiver);
        restoreSoundMode();
        getWindow().clearFlags(
                WindowManager.LayoutParams.FLAG_KEEP_SCREEN_ON);
        super.onDestroy();
    }
}
```

15.6 对实物模块的监控

I-7013D 模块的实物产品为 RS-485 接口,实物形态如图 14-1 的下半部分所示,使用场景相当于图 15-4 中的"设备",计算机通过扩展卡或者 USB 转换器得到 RS-485 接口。本章的监测软件只是一个具体化的 TCP 客户机,由于与 RS-485 接口不一致,并不能直接监测 I-7013D 模块的温度。借助"RS-485/TCP"转换软件,计算机将手机发送的命令转发给设备,将设备的响应转发给手机,即可实现本文的监测软件对实物模块的监测。

图 15-4 TCP 客户机用于测试设备

I-7065D 模块有 4 路开关量输入,5 路开关量输出,如果将图 15-4 中的设备替换成 I-7065D,即可用第 13 章的通用 TCP 客户机对实物 I-7065D 模块进行监控测试,即读取输入与输出开关的状态,或直接发送命令控制输出开关。关于 I-7013D 和 I-7065D 的详细应用请参考《计算机监控系统的仿真开发》(国防工业出版社出版)。

随着移动互联网的推广和普及，越来越多的产品直接支持 WiFi 接口，如有的淋浴热水器可以直接用手机调节温度。采用第 13 章的通用 TCP 客户机对该类设备进行测试后，即可获取相关参数，然后进行界面设计，融入功能，通用 TCP 客户机就可转换成一个专用的手机监控软件。

15.7 本章小结

本章的温度监测软件可以定时查询仿真模块的温度，并利用实时趋势线显示。查询命令的时间间隔通过监测软件自身测试所得，接收数据的延迟也实现了自适应调整的功能，使得监测软件灵敏可靠。对于温度超限，实现了声音和短信报警两种手段。转屏时可以顺利恢复历史数据，让温度趋势线连续显示。在参数设置方面，还可以添加报警温度、操作员等信息，以增加更多的灵活性。如果增加一个接口转换器，将 RS-485 接口的串行通信协议转换为 TCP/IP，则本监测软件不需要修改任何代码，即可完成对此实物模块的监测。

参 考 文 献

[1] 王小科，寇长梅. Android 入门经典 [M]. 北京：机械工业出版社，2013.
[2] 陈会安. Java 和 Android 开发实战详解 [M]. 北京：人民邮电出版社，2014.
[3] 刘雍，孙冰，马玉春. Android 平台下的通用 SQLite 模型的设计与实现 [J]. 电脑编程技巧与维护，2017.
[4] 刘雍，孙冰，马玉春. 基于消息驱动的 Android TCP 服务器类的设计 [J]. 海南热带海洋学院学报，2017.
[5] 刘雍，汪文彬，马玉春. Android 环境下的数据编码与处理技术 [J]. 电脑编程技巧与维护，2018.
[6] 马玉春，刘雍，乔丽娟，等. Android 平台下的 TCP 客户机教学设计 [J]. 软件，2018.
[7] 刘雍，汪文彬，马玉春. Android 环境下的数据包的校验技术 [J]. 电脑编程技巧与维护，2019.
[8] 刘雍，马玉春. Android 课堂随机点名软件：2017SR288840[P]. 2017-06-20.
[9] 马玉春. Android 版 I-7013D 模块仿真软件：2019SR0205622[P].2019-03-04.
[10] 马玉春. Android 版 I-7013D 模块监测软件：2019SR0856436[P].2019-08-19.

推荐阅读

交互式系统设计：HCI、UX和交互设计指南（原书第3版）

作者：David Benyon 译者：孙正兴 等 ISBN：978-7-111-52298-0 定价：129.00元

本书在人机交互、可用性、用户体验以及交互设计领域极具权威性。
书中囊括了作者关于创新产品及系统设计的大量案例和图解，
每章都包括发人深思的练习、挑战点评等内容，适合具有不同学科背景的人员学习和使用。

以用户为中心的系统设计

作者：Frank E. Ritter 等 译者：田丰 等 ISBN：978-7-111-57939-7 定价：85.00元

本书融合了作者多年的工作经验，阐述了影响用户与系统有效交互的众多因素，
其内容涉及人体测量学、行为、认知、社会层面等四个主要领域，
介绍了相关的基础研究，以及这些基础研究对系统设计的启示。

交互设计：超越人机交互（原书第4版）

作者：Jenny Preece 等 ISBN：978-7-111-58927-3 定价：119.00元

本书由交互设计界的三位顶尖学者联袂撰写，是该领域的经典著作，被全球各地的大学选作教材。
新版本继承了本书一贯的跨学科特色，并与时俱进地更新了大量实例，
涉及敏捷用户体验、社会媒体与情感交互、混合现实与脑机界面等。

推荐阅读

分布式机器学习：算法、理论与实践

作者：刘铁岩 等 ISBN：978-7-111-60918-6 定价：89.00元

机器学习：算法视角（原书第2版）

作者：史蒂芬·马斯兰 译者：高阳 等 ISBN：978-7-111-62226-0 定价：99.00元

当计算机体系结构遇到深度学习

作者：布兰登·里根 等 译者：杨海龙 等 ISBN：978-7-111-62248-2 定价：69.00元

机器学习精讲：基础、算法及应用

作者：杰瑞米·瓦特 等 译者：杨博 ISBN：978-7-111-61196-7 定价：69.00元

基于复杂网络的机器学习方法

作者：迪亚戈·克里斯蒂亚诺·席尔瓦 等 译者：李泽荃 等 ISBN：978-7-111-61149-3 定价：79.00元

卷积神经网络与视觉计算

作者：拉加夫·维凯特森 等 译者：钱亚冠 ISBN：978-7-111-61239-1 定价：59.00元